LangChain与新时代生产力
AI应用开发之路

陆梦阳　朱　剑　孙罗庚　韩中俊◎编著

跟我一起学 人工智能

清华大学出版社
北京

内 容 简 介

本书全面深入地探讨了AI生成大模型在多个领域的应用,从AI的历史探源、生成式模型在不同领域的应用(文本、图片、声乐、视频及多模态模型),到大型语言模型(如GPT)的运作机制、预训练、可扩展架构及其适应性。通过详细介绍LangChain的概念、应用、安装方法及如何利用LangChain进行软件开发和数据科学研究,本书为读者提供了一个关于AI生成大模型应用的全景视图。

本书共11章,覆盖了AI生成大模型的理论基础、LangChain的入门与进阶使用、LangChain在软件开发和数据科学中的应用及生成式AI的未来展望等方面。第1~3章介绍AI生成大模型的基础知识,如AI历史、生成式模型在不同领域的应用,以及大型语言模型,如GPT的基础和LangChain的初识与入门。第4~6章深入LangChain的进阶使用,探讨Agent构建、文档查询、聊天机器人开发,提供实战案例和技术细节。第7~9章聚焦LangChain在软件开发和数据科学的应用及定制LLM输出的策略,介绍编程、自动化软件开发、数据探索技术方法及LLM输出优化。第10章和第11章讨论LLM在生产环境的应用、监控、回调处理,以及生成式AI的未来展望,包括其潜力、社会影响和挑战。本书旨在为读者提供LangChain从入门到进阶的全面知识,适合不同背景的读者深入学习和应用。

本书特色在于其理论与实践相结合的方法论,适合对AI生成大模型感兴趣的学生、研究人员和软件开发者阅读。书中不仅详细介绍了AI生成大模型的基础知识和前沿技术,还提供了丰富的实践指导和案例分析。此外,配套资源丰富,包括示例代码、工具安装指南和案例研究,极大地增强了本书的实用性和指导性。无论是AI领域的新手还是有经验的开发者都能从中获得必要的知识和灵感。

版权所有,侵权必究。举报: 010-62782989,beiqinquan@tup.tsinghua.edu.cn。

图书在版编目(CIP)数据

LangChain与新时代生产力:AI应用开发之路/陆梦阳等编著. -- 北京:清华大学出版社,2024.11. -- (跟我一起学人工智能). -- ISBN 978-7-302-67615-7

Ⅰ.TP311.561

中国国家版本馆CIP数据核字第2024RT3045号

责任编辑: 赵佳霓
封面设计: 吴 刚
责任校对: 时翠兰
责任印制: 宋 林

出版发行: 清华大学出版社
网　　址: https://www.tup.com.cn,https://www.wqxuetang.com
地　　址: 北京清华大学学研大厦A座　　　邮　编: 100084
社 总 机: 010-83470000　　　　　　　　　邮　购: 010-62786544
投稿与读者服务: 010-62776969,c-service@tup.tsinghua.edu.cn
质量反馈: 010-62772015,zhiliang@tup.tsinghua.edu.cn
课件下载: https://www.tup.com.cn,010-83470236

印 装 者: 三河市君旺印务有限公司
经　　销: 全国新华书店
开　　本: 186mm×240mm　　印　张: 23.5　　字　数: 530千字
版　　次: 2024年12月第1版　　　　　　　印　次: 2024年12月第1次印刷
印　　数: 1~2000
定　　价: 89.00元

产品编号: 104883-01

前言
PREFACE

随着人工智能技术的突飞猛进，生成式 AI 大模型，特别是像 GPT 这样的语言模型和图像生成模型，在众多领域的应用已经成为一种革命性的力量。这些模型不仅改变了我们与文本、图像和视频进行交互的方式，还为自动化、创造性解决方案和智能系统的开发开辟了新的可能性。从业界到学术界，对这些技术的探索和应用已经超越了初期的好奇，步入了深度整合和创新应用的新阶段。

目前笔者在一家全球 500 强外企担任 AI 工程师，作为一名早期开始研究 AI 的工程师，笔者亲历了生成式 AI 从理论概念到实际应用的全过程。在这个过程中，笔者目睹了 LangChain 框架的诞生和发展，它不仅解决了大型语言模型（Large Language Model，LLM）的局限性，还开启了通过 AI 增强软件工具的新纪元。LangChain 的出现，标志着构建和利用 AI 模型的重大转变，尤其是在提升 AI 应用的复杂度和效率方面。

本书旨在为读者提供从生成式 AI 的基础知识到 LangChain 框架的高级应用的全面指南，通过深入浅出的解释、实际案例分析及详细的步骤指导，帮助读者理解并掌握 LangChain 的强大功能，进而开发出更智能、更高效的应用。

在编写本书的过程中，笔者不仅回顾了自己在 AI 研究和应用开发中的经验，也深入研究了最新的技术进展和实际案例。这一过程不仅加深了笔者对生成式 AI 技术潜力的认识，也让笔者对其未来的社会影响和发展前景有了更深刻的思考。笔者希望通过分享这些知识和见解，能够激发读者对 AI 技术的热情，推动更多的技术创新和应用，共同探索生成式 AI 的未来。

无论是对 AI 技术充满好奇的初学者，还是希望深入理解和应用 LangChain 框架的开发者，笔者相信本书都能为读者提供宝贵的信息和启发。让我们在这个充满无限可能的新时代一起探索 AI 技术。

本书主要内容

第 1 章介绍 AI 生成大模型的基础，涉及生成式模型在多个领域的应用，如文本、图像和视频等。本章还解释了 LLM（如 GPT）的核心机制和图像模型的工作原理，为读者提供 AI 生成技术的基础知识。

第 2 章讲解 LangChain 框架，包括其背景、必要性及工作机制。介绍了 LangChain 如何突破 LLM 的局限性，其支持的应用，以及核心组件（如 Agent、Chain 和 Memory）的功能。

第 3 章介绍安装 LangChain 的步骤、模型导入方法，以及模型输出解析技巧。讲解 LangChain 表达式语言和链的创建，结合客户服务助手案例展示了 LangChain 的实际应用。

第 4 章深入探讨 LangChain 中 Agent 的构建和运用，包括常见 Agent 类型及其实际应用。讨论了如何使 Agent 返回结构化输出和如何构建为图，提升 LangChain 应用的复杂度和效率。

第 5 章专注于使用 LangChain 进行高效文档查询，包括文档总结、信息提取等。本章解释了 LLM 推理的策略，为文档处理提供实用指南。

第 6 章探讨聊天机器人的开发过程，从基础概念到检索和向量的具体实现。介绍了 LangChain 在聊天机器人记忆管理中的应用，以及内容监管的重要性。

第 7 章讲解 LangChain 在软件开发领域的应用，从 AI 的最新进展到使用 LLM 进行编程和自动化软件开发的方法，提供了将 LangChain 集成到开发流程中的指南。

第 8 章介绍 LangChain 在数据科学中的应用，包括数据收集、处理和自动化机器学习。探讨了使用 Agent 进行数据探索的方法，展示了生成式 AI 如何改变数据科学。

第 9 章深入讨论如何调整和微调 LLM 输出，包括对齐方法和效果评估。介绍了 LangChain 微调 LLM 的技巧，以及提示词工程的高级应用。

第 10 章关注于 LLM 在生产环境的应用，包括部署、监测和 LangChain 回调使用。探讨了提高 LLM 应用可观测性的方法，以及 LangSmith 工具的实战应用。

第 11 章展望生成式 AI 的未来，讨论其潜力、社会影响及面临的挑战。本章为读者提供了对生成式 AI 未来走向的全面理解，探索了 AI 在多个领域的应用难题。

阅读建议

为了最大程度从本书中获益，笔者建议读者按照以下几点阅读建议进行学习。

（1）基础知识建立：对于初学者来讲，第 1 章是构建生成式 AI 技术基础知识的关键。通过理解 AI 生成大模型的基础概念，包括文本、图像和视频等多个领域的应用，以及 LLM（如 GPT）和图像模型的工作原理，读者可以获得对 AI 生成技术的初步了解。

（2）框架理解：第 2 章对 LangChain 框架进行了全面介绍，包括背景、必要性及工作机制。深入理解 LangChain 如何解决 LLM 的局限性及其支持的应用，对于想要在 AI 领域进一步深造的读者来讲至关重要。

（3）实操演练：从第 3 章开始，书中逐渐转向实操，介绍了安装 LangChain、模型导入方法及模型输出解析技巧。笔者建议读者跟随书中的步骤进行实际操作，通过实践来加深对理论知识的理解和应用。

（4）专题深化：第 4～10 章覆盖了 LangChain 的高级应用，包括 Agent 的构建和运用、高效文档查询、聊天机器人开发、软件开发和数据科学应用等。读者应根据个人兴趣和专业需求选择重点阅读和深入研究的章节。

（5）未来趋势与挑战：第 11 章对生成式 AI 的未来进行了展望，讨论了其潜力、社会影响及面临的挑战。笔者建议所有读者都阅读本章，以获得对 AI 技术未来走向的全面理解，同时思考如何在未来的技术浪潮中定位自己。

（6）动手实践与反复阅读：AI 是一个快速发展的领域，理论与实践紧密相连。笔者鼓励读者在阅读的同时，积极动手实践书中的案例和练习，甚至尝试自己的项目。同时，对于一些复杂的概念和技术，反复阅读和实践会有助于深化理解。

资源下载提示

素材（源码）等资源：扫描目录上方的二维码下载。

由于时间仓促，书中难免存在不妥之处，请读者见谅，并提宝贵意见。

陆梦阳

2024 年 9 月

目 录
CONTENTS

本书源码

第 1 章　什么是 AI 生成大模型 ⋯⋯⋯⋯⋯⋯⋯⋯⋯⋯⋯⋯⋯⋯⋯⋯⋯⋯⋯⋯⋯⋯⋯⋯ 1

　1.1　AI 历史探源 ⋯⋯⋯⋯⋯⋯⋯⋯⋯⋯⋯⋯⋯⋯⋯⋯⋯⋯⋯⋯⋯⋯⋯⋯⋯⋯⋯⋯⋯ 1
　　1.1.1　生成式模型 ⋯⋯⋯⋯⋯⋯⋯⋯⋯⋯⋯⋯⋯⋯⋯⋯⋯⋯⋯⋯⋯⋯⋯⋯⋯⋯ 2
　　1.1.2　生成式模型在文本领域 ⋯⋯⋯⋯⋯⋯⋯⋯⋯⋯⋯⋯⋯⋯⋯⋯⋯⋯⋯⋯ 3
　　1.1.3　生成式模型在图片领域 ⋯⋯⋯⋯⋯⋯⋯⋯⋯⋯⋯⋯⋯⋯⋯⋯⋯⋯⋯⋯ 4
　　1.1.4　生成式模型在声乐领域 ⋯⋯⋯⋯⋯⋯⋯⋯⋯⋯⋯⋯⋯⋯⋯⋯⋯⋯⋯⋯ 5
　　1.1.5　生成式模型在视频领域 ⋯⋯⋯⋯⋯⋯⋯⋯⋯⋯⋯⋯⋯⋯⋯⋯⋯⋯⋯⋯ 6
　　1.1.6　多模态模型 ⋯⋯⋯⋯⋯⋯⋯⋯⋯⋯⋯⋯⋯⋯⋯⋯⋯⋯⋯⋯⋯⋯⋯⋯⋯ 6
　1.2　LLM 简介 ⋯⋯⋯⋯⋯⋯⋯⋯⋯⋯⋯⋯⋯⋯⋯⋯⋯⋯⋯⋯⋯⋯⋯⋯⋯⋯⋯⋯⋯ 6
　　1.2.1　潮流涌现 ⋯⋯⋯⋯⋯⋯⋯⋯⋯⋯⋯⋯⋯⋯⋯⋯⋯⋯⋯⋯⋯⋯⋯⋯⋯⋯ 8
　　1.2.2　GPT 的运作机制 ⋯⋯⋯⋯⋯⋯⋯⋯⋯⋯⋯⋯⋯⋯⋯⋯⋯⋯⋯⋯⋯⋯ 10
　　1.2.3　模型的预训练 ⋯⋯⋯⋯⋯⋯⋯⋯⋯⋯⋯⋯⋯⋯⋯⋯⋯⋯⋯⋯⋯⋯⋯ 12
　　1.2.4　可扩展架构 ⋯⋯⋯⋯⋯⋯⋯⋯⋯⋯⋯⋯⋯⋯⋯⋯⋯⋯⋯⋯⋯⋯⋯⋯ 12
　　1.2.5　模型的适应性 ⋯⋯⋯⋯⋯⋯⋯⋯⋯⋯⋯⋯⋯⋯⋯⋯⋯⋯⋯⋯⋯⋯⋯ 13
　　1.2.6　上手 GPT ⋯⋯⋯⋯⋯⋯⋯⋯⋯⋯⋯⋯⋯⋯⋯⋯⋯⋯⋯⋯⋯⋯⋯⋯⋯ 13
　1.3　图像模型 ⋯⋯⋯⋯⋯⋯⋯⋯⋯⋯⋯⋯⋯⋯⋯⋯⋯⋯⋯⋯⋯⋯⋯⋯⋯⋯⋯⋯⋯ 13
　　1.3.1　图像模型工作原理 ⋯⋯⋯⋯⋯⋯⋯⋯⋯⋯⋯⋯⋯⋯⋯⋯⋯⋯⋯⋯⋯ 14
　　1.3.2　图像模型的不同版本 ⋯⋯⋯⋯⋯⋯⋯⋯⋯⋯⋯⋯⋯⋯⋯⋯⋯⋯⋯⋯ 15
　　1.3.3　图形模型的调节性 ⋯⋯⋯⋯⋯⋯⋯⋯⋯⋯⋯⋯⋯⋯⋯⋯⋯⋯⋯⋯⋯ 16
　1.4　总结 ⋯⋯⋯⋯⋯⋯⋯⋯⋯⋯⋯⋯⋯⋯⋯⋯⋯⋯⋯⋯⋯⋯⋯⋯⋯⋯⋯⋯⋯⋯⋯ 16

第 2 章　初识 LangChain ⋯⋯⋯⋯⋯⋯⋯⋯⋯⋯⋯⋯⋯⋯⋯⋯⋯⋯⋯⋯⋯⋯⋯⋯⋯ 17

　2.1　LLM 的局限性 ⋯⋯⋯⋯⋯⋯⋯⋯⋯⋯⋯⋯⋯⋯⋯⋯⋯⋯⋯⋯⋯⋯⋯⋯⋯⋯⋯ 17
　2.2　LLM 应用介绍 ⋯⋯⋯⋯⋯⋯⋯⋯⋯⋯⋯⋯⋯⋯⋯⋯⋯⋯⋯⋯⋯⋯⋯⋯⋯⋯⋯ 19
　2.3　LangChain 介绍 ⋯⋯⋯⋯⋯⋯⋯⋯⋯⋯⋯⋯⋯⋯⋯⋯⋯⋯⋯⋯⋯⋯⋯⋯⋯⋯ 21

2.3.1 LangChain 的必要性 23
2.3.2 LangChain 支持的应用 23
2.4 LangChain 的工作机制 24
2.4.1 初识 Agent 24
2.4.2 初识链 27
2.4.3 初识记忆 28
2.4.4 LangChain 中的工具 28
2.5 总结 29

第 3 章 LangChain 入门 30

3.1 安装 LangChain 方法 30
3.1.1 安装 Python 30
3.1.2 Jupyter Notebook 和 JupyterLab 30
3.1.3 环境管理 32
3.2 导入模型 34
3.2.1 虚拟 LLM(Fake LLM) 35
3.2.2 OpenAI 36
3.2.3 HuggingFace 37
3.2.4 微软云 38
3.2.5 谷歌云 38
3.2.6 Jina AI 40
3.2.7 Replicate 42
3.2.8 本地模型 42
3.3 模型输出解析 45
3.3.1 列表解析器 47
3.3.2 日期解析器 48
3.3.3 自动修复解析器 49
3.3.4 Pydantic(JSON)解析器 51
3.3.5 重试解析器 53
3.3.6 结构化输出解析器 55
3.3.7 XML 解析器 57
3.4 LangChain 表达式语言 59
3.4.1 LCEL 接口简介 60
3.4.2 绑定运行时参数 75
3.4.3 运行自定义函数 77
3.4.4 流式传输自定义生成器函数 78

		3.4.5 并行化步骤 ········· 80
		3.4.6 根据输入的动态路由逻辑 ········· 82
	3.5	链 ········· 86
		3.5.1 链接口中的方法调用 ········· 88
		3.5.2 自定义链的创建 ········· 89
		3.5.3 几种常见的链 ········· 92
	3.6	实战案例：客户服务助手应用程序开发 ········· 109
	3.7	总结 ········· 113

第 4 章 LangChain 进阶：Agent ········· 114

4.1	构建自己的第 1 个 Agent ········· 115
4.2	LangChain 中的常见 Agent 类型 ········· 121
	4.2.1 Zero-shot ReAct ········· 122
	4.2.2 Structured Input ReAct ········· 124
	4.2.3 OpenAI Functions ········· 128
	4.2.4 Conversational ········· 130
	4.2.5 ReAct Document Store ········· 133
4.3	迭代器运行 Agent ········· 135
4.4	让 Agent 返回结构化输出 ········· 138
4.5	处理 Agent 解析错误 ········· 143
4.6	将 Agent 构建为图 ········· 144
	4.6.1 快速开始 ········· 144
	4.6.2 流式输出 ········· 147

第 5 章 使用 LangChain 工具进行文档查询 ········· 153

5.1	幻觉现象 ········· 153
5.2	文档总结 ········· 156
5.3	信息提取 ········· 159
5.4	使用工具 ········· 161
5.5	解剖 LLM 推理的底层策略 ········· 165
5.6	总结 ········· 168

第 6 章 聊天机器人 ········· 170

6.1	聊天机器人简介 ········· 170
	6.1.1 历史溯源 ········· 171
	6.1.2 上下文和记忆 ········· 173

6.1.3　意识性与主动性 …… 174
　6.2　检索和向量 …… 174
　　6.2.1　嵌入 …… 176
　　6.2.2　存储嵌入的方式 …… 179
　　6.2.3　索引 …… 179
　　6.2.4　向量库 …… 180
　　6.2.5　向量数据库 …… 181
　　6.2.6　文档加载器 …… 184
　　6.2.7　LangChain 中的检索器 …… 184
　6.3　实战案例：实现一个聊天机器人 …… 187
　6.4　LangChain 中的记忆机制 …… 191
　　6.4.1　快速开始 …… 192
　　6.4.2　LangChain 中基础的记忆类型 …… 194
　　6.4.3　其他高级记忆类型 …… 211
　　6.4.4　记忆和 LLM 链 …… 225
　　6.4.5　记忆和 Agent …… 228
　　6.4.6　自定义会话记忆 …… 233
　　6.4.7　自定义记忆 …… 237
　　6.4.8　聊天机器人的记忆 …… 240
　6.5　内容监管 …… 245
　6.6　总结 …… 246

第 7 章　LangChain 和软件开发 …… 248

　7.1　步入新时代 …… 248
　　7.1.1　AI 在软件领域的最新进展 …… 249
　　7.1.2　代码生成 LLM …… 250
　　7.1.3　未来展望 …… 253
　7.2　使用 LLM 编程 …… 254
　7.3　LLM 自动化软件开发 …… 260
　7.4　总结 …… 265

第 8 章　LangChain 和数据科学 …… 266

　8.1　自动化数据科学简介 …… 266
　　8.1.1　数据收集 …… 268
　　8.1.2　可视化和探索性数据分析 …… 270
　　8.1.3　数据预处理和特征提取 …… 270

8.1.4　自动化机器学习 ········· 271
　　8.1.5　生成式 AI 对数据科学的变革 ········· 274
8.2　使用 Agent ········· 276
8.3　数据探索和 LLM ········· 279
8.4　总结 ········· 282

第 9 章　绽放 LangChain 的魅力：定制 LLM 输出 ········· 283

9.1　调整与对齐 ········· 283
　　9.1.1　对齐的方法 ········· 284
　　9.1.2　变革者：InstructGPT ········· 286
　　9.1.3　LLM 推理过程的调整方法 ········· 287
　　9.1.4　效果评估 ········· 288
9.2　实战案例：LangChain 微调 LLM ········· 288
9.3　提示词工程 ········· 295
　　9.3.1　提示词的结构 ········· 295
　　9.3.2　提示模板 ········· 297
　　9.3.3　高级提示词工程 ········· 298
9.4　总结 ········· 301

第 10 章　生产环境 LLM ········· 303

10.1　引言 ········· 303
10.2　LLM 应用评估 ········· 305
　　10.2.1　比较两个输出 ········· 306
　　10.2.2　基于标准的比较 ········· 307
　　10.2.3　字符串和语义比较 ········· 308
　　10.2.4　基准数据集 ········· 309
10.3　部署 LLM 应用 ········· 312
　　10.3.1　FastAPI ········· 314
　　10.3.2　Ray ········· 316
10.4　监测 LLM 应用 ········· 320
　　10.4.1　跟踪和追踪 ········· 321
　　10.4.2　可观测性工具 ········· 323
10.5　LangChain 回调 ········· 324
　　10.5.1　异步回调 ········· 325
　　10.5.2　自定义回调处理程序 ········· 327
　　10.5.3　记录到文件 ········· 328

10.5.4　多个回调处理程序 ……………………………………………… 330
　　　10.5.5　Token 计算 …………………………………………………… 333
　10.6　LangSmith ………………………………………………………… 334
　　　10.6.1　LangSmith 调试 ……………………………………………… 334
　　　10.6.2　LangSmith 样本收集 ………………………………………… 335
　　　10.6.3　LangSmith 测试评估 ………………………………………… 336
　　　10.6.4　LangSmith 人工评估 ………………………………………… 336
　　　10.6.5　LangSmith 监控 ……………………………………………… 336
　　　10.6.6　LangSmith 实战演示 ………………………………………… 337
　10.7　总结 ………………………………………………………………… 344

第 11 章　生成式 AI 的未来展望 …………………………………………… 346

　11.1　当前的生成式 AI …………………………………………………… 346
　11.2　未来的能力 ………………………………………………………… 349
　11.3　AI 的社会影响 ……………………………………………………… 351
　　　11.3.1　AI 和创意行业 ………………………………………………… 353
　　　11.3.2　AI 和社会经济 ………………………………………………… 354
　　　11.3.3　AI 和教育 ……………………………………………………… 354
　　　11.3.4　AI 和就业 ……………………………………………………… 355
　　　11.3.5　AI 和其他行业 ………………………………………………… 357
　　　11.3.6　AI 和网络安全 ………………………………………………… 357
　11.4　应用难题探索 ……………………………………………………… 358
　11.5　写在最后 …………………………………………………………… 359

参考文献 …………………………………………………………………… 361

第 1 章 什么是 AI 生成大模型

人工智能（Artificial Intelligence，AI）已经取得了重大进步，影响着企业、社会和个人。大约在过去的十年中，深度学习已经发展到可以处理和生成文本、图像、视频等非结构化数据。这些基于深度学习的先进人工智能模型已在各个行业中广受欢迎，其中包括大型语言模型（LLM）。目前媒体和各行各业对人工智能的吹捧程度很高。这是由多种因素驱动的，包括技术进步、备受瞩目的应用及跨多个部门产生变革性影响的潜力。在本章中将讨论生成模型，特别是 LLM，以及其在文本、图像、音频、视频等领域的应用。本书将介绍一些使它们发挥作用的技术背景，以及它们是如何训练的。本书将从简介开始，阐明目前在技术方面所处的阶段，以及未来的发展机遇。

1.1 AI 历史探源

在媒体上，有大量关于人工智能相关突破及其潜在影响的报道，主要包括从自然语言处理和计算机视觉的进步到 GPT-3 等复杂语言模型的开发。媒体经常强调人工智能的能力及其彻底改变医疗保健、金融、运输等行业的潜力。特别是生成模型因其生成文本、图像和其他创造性内容的能力而受到了很多关注，这些内容通常与人类生成的内容无法区分。这些相同的模型还提供了广泛的功能，包括语义搜索、内容操作和分类等。这可以通过自动化节省成本，并允许人类以前所未有的水平利用它们的创造力。

特别需要强调的是 OpenAI 通过公共用户界面提供的模型的进展，特别是版本之间的改进，如从 GTP-2 到 GPT-3 和从 GPT-3.5 到 GPT-4。这些模型直到最近才开始比普通的人类评分员表现得更好，但仍然没有达到人类的最高表现水平。人类工程学的这些成就令人印象深刻，但是，应该注意的是，这些模型的性能取决于现实测试的情况。大多数模型在 GSM8K 小学数学推理问题的基准测试中仍然表现不佳。

OpenAI 是一家总部位于美国的人工智能研究实验室，旨在促进和发展对人类友好的人工智能。该公司成立于 2015 年，得到了几位有影响力的人物和公司的支持，他们承诺向该合资企业提供超过 10 亿美元的资金。该组织最初致力于非营利，通过向公众开放其专利和研究来与其他机构和研究人员合作。2018年,埃隆·马斯克(Elon Musk)辞去了董事会职务,理由是与他在特斯拉的角色存在潜在的利益冲突。2019年,OpenAI转型为营利性组织,随后微软公司对OpenAI进行了大量投资,从而将OpenAI系统与微软公司基于Azure的超级计算平台集成到Bing搜索引擎中。该公司最重要的成就包括用于训练强化算法的OpenAI Gym,以及最近的GPT系列模型和DALL-E等。

GPT模型(如最近推出的OpenAI的Chat)是LLM领域AI进步的主要例子。ChatGPT通过更大规模的训练和比以前大得多的模型,大大提高了聊天机器人的性能。这些基于人工智能的聊天机器人可以生成类似人类的响应,作为对客户的实时反馈,并可应用于从软件开发和测试到诗歌和商业沟通等广泛用例。在行业内,人们对人工智能的能力及其对业务运营的潜在影响越来越感到兴奋。随着像OpenAI的GPT这样的人工智能模型的不断改进,它们可能会成为需要多样化知识和技能的团队不可或缺的资产。例如,GPT-4可以被认为是一个精通多方面知识的专家,而且它可以不知疲倦地工作而不要求工资(除了订阅或API费用),在数学、语言、统计、宏观经济学、生物学等科目上提供帮助,甚至可以辅导通过律师考试。随着这些人工智能模型变得更加成熟和易于访问,它们可能会在工作和学习方面发挥重要作用。通过使知识更容易获得和适应性,这些模式有可能创造公平的竞争环境,并为各行各业的人们创造新的机会。这些模型在需要更高水平推理和理解的领域显示出潜力,尽管进展因所涉及的任务的复杂性而异。至于带有图像的生成模型,可以期待具有更好功能的模型来帮助创建视觉内容,并可能改进计算机视觉任务,如对象检测、分割、添加字幕等。接下来引入一些人工智能方面的术语,以便向读者更详细地解释生成式模型、人工智能、深度学习和机器学习的含义。

1.1.1 生成式模型

在媒体中,在提到这些新模型时,人工智能一词一直被大量使用,但是有必要更清楚地区分术语生成模型与人工智能、深度学习、机器学习和语言模型的不同之处:

人工智能是计算机科学的一个广泛领域,涉及智能代理(Agent)的创建,智能Agent是可以推理、学习和自主行动的系统。

机器学习(Machine Learning,ML)是人工智能的一个子集,用于开发可以从数据中学习的算法。机器学习算法在一组数据上进行训练,然后它们可以使用该数据进行预测或决策。

深度学习(Deep Learning,DL)是机器学习的一个子集,它使用人工神经网络从数据中学习。神经网络受到人脑的启发,它们能够从数据中学习复杂的模式。

生成式模型(Generative Models,GM)是一种可以生成新数据的机器学习模型。生成模型在一组数据上进行训练,然后它们可以使用该数据创建类似于训练数据的新数据。

语言模型(Language Models)是预测序列中的标记(通常是单词)的统计模型,其中一些能够执行更复杂的任务的模型由许多参数(数十亿甚至数万亿的数量级)组成,因此它们被称为LLM。

生成模型与其他类型的机器学习模型之间的主要区别在于,生成模型不仅可以做出预测,还可以做出决策。它们实际上可以创建新数据。这使生成模型非常强大,可以用于各种任务,例如生成图像、文本、音乐和视频。人工智能、机器学习、深度学习、生成式模型和语言模型之间的差异见表 1-1。

表 1-1　人工智能、机器学习、深度学习、生成式模型和语言模型之间的差异

术　　语	定　　义
人工智能	一个广泛的计算机科学领域,涉及智能 Agent 的创建
机器学习	人工智能的一个子集,处理可以从数据中学习的算法的开发
深度学习	机器学习的一个子集,它使用人工神经网络从数据中学习
语言模型	一种模型,现在主要是深度学习模型,用于预测上下文中的 Token(令牌)
生成式模型	一种可以生成新数据的机器学习模型

生成式模型是一种功能强大的人工智能,可以生成类似于训练数据的新数据样本。生成式模型已经有了一段发展历史,可以使用数据中的模式从头开始生成新示例。这些模型可以处理不同类型的数据,并用于各个领域,包括文本生成、图像生成、音乐生成和视频生成。对于语言模型,重要的是要注意其中一些特性,特别是新一代语言模型是生成性的,从某种意义上讲,它们可以产生语言(文本),而另一些则不是。这些生成式模型有助于创建合成数据,以便在真实数据稀缺或受限时训练人工智能模型。这种类型的数据生成降低了标记成本,提高了训练效率。微软公司的研究人员采用这种方法[1]来训练他们的 phi-1 模型,他们使用 GPT-3.5 作为训练数据集创建了合成教科书和练习。在以下部分中,本书将介绍生成式模型的不同应用领域,例如文本、图像、音频、视频。应用程序主要围绕内容生成、编辑和处理(识别)。

1.1.2　生成式模型在文本领域

OpenAI 的 GPT-4 可以生成连贯且语法正确的诗歌,或不同语言的代码,并可以提取关键字和主题等特征。这些模型在内容创建和自然语言处理(Natural Language Processing,NLP)等领域具有实际应用,其最终目标是创建能够解释人类语言的算法。语言建模旨在根据序列中的前一个单词、字符甚至句子预测下一个单词、字符甚至句子。从这个意义上讲,语言建模是一种以机器可以理解的方式对语言的规则和结构进行编码的方法。大型语言模型从语法、句法和语义方面捕获人类语言的结构。这些模型很重要,因为它们构成了许多较大的 NLP 任务,例如内容创建、翻译、摘要、机器翻译和文本编辑任务,以及拼写更正的主干。语言建模的核心,以及更广泛的自然语言处理,在很大程度上依赖于表征学习的质量。训练良好的语言模型对有关其训练的文本的信息进行编码,并根据这些学习生成新文本,从而承担文本生成任务。最近,大型语言模型已经应用于论文生成、代码开发、翻译和理解基因序列等任务。语言模型的更广泛应用涉及多个领域。

（1）问答：人工智能聊天机器人和虚拟助手可以提供个性化和高效的帮助,减少客户支持的响应时间,从而增强客户体验。这些系统可用于解决餐厅预订和票务预订等特定问题。

（2）自动摘要：语言模型可以创建文章、研究论文和其他内容的简明摘要,使用户能够快速理解信息。

（3）情感分析：分析文本中的观点和情感,语言模型可以帮助企业更有效地理解客户反馈和意见。

（4）主题建模和语义搜索：这些模型可以识别及按主题分类,并将文档压缩为简洁的向量,使组织更容易管理和发现内容。

（5）机器翻译：人工智能驱动的语言模型可以将文本从一种语言翻译成另一种语言,支持企业的全球扩张工作。新的生成模型可以与商业产品（例如谷歌翻译）竞争。

尽管取得了显著的成就,但语言模型在处理复杂的数学或逻辑推理任务时仍然面临局限性。不断增加的语言模型规模是否将不可避免地提高新的推理能力,目前还不确定。如前所述,我们还必须考虑数据质量和规模的重要性,因为这些因素在不同任务和领域的语言模型性能提高方面发挥着重要作用。接下来将阐述生成模型的另外一个重要领域——图像生成。

1.1.3 生成式模型在图片领域

生成式模型被广泛地用于生成三维图像、头像、视频、图形、虚拟或增强现实插图、视频游戏图形设计、徽标创建、图像编辑或增强。Stable Diffusion 模型使用文本提示生成图像的过程如图 1-1 所示。

图 1-1　Stable Diffusion 从文本提示"玻璃制成的鸭子的透明雕塑"生成图像[2]

使用 Stable Diffusion 模型,只需对模型的初始设置,或者（在本例中）数值求解器和采样器进行更改,便可看到各种各样的结果。尽管它们有时会产生惊人的结果,但这种不稳定性和不一致性是更广泛地应用这些模型的重大挑战。Midjourney、DALL-E 2 和 Stable Diffusion 等服务提供源自文本输入或其他图像的创意和逼真的图像。DreamFusion、

Magic3D 和 Get3D 等服务使用户能够将文本描述转换为三维模型和场景，从而推动设计、游戏和虚拟体验的创新。这里面有 3 种主要应用。

（1）图像生成：模型可以生成图像，例如绘画、照片和草图。这可用于多种目的，例如创建艺术、设计产品和生成逼真的视觉效果。

（2）图像编辑：模型可以执行删除对象、更改颜色和添加效果等任务。这可用于提高图像质量，并使它们更具视觉吸引力。

（3）图像识别：大型基础模型可用于识别图像，包括分类场景，以及对象检测，例如检测人脸。

生成对抗网络（Generative Adversarial Network，GAN）和 DALL-E 等模型的区别在于，GAN 生成具有众多业务应用程序的逼真图像，而 DALL-E 从文本描述创建图像，这有助于创意产业设计广告、产品和时尚。图像编辑涉及通过面部属性编辑或图像变形等技术更改内容或样式属性来修改图像的语义。基于优化和学习的方法能够创造出具有独特风格的图像。这是通过使用预先训练好的 GAN 模型实现的，例如 StyleGAN。这些模型中的"潜在表示"能够帮助用户捕捉和模仿不同的图像风格。Stable Diffusion 模型最近已用于高级图像编辑任务，例如无缝连接手动设计的遮罩区域或通过文本引导生成三维对象操作。这些技术可实现灵活的图像生成，但面临有限的多样性问题，可以通过将其他文本输入合并到流程中来缓解这些问题。图像编辑类别还包括图像恢复等任务，这意味着从降级版本中恢复干净的图像，涉及图像超分辨率、修复、去噪、去雾和去模糊等任务。与传统方法相比，使用 CNN 和转换器架构的基于深度学习的方法由于具有优越的视觉质量而被广泛应用。像 GAN 和 Stable Diffusion 模型这样的生成式模型用于恢复，但可能会遭受复杂的训练过程和模式崩溃的影响。具有注意力模块或引导子网络的多失真数据集和单网络方法提高了处理多种退化类型的有效性。在接下来的章节里将看到模型可以对声音和音乐做什么。

1.1.4　生成式模型在声乐领域

生成式模型可以根据文本输入创作歌曲和对音频进行剪辑，识别视频中的对象并创建伴随音频，以及创建自定义音乐，其中可以再次将应用程序大致分为音乐生成、声音编辑和声音识别三类。

（1）音乐生成：生成模型可用于生成音乐，例如歌曲、节拍和旋律。这可用于多种目的，例如创作新音乐、创作配乐和生成个性化播放列表。

（2）声音编辑：生成模型可用于编辑声音，例如去除噪声、改变音高和添加效果。这可用于提高声音质量，并使其更具吸引力。

（3）声音识别：生成模型可用于声音识别，例如识别乐器、分类流派和检测语音。这可用于多种用途，例如音乐分析、搜索和推荐系统。

音乐生成算法始于 20 世纪 50 年代的算法创作，最近可以看到谷歌的 WaveNet 和 OpenAI 的点唱机等创新。这些模型催生了人工智能作曲家助手，它可以生成各种风格的音乐，并启用语音合成等较新的应用程序。作为一种特殊情况，语音转文本生成，也称为自动语音识别（ASR），是将口语转换为文本的过程。它们接受过声音和文本方面的训练。ASR 系统变得越来越精确，现在已用于各种应用，然而，仍有一些挑战需要解决，例如提高处理嘈杂环境和不同口音的能力。随着许多潜在的应用，如语音拨号、计算机辅助、个人协助、Alexa 和 Siri，ASR 背后的技术从马尔可夫模型演变为依赖 GPT。接下来探讨生成式模型在视频领域的应用。

1.1.5　生成式模型在视频领域

像 DeepMind 的 Motion to Video 和 NVIDIA 的 Vid2Vid 这样的视频生成模型依靠 GAN 实现高质量的视频合成。它们可以在不同域之间转换视频，修改现有视频，显示出视频编辑和媒体制作的巨大潜力。Make-a-Video 和 Imagen Video 等工具将自然语言提示转换为视频剪辑，从而简化视频制作和内容创建过程。广泛的应用类别如下。

（1）视频生成：生成模型可用于生成视频，例如短片、动画和商业广告。这可以用于创建新内容、广告产品和生成逼真的视觉效果。

（2）视频编辑：可以编辑视频，例如删除对象、更改颜色和添加效果。这有助于提高视频的质量，并使它们更具视觉吸引力。

（3）视频识别：模型可以识别视频，例如识别对象、对场景进行分类和检测人脸。这对于安全、搜索和推荐系统等应用程序很有帮助。

某些模型可以在多个域或模态中生成内容，这种模型称为多模态模型。

1.1.6　多模态模型

多模态模型可以生成文本、图像、声音和视频，能够提供更逼真和身临其境的体验。多模态模型仍处于发展的早期阶段，但它们有可能彻底改变我们与计算机交互的方式及体验世界的方式。例如，这些进步显著地提高了图像字幕任务（通过自然语言描述图像内容的过程）的性能。多模态模型采用生成式架构，将图像和标题融合到一个模型中，以实现共享学习空间。该过程涉及编码器-解码器架构：视觉编码和语言解码。可以区分这些潜在的用例。

（1）虚拟现实：这些模型可用于创建更逼真和身临其境的虚拟现实体验。这可以被广泛地应用于游戏、教育和培训等行业。

（2）增强现实：可以提高增强现实体验，将数字内容叠加在现实世界中。这对于导航、购物和娱乐非常有用。

1.2 节中将讨论 LLM 的技术背景。

1.2　LLM 简介

LLM 是经过深度训练的神经网络，擅长理解和生成人类语言。当前一代的 LLM（如 ChatGPT）是深度神经网络架构，它利用 Transformer 模型，对大量文本数据的无监督学习进行预训练，使其能够学习语言模式和结构。最新一代的 LLM 应用于对话聊天机器人（ChatBot）的显著优势在于它们能够产生连贯且上下文适当的文本，即使在开放式对话中也是如此。该模型能够产生流畅且连贯的文本，通常与人类生成的文本无法区分，然而，正如 OpenAI 的免责声明中所表达的那样，ChatGPT 有时会写出听起来合理但不正确或荒谬的答案。这被称为幻觉，只是 LLM 的问题之一。Transformer 是一种深度学习架构，由谷歌和多伦多大学的研究人员于 2017 年首次推出（在一篇名为"注意力是你所需要的"的文章中），它包括自我注意和前馈神经网络，使其能够有效地捕获句子中的单词关系。注意力机制使模型能够专注于输入序列的不同部分。生成式预训练转换器（GPT）由 OpenAI 的研究人员于 2018 年与他们的第 1 个同名 GPT 模型 GPT-1 一起推出，并作为"通过生成式预训练提高语言理解"发布。预训练过程包括预测文本序列中的下一个单词，增强模型对语言的掌握，以输出质量进行衡量。预训练后，可以针对特定的语言处理任务（如情绪分析、语言翻译或聊天）对模型进行微调。这种无监督和监督学习的结合使 GPT 模型能够在一系列 NLP 任务中表现得更好，并减少与训练 LLM 相关的挑战。LLM 培训语料库的规模一直在急剧增加。GPT-1 由 OpenAI 于 2018 年推出，在 BookCorpus 上进行了 9.85 亿字的训练。同年发布的 BERT 在书语料库和英语维基百科的组合语料库上进行了训练，总计 33 亿字。现在，LLM 的训练语料库达到了数万亿个 Token。

　　GPT-3 在 3000 亿个 Token 上进行训练，具有 1750 亿个参数，这对于深度学习模型来讲是前所未有的规模。GPT-4 是该系列中最新的，尽管由于竞争和安全问题，其规模和训练细节尚未公布，但多方猜测估计参数数量在 2000 亿到 5000 亿个。OpenAI 首席执行官 Sam Altman 表示，训练 GPT-4 的成本超过 1 亿美元。对话模型 ChatGPT 于 2022 年 11 月由 OpenAI 发布。基于以前的 GPT 模型（特别是 GPT-3）并针对对话进行了优化，它结合了人工生成的角色扮演对话和所需模型行为的人工标记器演示数据集。该模型在多回合对话中表现出色，例如广泛的知识保留和精确的上下文跟踪。GPT-4 在 2023 年 3 月取得了另一项重大进展，它超越了文本输入，包括多模式信号。GPT-4 在各种评估任务上具有卓越的性能，并且由于在训练期间进行了 6 个月的迭代对齐，因此可以更好地避免恶意或挑衅性查询的响应。除了 OpenAI 之外，其他值得注意的基础 GPT 模型包括谷歌公司的 PaLM2。尽管 GPT-4 在性能方面领先于大多数模型，但其他模型在某些任务中表现出比 GPT4 更加强大的性能，并为基于生成转换器的语言模型的进步做出了贡献。Meta 的 LLaMA 是在 1.4 万亿个 Token 上训练的，而谷歌聊天机器人 Bard 背后的模型 PaLM2 由 3400 亿个参数组成，这比以前的 LLM 小，初步预计具有至少 100 种语言的更大规模的训练数据。

　　有相当多的公司和组织正在开发 LLM，他们以不同的条件发布它们。OpenAI 已经开源了 GPT-2，后续模型是闭源的，但可以在他们的网站上或通过 API 开放使用。Meta 正在发布从 Roberta、BART 到 LLaMA 的模型，包括模型的参数（权重），以及用于设置和训练模型的源代码。Google AI 及其 DeepMind 部门开发了许多 LLM，包括 BERT、GPT-2、Lambda、Chinchilla、Gopher、PaLM 和 PaLM2。他们一直在开源许可下发布一些模型的代码和权重，但是最近在开发中已经朝着更加保密的方向发展。微软公司已经开发了许多

LLM，包括图灵 NLG 和威震天-图灵 NLG，但是，他们已经将 OpenAI 模型集成到 Microsoft 365 和 Bing 中。技术创新研究所（TII）是阿布扎比政府资助的研究机构，拥有开源的 Falcon LLM 用于研究和商业用途。

GPT 模型还可以使用文本以外的模式进行输入和输出，如 GPT-4 具有将图像输入与文本一起处理的能力。此外，它们还是扩散和并行解码等文本到图像技术的基础，能够为处理图像的系统开发视觉基础模型（Visual Foundation Models，VFM）。总之，GPT 模型发展迅速，能够创建适用于各种下游任务和模式的多功能基础 AI 模型，最终推动各种应用程序和行业的创新。1.2.1 节中将回顾深度学习和生成式模型近年来取得的进展。

1.2.1 潮流涌现

生成式模型在 2022 年成为公众关注的焦点，这可归因于几个相互关联的驱动因素。生成式模型的开发和成功依赖于改进的算法、计算能力和硬件设计的巨大进步、大型标记数据集的可用性及活跃和协作的研究社区，帮助发展一套工具和技术。更复杂的数学和计算方法的发展在生成式模型的发展中发挥了至关重要的作用。Geoffrey Hinton、David Rumelhart 和 Ronald Williams[3]在 20 世纪 80 年代引入的反向传播算法就是这样一个例子。它提供了一种有效训练多层神经网络的方法。2000 年，随着研究人员开发出更复杂的架构，神经网络开始重新流行起来。然而，深度学习是一种具有多层的神经网络，这标志着这些模型的性能和功能的重大转折点。有趣的是，尽管深度学习的概念已经存在了一段时间，但生成式模型的发展和扩展与硬件的重大进步相关，特别是图形处理单元（Graphical Processing Units，GPU），这些进步有助于推动该领域向前发展。如前所述，更便宜、更强大的硬件的可用性一直是开发更深层次模型的关键因素。这是因为深度学习模型需要强大的计算能力来进行训练和运行。这涉及处理能力、内存和磁盘空间等方面。不同介质（如硬件存储、固态存储、闪存和内存）随时间推移的计算机存储成本如图 1-2 所示[4]。

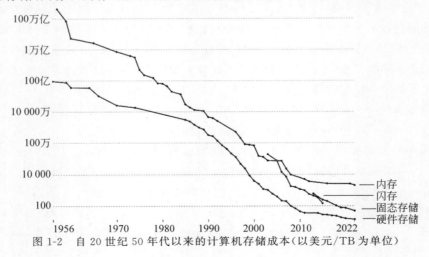

图 1-2　自 20 世纪 50 年代以来的计算机存储成本（以美元/TB 为单位）

虽然训练深度学习模型非常昂贵，但随着硬件成本的下降，在更大的数据集上训练更大的模型成为可能。模型大小是决定模型对训练数据集的近似程度（以困惑度为衡量）的因素之一。

模型具有的参数越多，其捕获单词和短语之间的关系作为知识的能力就越高。作为这些高阶相关性的简单示例，LLM 可以理解到如果单词 cat 前面有单词 chase，则它后面更有可能是单词 dog，即使中间还有其他单词。通常，模型的困惑度越低，它的性能就越好，例如在回答问题方面。特别是，似乎在由 20 亿到 70 亿个参数组成的模型中出现了新功能，例如生成不同创意文本格式的能力，如诗歌、代码、脚本、音乐作品、电子邮件、信件，并以信息丰富的方式回答问题，即使它们是开放式和具有挑战性的。

这种向大型模型发展的趋势始于 2009 年左右，当时英伟达公司催化了通常被称为深度学习的大爆炸。GPU 特别适合训练深度学习神经网络所需的矩阵/向量计算，因此将这些系统的速度和效率显著地提高了几个数量级，并将运行时间从几周缩短到几天。特别是，英伟达公司的 CUDA 平台允许直接对 GPU 进行编程，使研究人员和开发人员比以往任何时候都更容易试验和部署复杂的生成模型。2010 年，不同类型的生成模型开始受到关注。自动编码器是一种神经网络，可以学习将数据从输入层压缩到表示，然后重建输入，这是 2013 年首次提出的变分自动编码器（Variational Auto-Encoders，VAE）等更高级模型的基础。与传统的自动编码器不同，VAE 使用变分推理来学习数据的分布，也称为输入数据的潜在空间。大约在同一时间，Ian Goodfellow 等在 2014 年提出了 GAN。训练 GAN 的设置如图 1-3 所示（摘自《使用 GAN 生成文本的调查》，G. de Rosa 和 J. P. Papa，2022 年）[4]。

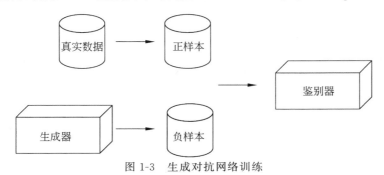

图 1-3　生成对抗网络训练

GAN 由两个网络组成，它们在类似游戏的环境中相互对抗生成新数据（通常是图像）的生成器和估计新数据真实概率的鉴别器。当它们相互竞争时，GAN 会更好地完成其任务，能够生成逼真的图像和其他类型的数据。在过去的十年中，深度学习中使用的基本算法取得了重大进展，例如更好的优化方法、更复杂的模型架构和改进的正则化技术。2017 年推出的 Transformer 模型建立在这一进展的基础上，并能够创建像 GPT-3 这样的大型模型。Transformer 依赖于注意力机制，并导致生成模型性能的进一步飞跃。这些模型，如谷歌公司的 BERT 和 OpenAI 公司的 GPT 系列，可以生成高度连贯和上下文相关的文本。迁移学习技术的发展也很重要，该技术允许在一项任务上预先训练的模型在另一项类似任务上进行微调。这些技术使训练大型生成模型变得更加高效和实用。

此外，生成模型的兴起部分可以归因于软件库和工具（TensorFlow、PyTorch、Keras）的开发，这些软件库和工具专门用于与这些人工神经网络一起工作，简化了构建、训练和部署它们的过程。为了进一步推动生成模型的发展，研究界定期举办像 ImageNet 这样的图像分类挑战，并开始对生成模型做同样的事情，例如生成对抗网络竞赛。除了更便宜和更强大的硬件的可用性外，标记数据的大型数据集的可用性也是生成模型开发的关键因素。这是因为深度学习模型和生成模型需要大量的文本数据才能进行有效训练。互联网上数据的爆炸式增长，特别是在过去十年中，为这种模型的蓬勃发展创造了合适的环境。随着互联网变得越来越流行，收集文本、图像和其他数据的大型数据集变得更加容易。这使在比过去大得多的数据集上训练生成模型成为可能。总之，生成建模领域是一个引人入胜且快速发展的领域。它有可能彻底改变我们与计算机进行交互的方式及我们创建新内容的方式。

1.2.2　GPT 的运作机制

Transformer 深度神经网络架构使 BERT 和 GPT 等模型成为可能，该架构改变了自然语言处理的游戏规则。旨在避免递归以允许并行计算，不同变体的 Transformer 架构正在继续推动自然语言处理和生成 AI 领域的发展。Transformer 的一个决定性特征是注意力机制。传统的序列间模型经常遇到处理长依赖性的问题。如果序列太长，则它们很难记住相关信息。Transformer 模型引入了注意力机制来解决这个问题。自我注意机制通常被称为 Transformer 模型的核心，它为序列中的每个单词分配一个分数，确定应该对该单词给予多少关注。Transformer 由可以堆叠的模块组成，从而可以创建学习大量数据集的非常大的模型。模型结构如图 1-4 所示[5]。

Transformer 模型具有以下几个重要组成部分。

（1）编码器-解码器结构：转换器模型遵循编码器-解码器结构。编码器获取输入序列并为每个单词计算一系列表示形式（上下文嵌入）。这些表示不仅要考虑单词的固有含义（它们的语义值），还要考虑它们在序列中的上下文，然后解码器使用此编码信息，使用先前生成的项目的上下文，一次生成一个项目的输出序列。

（2）位置编码：由于 Transformer 不按顺序处理单词，而是同时处理所有单词，因此它缺乏单词顺序的任何概念。为了解决这个问题，使用位置编码将有关单词在序列中的位置信息注入模型中。这些编码被添加到表示每个单词的输入嵌入中，从而允许模型考虑序列中单词的顺序。

（3）规范化层：为了稳定网络的学习，Transformer 使用了一种称为层规范化的技术。此技术跨特征维度（而不是批处理归一化中的批处理维度）规范化模型的输入，从而提高学习的整体速度和稳定性。

（4）多头注意力：Transformer 不是一次性应用注意力，而是并行多次应用它，提高了模型专注于不同类型信息的能力，从而可以捕获更丰富的特征组合。

另一个不是特定于 Transformer 的可选架构功能是跳过连接（也称为残差连接）。为了

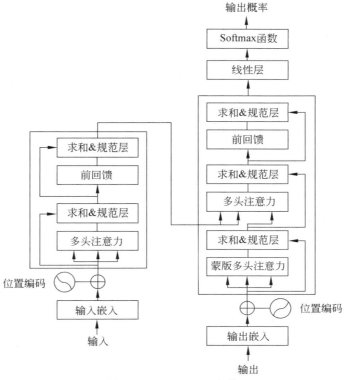

图 1-4 Transformer 架构

缓解随着网络深度的加深而出现的退化问题,所以使用跳过连接。这允许梯度通过将快捷方式输入更深的层来在层间保持不变的流动。Transformer 在 NLP 领域突破了极限,尤其是在翻译和语言理解方面。神经机器翻译(Neural Machine Translation,NMT)是一种主流的机器翻译方法,它使用深度学习来捕获句子中的长期依赖关系。基于 Transformer 的 NMT 优于以前的方法,例如使用递归神经网络,特别是长短期记忆(Long Short Term Memory,LSTM)网络。这可以归因于这种强大的架构,包括首要的关注,它允许 Transformer 模型以灵活的方式处理词序,无论它们相距多远,对于每种特定情况都是最佳的。此外,这些架构特征的结合使其能够成功地处理涉及理解和生成人类语言及其他领域的任务。OpenAI 强大的语言生成 GPT 模型也是一个转换器,DeepMind 的 AlphaFold 2 也是如此,这是一个从基因序列预测蛋白质结构的模型。与其他模型(例如递归神经网络)相比,Transformer 能够在较长的序列上更好地保持性能。这为它们的成功做出了贡献,然而,变压器架构意味着它们只能捕获固定输入宽度内的依赖关系。早期的注意力机制与数据点的数量呈二次缩放,使它们不适用于具有大量输入的设置。已经提出了许多方法来提高效率,例如稀疏、低秩的自我关注和潜在的瓶颈,此处仅举几例,其他工作试图将序列扩展到固定输入大小之外,诸如 Transformer-XL 之类的架构通过存储已编码句子的隐藏状态来重新引入递归，以便在下一个句子的后续编码中利用它们。GPT 模型之所以特殊是因为它的预训练方式。接下来将探讨 LLM 是如何预训练的。

1.2.3 模型的预训练

训练 LLM 的第 1 步是标记化。这个过程涉及将单词转换为数字，以便模型可以处理它们，因为 LLM 是需要数字输入和输出的数学函数。为了执行这种标记化，LLM 使用唯一的标记器。分词器将文本中的单词映射到相应的整数列表。在训练 LLM 之前，分词器通常会被拟合到整个训练数据集，然后冻结。采用的常见分词器类型是字节对编码。务必注意，分词器不会生成任意整数。相反，它们会输出特定范围内的整数。现在，考虑输出，当 LLM 接收到文本时，它主要产生一个向量并落在输出向量，然后此输出向量通过 Softmax 函数以产生另一个向量，该向量称为概率向量。由于它的条目是非负的，并且总和为 1，这个向量可以解释为 LLM 词汇表上的概率分布。此外，有必要指出，LLM 只能基于不超过其上下文窗口的 Token 序列生成令牌。此上下文窗口是指 LLM 可以使用的最长令牌序列的长度。如果显示的序列长于此窗口，则 LLM 将需要截断序列或采用算法来处理它。LLM 的典型上下文窗口大小范围约为 1000～10 000 个 Token。LLM 的训练包括一个特定的过程：首先标记输入数据，然后将这些数据输入模型中，最后生成一个输出，这个输出基于模型词汇表上的概率分布。此过程中的特定机制，例如 Softmax 函数和上下文窗口，有助于促进 LLM 对输入数据的理解和响应。负对数似然（Negative Log Likelihood，NLL）和困惑度（Perplexity，PPL）是在训练和评估语言模型的过程中使用的重要指标。NLL 是机器学习算法中使用的损失函数，旨在最大化正确预测的概率。较低的 NLL 表示网络已成功地从训练集中学习到了预测模式，因此它将能够准确预测训练样本的标签。值得一提的是，NLL 是一个受正区间约束的值。另外，PPL 是 NLL 的幂，提供了一种更直观的方法来理解模型的性能。较小的 PPL 值表示训练有素的网络可以准确预测，而较高的值表示学习表现较差。直观地说，低困惑度意味着模型对下一个单词的惊讶程度较低，因此，预训练的目标是最大程度地减少困惑，这意味着模型的预测与实际结果更加一致。在比较不同的语言模型时，困惑度通常被用作各种任务的基准指标。它给出了语言模型性能如何的想法，其中较低的困惑度表示模型对其预测的确定性更高，因此，与其他困惑度较高的模型相比，具有较低困惑度的模型将被认为表现更好。

1.2.4 可扩展架构

本节简要讨论一下架构的选择，以及为什么这些模型如此之大。2020 年在 OpenAI 研究人员发表的一篇论文中，卡普兰和其他人讨论了缩放定律和参数选择。有趣的是，他们比较了许多不同的架构选择，结果表明 Transformer 模型在困惑度方面优于 LSTM 作为语言模型，这在很大程度上是由于改进了对字数较多的上下文的使用。虽然这些循环网络在不到 100 个 Token 后停滞不前，但 Transformer 模型在整个上下文中得到了改进，因此，Transformer 模型不仅具有相对于 LSTM 更好的训练和推理速度，而且在查看相关上下文时也具有更好的性能。此外，他们发现了数据集大小、模型大小（参数数量）和用于训练的计算量的幂律关系，从某种意义上讲，为了通过某个因素提高性能，其中一个因素必须作为因子的幂放大。但是，为了获得最佳性能，所有3个因素必须串联缩放，以免产生瓶颈效应。DeepMind的研究人员[6]分析了LLM的训练计算和数据集大小，并得出结论，正如缩放定律所建议的那样，LLM在计算预算和数据集大小方面训练不足。他们预测，如果大型模型越小，训练时间越长，则大型模型的表现会更好，事实上，他们验证了他们的预测，将基准上的700亿参数龙猫模型与他们的Gopher模型进行比较，Gopher模型由2800亿个参数组成。最近，该团队发现，在epoch方面进行更长时间的训练或以petaflops计算的更多计算似乎不再能提高性能，而较小的网络和更高质量的数据集可以提供非常有竞争力的性能。

1.2.5 模型的适应性

可适应性LLM是指使模型适应特定任务。不同的调节方法包括微调、提示、指令调整和强化学习。

微调涉及通过监督学习在特定任务上训练预先训练的语言模型来修改它。例如，为了使模型更适合与人类聊天，该模型根据表述为自然语言指令（指令调整）的任务示例进行训练。

提示技术将问题呈现为文本提示，并期望模型完成。

对于微调，强化学习通常将监督微调与使用人类反馈的强化学习相结合，以根据人类偏好训练模型。LLM可以在一组训练示例上进行训练，这些示例集本身由LLM生成（从一小组初始人工生成的示例集引导），例如在微软研究所的phi-1训练集中[7]。通过提示技术，将介绍类似问题的文本示例及其解决方案。零镜头提示不涉及已解决的示例，而少数镜头提示包括少量相似（问题、解决方案）对的示例。这些调节方法不断发展，在广泛的应用中变得更加有效和有用。

1.2.6 上手GPT

用户可以通过OpenAI网站或他们的API访问OpenAI的GPT家族模型。如果想在笔记本电脑上尝试其他LLM，则开源LLM是一个不错的起点。可以通过HuggingFace或其他提供商访问这些模型。在HuggingFace上甚至可以下载它们，微调它们，或者从头开始训练一个模型。接下来将介绍Stable Diffusion模型和它的工作原理。

1.3 图像模型

图像生成模型是一种可用于生成图像的生成模型。图像生成模型是一种强大的工具，可用于生成逼真和创造性的图像。它们仍处于开发的早期阶段，但有可能彻底改变创建和消费图像的方式。最流行的图像生成模型之一是Stable Diffusion；另一个是中途。简单来讲，这些是一种深度学习模型，可以在给定文本提示的情况下创建图像。谷歌大脑宣布在

2022年创建了两个文本到图像模型Imagen和Parti。

1.3.1 图像模型工作原理

  Stable Diffusion模型是由慕尼黑路德维希马克西米利安大学和跑道分校CompVis小组的研究人员开发的深度学习文本到图像模型。它以文本描述为条件生成详细的图像，并利用潜在的扩散模型架构。该模型的源代码和权重都在CreativeML OpenRAIL-M许可证下公开发布，该许可证不对重用、分发、商业化、改编施加任何限制。该型号可以在配备适度GPU的消费类硬件上运行（例如GeForce 40系列）。Stable Diffusion是一种扩散模型，它使用Gumbel分布向图像添加噪声。Gumbel分布是一种连续概率分布，通常用于机器学习，因为它易于采样并且具有更稳定的特性。稳定性意味着模型不太可能卡在局部最小值中，而其他类型的扩散模型可能会发生这种情况。该模型由VAE、U-Net和文本编码器组成。VAE由编码器和解码器两部分组成，首先将原始高维图像压缩到低维潜在空间中，然后将其重建回图像空间。潜在空间显著地降低了计算复杂度，使扩散过程更快。

  VAE编码器将图像压缩到潜在空间中，而U-Net则执行前向扩散进行去噪以获得潜在表示，然后VAE解码器生成最终图像。该模型可以灵活地以各种模式（包括文本）为条件，并利用交叉注意力机制来合并条件信息。U-Net是一种流行的卷积神经网络（Convolutional Neural Network，CNN），具有对称编码器-解码器结构。它通常用于图像分割任务，但在Stable Diffusion的上下文中，它用于预测图像中的噪声。U-Net将噪声图像作为输入，并通过一系列卷积层对其进行处理，以提取特征并学习表示。这些卷积层通常以收缩路径组织，在增加通道数量的同时减少了空间维度。一旦收缩路径到达U-Net的瓶颈，它就会通过对称的扩展路径进行扩展。在扩展路径中，应用转置卷积（也称为上采样或反卷积）以逐步上采样空间维度，同时减少通道数。在扩散过程中，U-Net的扩展路径采用噪声图像，并从前向扩散重建潜在表示。通过将重建的潜在表示与真实的潜在表示进行比较，U-Net便可预测原始图像中噪声的估计。这种噪声预测有助于反向扩散过程恢复原始图像。扩散模型通过类似于物理学中的扩散过程运行。它遵循前向扩散过程，向图像添加噪声，直到它变得不典型和嘈杂。这个过程类似于墨滴落入一杯水中并逐渐扩散。这里的独特之处在于反向扩散过程，其中模型试图从嘈杂、无意义的图像中恢复原始图像。该结果是通过逐步从噪声图像中减去估计噪声实现的，最终恢复类似于原始图像的图像。降噪过程如图1-5所示。

  在图1-5中可以看到图像的生成步骤，以及使用DDIM采样方法的U-Net去噪过程，该方法反复去除高斯噪声，然后将去噪输出解码为像素空间。Stable Diffusion是一种深度学习模型，它利用扩散过程通过几个明确的步骤从文本提示生成图像。

  首先在潜在空间中产生一个随机张量（随机图像），作为初始图像的噪声。噪声预测器（U-Net）接收潜在噪声图像和提供的文本提示并预测噪声，然后模型从潜在图像中减去潜在噪声。

  步骤2和步骤3重复一定数量的采样步骤，例如，图1-5所示的40次。最后，VAE的

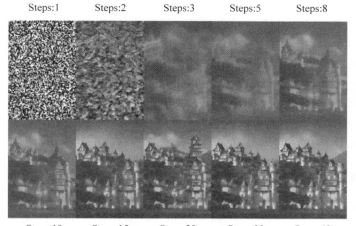

图 1-5 使用 Stable Diffusion v1～5 AI 扩散模型创建的
日本欧式城堡。仅显示 40 个步骤的生成过程

解码器组件将潜在图像转换回像素空间，提供最终的输出图像。在图像生成模型的训练过程中，使用损失函数来评估生成的图像的质量。一种常用的损失函数是均方误差（Mean Square Error，MSE）损失，它量化了生成的图像和目标图像之间的差异。该模型经过优化以最大程度地减少这种损失，鼓励其生成与所需输出非常相似的图像。该模型在一个名为 LAION-5B 的数据集上进行训练，该数据集源自 Common Crawl 数据，包含数十亿个图像文本对。根据语言、分辨率、水印可能性和美学分数对训练数据集进行分类。Stable Diffusion 是在这个数据集的子集上训练的。该模型的训练数据有多种来源，其中很大一部分来自 Pinterest、WordPress、Blogspot、Flickr、DeviantArt 等网站。总体而言，Stable Diffusion 和 Midjourney 等图像生成模型将文本提示处理成生成的图像，利用正向和反向扩散过程的概念，并在较低维的潜在空间中运行以提高效率。到目前为止，Stable Diffusion 有两个主要模型版本，版本 1 和版本 2。让我们看一看它们有何不同。

1.3.2 图像模型的不同版本

Stable Diffusion v1 和 v2 在文本处理、训练数据及其结果方面有所不同。在文本处理方面，Stable Diffusion v2 使用 OpenClip 进行文本嵌入，而 v1 使用 Open AI 的 CLIP ViT-L/14 进行文本嵌入。OpenClip 比 CLIP 大 5 倍，这提高了图像质量，也为研究人员在研究和优化模型方面提供了更多的透明度。关于训练数据，Stable Diffusion v1.4 使用 3 个不同的数据集进行训练，而 Stable Diffusion v2 在 LAION-5B 的一个子集上进行训练，过滤出露骨的色情材料（NSFW 过滤器）和高于阈值的美学分数。LAION-5B 数据集是一个由 58.5 亿个 CLIP 过滤图像文本对组成的大规模数据集。数据集中超过 23 亿个样本包含英语，而 22 亿个样本来自 100 多种其他语言，其余 10 亿个样本不允许特定语言分配，例如名称。数据集的采集管道非常复杂，需要大量处理。它包括 PB 级通用爬网数据集的分布式处理、图像的分布式下载及少量 GPU 节点对数据进行后处理,从而生成最终数据集。过滤并删除了重复样本,将数据集从 500 亿个候选者修剪到略低于 60 亿个 CLIP 过滤的图像文本对。就结果而言,Stable Diffusion v2 更难用于控制风格和生成名人照片。这种差异可能是由于训练数据的差异,因为 OpenAI 的专有数据可能有更多的艺术品和名人照片,这些照片不包括在 Stable Diffusion v2 训练数据中。总之,Stable Diffusion v2 使用不同的文本嵌入模型,并在不同的数据子集上进行训练,与 Stable Diffusion v1 相比,结果不同。虽然 Stable Diffusion v2 可能更透明,更适合长期开发,但因为它的训练数据的关系,Stable Diffusion v1 对于特定用例(例如控制样式或生成名人照片)可能会有更好的结果。

1.3.3　图形模型的调节性

调节过程允许这些模型受到输入文本提示或其他输入类型(如深度图或轮廓)的影响,以提高精度,以便创建相关图像。在调节期间,提示被标记化,每个标记被转换为嵌入,一定长度的向量,有时是 768 个值。这些嵌入解释了单词之间的语义关系,然后由文本转换器处理并馈送到噪声预测器,引导它生成与文本提示一致的图像。在文本到生成图像的过程中,模型使用文本提示来生成全新的图像。文本提示被编码到潜在空间中,扩散过程逐渐添加噪声(由去噪强度控制),以使初始图像向输出图像演化。

1.4　总结

像 LLM 这样的生成模型因其彻底改变了众多行业并且具有强大的潜力,因而受到相当大的关注。特别是它们在文本生成和图像合成中的应用获得了媒体的大量炒作。OpenAI 等领先公司正在突破 LLM 的界限,其 GPT 模型系列因其令人印象深刻的语言生成功能而受到广泛关注。本章讨论了媒体对最新突破的关注、深度学习和人工智能的近期历史、生成式模型、LLM 及支撑它们的理论思想,尤其是 Transformer 架构。本章还讨论了图像的 Stable Diffusion 模型,以及文本、图像、声音和视频的应用。第 2 章将探讨使用 LangChain 框架在 LLM 优化和增强 LLM 中的基础知识、实现和使用。

第 2 章 初识 LangChain

本章讨论 LLM 的局限性,以及 LLM 与工具相结合如何克服这些挑战,从而构建基于语言的创新应用程序。有一些强大的框架通过为快速工程、链接、数据检索等提供强大的工具来增强开发人员的能力。无论您是开发人员、数据科学家,还是只对 NLP 或生成式模型的技术进步感到好奇,您都应该了解这些框架中最强大和最受欢迎的 LangChain。LangChain 解决了与 LLM 相关的痛点,并提供了一个直观的框架来创建定制的 NLP 解决方案。在 LangChain 中,LLM、互联网搜索和数据库查找等组件可以链接在一起,这是指根据数据或任务的要求,按顺序执行不同的任务。通过利用其功能,开发人员可以构建动态和数据感知应用程序。本章将包括一些用例来说明该框架如何帮助不同领域的企业和组织。LangChain 对 Agent 和记忆的支持使构建各种应用程序成为可能,这些应用程序比那些通过 API 调用语言模型构建的应用程序更强大、更灵活。本章将讨论与框架相关的重要概念,例如 Agent、链(Chain)、行动计划生成和记忆(Memory)。所有这些概念对于理解 LangChain 的工作原理都很重要。

2.1 LLM 的局限性

LLM 因其能够生成类似人类的文本和理解自然语言而获得了极大的关注和普及,这使它们在围绕内容生成、文本分类和摘要的场景中非常有用。虽然 LLM 提供了令人印象深刻的功能,但它们存在局限性,可能会阻碍它们在某些情况下的有效性。在开发应用程序时,了解这些限制至关重要。目前与 LLM 相关的一些痛点主要包括以下几项。

(1)过时的知识:LLM 无法提供实时或最近的数据,因为它们完全依赖于提供给它们的训练数据。

(2)无法执行外部动作:LLM 无法执行操作或与外部系统交互,从而限制了其功能。例如,LLM 无法启动 Web 搜索、实时查询数据库或使用计算器进行运算。

(3)缺乏上下文和附加信息:LLM 可能难以理解和整合先前提示或对话中的上下文。它们可能不记得前面提到的细节,或者在给定的提示之外没有提供其他相关信息。

（4）复杂性和学习曲线：使用大型语言模型开发应用程序通常需要对 AI 概念、复杂算法和 API 有深入的了解。对于可能不具备这些领域专业知识的开发人员来讲，这可能会带来挑战。

（5）幻觉：LLM 的权重中隐含着很多关于世界的一般知识，然而，这导致 LLM 可能对某些主题的理解不足，并产生与事实不符或不连贯的回答。例如，LLM 可能会生成不存在的信息或提供不准确的详细信息。

（6）偏见和歧视：根据接受训练的数据，LLM 可能会表现出偏见，这些偏见可能是宗教、意识形态、政治和其他性质的。LLM 没有关于时事的信息，因为它们与外界没有联系，它们不会知道它们没有接受过训练的任何事情，例如截止日期之后的任何事情，即生成训练数据的时间。不仅如此，它们还在训练数据限制之外的上下文理解方面出现了很大的问题。例如，由于模型无法直接执行操作或与外部系统交互，因此它们不会知道天气情况，也无法访问文档。ChatGPT 聊天界面中说明了这个截止日期的问题，例如询问 ChatGPT 什么是 LangChain，结果如图 2-1 所示。

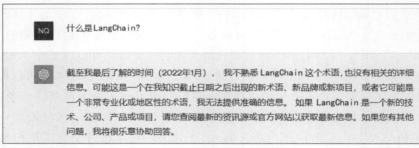

图 2-1　ChatGPT 缺乏最新信息

在这种情况下，模型能够正确地捕获问题并提供正确的反馈，但是，如果在 GPT-3 游乐场中提出同样的问题，则会得到这样的回答，如图 2-2 所示。

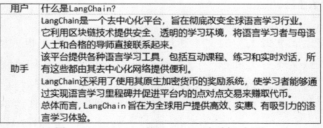

图 2-2　GPT-3.5 的 OpenAI 游乐场-幻觉

在这个例子中可以看到该模型虚构了该术语，并发明了一个去中心化的平台。这是一种幻觉。注意这些问题很重要。此类问题可以通过访问外部数据（例如天气 API、用户首选项或来自 Web 的相关信息）来解决，这对于创建个性化且准确的语言驱动应用程序至关重要。LLM 擅长生成文本，但缺乏真正的理解和推理能力，但是，它们可能会在逻辑推理方面遇到困难。例如，即使是高级 LLM 在小学水平的数学中表现仍然不佳，也无法执行它们以前从未见过的简单数学运算。在这里可以通过一个简单的演示来说明这一点，如图 2-3 所示。

图 2-3　ChatGPT 解决数学问题

因此，该模型对第 1 个问题给出了正确的答案，但对第 2 个问题却给出了错误的答案。使用计算机得出 2555×2555 的答案是 6 528 025，而不是模型给出的答案 6 527 025。

LLM 没有将计算结果存储在训练数据中，在训练数据中没有频繁地遇到它，因此没有记住答案，就像在其权重中编码一样，从而导致它无法正确地提出解决方案。在这种情况下，基于 Transformer 的 LLM 不适合这项工作。在客户服务、教育和营销等领域部署应用之前，可能需要监控和纠正 LLM 的输出，以确保准确性及避免偏见和不恰当的语言。聊天机器人中存在严重偏见的例子是微软公司的 Tay 聊天机器人，它因为种族诽谤和其他仇外言论而为微软公司带来了一场公关灾难。对于所有这些问题，LLM 需要与外部数据源、内存记忆和功能集成，以便与其环境进行动态交互，并根据提供的数据做出适当的响应。然而，将 LLM 与不同的数据源和计算连接起来可能很棘手，需要开发和仔细测试特定的定制工具，因此，使用生成式模型构建数据响应式应用程序可能很复杂，并且可能需要大量的编码和数据处理。最后，直接使用 LLM 模型可能具有挑战性且耗时。这从提示词工程开始，但延伸到更远的其他地方。这里面固有的挑战在于驾驭这些复杂的模型，提供有效的提示，并解析它们的输出。

2.2　LLM 应用介绍

为了解决上述挑战问题和局限性问题，LLM 可以与对其他程序或服务的调用相结合。主要思想是，通过工具将它们连接在一起，以此来增强 LLM 的能力。LLM 驱动的应用程序将 LLM 与其他工具结合到使用专用工具的应用程序中，有可能改变我们的数字世界。这通常是通过对 LLM 的一个或多个提示调用链来完成的，但也可以利用其他外部服务（例如 API 或数据源）实现特定任务。

LLM 应用程序是一种使用 ChatGPT 等 LLM 来协助完成各种任务的应用程序。它通过向语言模型发送提示来生成响应并运行，还可以与其他外部服务（如 API 或数据源）集成以实现特定目标。

为了说明LLM应用程序的外观,这里有一个非常简单的LLM应用程序,它包括一个提示和一个LLM,如图2-4所示。

图2-4　一个简单的LLM应用程序,将提示与LLM相结合

LLM应用程序对人类具有巨大的潜力,因为它们增强了生成式模型的能力、简化了流程并在各个领域提供了宝贵的帮助。以下是LLM应用程序很重要的一些关键原因。

(1) 效率和生产力:LLM应用程序可自动执行任务,从而更快、更准确地完成重复或复杂的操作。它们可以进行数据处理、分析、模式识别和决策,其速度和准确性超过了人类的能力。这提高了数据分析、客户服务、内容生成等领域的效率和生产力。

(2) 任务简化:LLM应用程序通过将复杂的任务分解为可管理的步骤或提供直观的界面供用户交互来简化复杂的任务。这些工具可以自动执行复杂的工作流程,使更广泛的用户无须专业知识即可访问它们。

(3) 增强的决策:LLM应用程序提供高级分析功能,可实现数据驱动的决策。它们可以快速分析大量信息,识别仅对人类来讲可能不明显的趋势或模式,并为战略规划或解决问题提供有价值的见解。

(4) 个性化:人工智能驱动的推荐系统根据个人偏好和行为模式提升用户体验。这些应用程序会考虑用户数据,以便在电子商务、娱乐和在线平台等各个领域提供量身定制的建议、推荐和个性化内容。

一个特定的增长领域是将公司数据(尤其是客户数据)与LLM一起使用,但是在使用过程中必须小心并考虑对隐私和数据保护的影响。绝不应将个人身份信息数据馈送到公共API端点。对于这些用例,在内部基础设施或私有云上部署模型至关重要,而微调甚至训练专用模型可以对此进行改进。本书在后面的章节将会涉及LLM应用程序在生产中的实际应用并将更加深入地进行探讨。下面比较一些可以帮助构建LLM应用程序的框架。

LLM应用程序框架的开发旨在提供专门的工具,可以有效地利用LLM的力量来解决复杂问题。目前已经出现了一些库,可以满足将生成式AI模型与其他工具有效结合的要求,并以此构建LLM应用程序。另外市面上也有几个开源框架能够用于构建动态LLM应用程序。它们都为开发尖端的LLM应用程序提供了价值。各个框架随时间推移的受欢迎程度如图2-5所示。

从图2-5中可以看出,Haystack是这些框架中最古老的,于2020年初启动(根据GitHub提交)。就GitHub上的星星而言,它也是最不受欢迎的。LangChain、LlamaIndex(以前称为GPTIndex)和SuperAGI于2022年底或2023年初启动,它们在很短的时间内很快流行起来,其中LangChain的增长最令人印象深刻。LlamaIndex专注于高级检索,而不

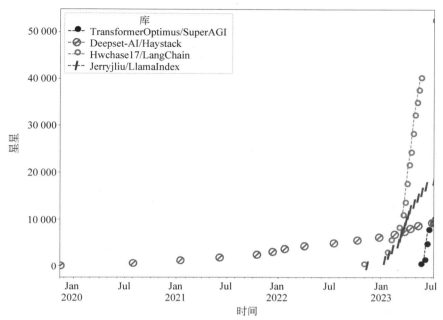

图 2-5　Python 中不同框架之间的受欢迎度比较，图中展示了每个项目的星星数量随时间的变化

是 LLM 应用程序的更广泛方面。同样，Haystack 专注于创建大规模搜索系统，其组件专门用于使用检索器、阅读器和其他数据处理程序进行可扩展的信息检索，并通过预先训练的模型进行语义索引。LangChain 擅长使用 Agent 将 LLM 链接在一起，以便将操作委托给模型。它的用例强调快速优化和上下文感知信息检索/生成，但凭借其高度模块化的 Pythonic 界面和庞大的工具集合，它是实现复杂业务逻辑的头号工具。SuperAGI 具有与 LangChain 相似的功能。它甚至带有一个市场、一个工具和 Agent 的存储库，然而，它不像 LangChain 那样得到很好支持，没有包括 AutoGPT（例如 AutoLlama 这样的类似工具）。这是一个分解任务的递归应用程序，因为它的推理能力基于人类和 LLM 的反馈，与 LangChain 相比非常有限。因此，它经常陷入逻辑循环，并经常重复步骤。另外此处还省略了一些专注于提示工程的库，例如 Promptify。还有其他 LLM 应用程序框架，例如 Rust、JavaScript、Ruby 和 Java。例如，用 Rust 编写的 Dust 专注于 LLM 应用程序的设计及其部署。接下来本书将深入分析 LangChain。

2.3　LangChain 介绍

LangChain 是一个用于开发由语言模型驱动的应用程序的框架，使用户能够更有效地使用 LLM 构建应用程序。它提供了一个标准接口，用于将语言模型连接到其他数据源，以及构建可以与其环境交互的 Agent。LangChain 被设计为模块化和可扩展的库，开发者可以轻松地构建适应各种领域的复杂应用程序。LangChain 是开源的，主要用 Python 编写，但是存在用 JavaScript 或更准确地说是 TypeScript（LangChain.js）实现的配套项目，以及 Ruby 的 LangChain.rb 项目，它带有用于代码执行的 Ruby 解释器。本书将重点介绍该框架的 Python 版本。

LangChain 是一个开源框架，允许 AI 开发人员将 ChatGPT 等 LLM 与其他计算和信息资源相结合。Harrison Chase 于 2022 年 10 月作为 GitHub 上的一个开源项目开始，它根据 MIT 许可证获得许可，这是一种通用许可证，允许商业使用、修改、分发和私人使用，但限制了责任和保证。LangChain 仍然很新，但是，它已经具有 100 多个集成工具。在 Discord 聊天服务器上有积极的讨论，有博客，并且在旧金山和伦敦都有定期的聚会。甚至还有一个聊天机器人 ChatLangChain，可以回答有关使用 LangChain 和 FastAPI 构建的 LangChain 文档的问题，该文档可通过文档网站在线获得。该项目吸引了来自红杉资本和 Benchmark 等公司的数百万美元风险投资资金，这些公司为苹果、思科、谷歌、WeWork、Dropbox 和许多其他成功的公司提供了资金。LangChain 带有许多扩展和围绕它开发的更大的生态系统。如前所述，它已经有大量的集成工具，每周都有许多新的集成工具。

例如，LangChainHub 是一个工件存储库，可用于处理 LangChain，例如提示、链和 Agent，它们组合在一起形成复杂的 LLM 应用程序。从 HuggingFace Hub（一个模型集合）中汲取灵感，它的目标是成为共享和发现高质量 LangChain 原语与应用程序的中心资源。目前 LangChainHub 仅包含一组提示，但未来会随着社区添加到此集合中，开发者有望很快就能找到链和 Agent。此外，LlamaHub 库扩展了 LangChain 和 LlamaIndex 更多的数据加载器和阅读器，例如用于 Google Docs、SQL 数据库、PowerPoints、Notion、Slack 和 Obsidian。LangFlow 是一个 UI，它允许通过将侧边栏组件拖到画布上并将它们连接在一起以创建管道，从而在可执行流程图中链接 LangChain 组件。这是试验和原型化管道的快速方法。

在 LangFlow 的浏览器界面的侧边栏中，开发者可以看到所有不同的 LangChain 组件，例如零样本提示、数据加载器和语言模型包装器。这些流可以直接导出并加载到 LangChain 中，也可以通过对本地服务器的 API 调用来进行使用。LangChain 和 LangFlow 可以部署在本地，例如使用 Chainlit 库，也可以部署在包括 Google Cloud 在内的不同平台上。LangChain-serve 库有助于通过单个命令将 LangChain 和 LangFlow 作为 LLM 应用（服务）部署在 Jina AI 云上。LangChain 提供了一个直观的框架，使开发人员、数据科学家，甚至 NLP 技术的新手都可以更轻松地使用 LLM 创建应用程序。需要注意的是，LangChain 既不是一个模型，也不是一个提供者，而本质上是一个促进与不同模型无缝交互的框架。开发者使用 LangChain 的时候无须成为 AI 或复杂算法方面的专家，从而大大地简化了流程并缩短了学习曲线。

通过数据感知和 Agent，LangChain 允许与各种数据源轻松集成，包括 Google Drive、Notion、Wikipedia、Apify Actors 等。这种数据感知使应用程序能够根据用户偏好或来自外部来源的实时信息生成个性化且与上下文相关的内容。接下来探讨为什么 LangChain 很重要，以及它的用途。

2.3.1　LangChain 的必要性

LangChain 解决了本书在 LLM 局限性章节中提出的问题。简而言之，它简化了应用程序的开发过程，提供了一种构建应用程序的方法，这些应用程序比通过 API 简单地调用语言模型所构建的应用程序更强大、更灵活。特别是 LangChain 对 Agent 和记忆的支持允许开发人员构建能够以更复杂的方式与环境交互的应用程序，并且可以随着时间的推移存储和重用信息。LangChain 可用于提高各种领域中应用程序的性能和可靠性。在医疗保健领域，它可用于构建聊天机器人，可以回答患者的问题并提供医疗建议，其中必须非常谨慎地遵守有关信息可靠性、机密性的监管和道德约束。在金融领域，该框架可用于构建可以分析财务数据并进行预测的工具，其中必须考虑这些模型的可解释性。在教育领域，LangChain 可以用来构建帮助学生学习新概念的工具，这可能是最令人兴奋的领域之一，完整的教学大纲可以按 LLM 分解，并在定制的互动课程中提供，针对个人学习者进行个性化设置。LangChain 的多功能性使其能够以多种动态方式使用，例如构建能够回忆以前交互的虚拟个人助理；提取分析结构化数据集；创建问答应用程序，提供与实时更新的 API 进行交互；执行代码理解，从 GitHub 提取交互式源代码，提高开发人员体验，增强编码性能。使用 LangChain 好处如下。

（1）高灵活性：它为构建功能强大的应用程序提供了广泛的工具和功能。此外，它的模块化设计使构建可适应各种领域的复杂应用程序变得容易。

（2）高性能：支持生成行动计划有助于提高应用程序的性能。

（3）高可靠性：LangChain 对记忆的支持可以通过随着时间的推移存储和重用信息来帮助提高应用程序的可靠性，并且通过访问外部信息它可以减少 LLM 出现幻觉现象的次数。

（4）开源：开放的商业友好型许可证与庞大的开发人员和用户社区相结合，意味着开发者可以根据自己的需求对其进行自定义，并获得广泛支持。

综上所述，使用 LangChain 的原因有很多，但是仍然应当注意的是，由于 LangChain 很新，可能有一些错误或问题尚未解决。不过截至本书撰写的时间点，LangChain 的文档已经相对全面。

2.3.2　LangChain 支持的应用

LangChain 支持各种 NLP 用例，例如虚拟助手、摘要或翻译的内容生成模型、问答系统等。它已被用于解决现实世界的各种问题。例如，LangChain 已被用于构建聊天机器人、问答系统和数据分析工具。它还被用于许多不同的领域，包括医疗保健、金融和教育。开发者可以使用 LangChain 构建各种各样的应用程序，主要包括以下几种。

（1）聊天机器人：它可用于构建可以以自然方式与用户互动的聊天机器人。

（2）问答：LangChain 可用于构建问答系统，可以回答有关各种主题的问题。

（3）数据分析：您可以使用它进行自动化数据分析和可视化，以提取见解。

（4）代码生成：您可以设置软件对编程助手，以帮助解决业务问题。

2.4 LangChain 的工作机制

借助 LangChain，开发者可以构建动态应用程序，从而利用 NLP 的最新突破。通过连接来自多个模块的组件创建围绕 LLM 定制的独特应用程序。从情感分析到聊天机器人，这里的可能性是巨大的。LangChain 框架的主要价值由以下几部分组成。

（1）模型 I/O：此组件提供 LLM 包装器作为用于连接到语言模型的标准化接口。

（2）提示模板（Prompt Template）：此组件允许开发者管理和优化提示。

（3）记忆（Memory）：此组件的索引用于在链/Agent 的调用之间存储和重用信息。

（4）Agent：Agent 允许 LLM 与其环境交互。它们决定要采取的行动并执行行动。

（5）链：这些链条将组件组装在一起以解决问题。它们可以由对语言模型和其他实用工具的调用序列组成。

关于这些部分，有很多内容需要展开详细解释。虽然 LangChain 本身不提供模型，但它支持通过 LLM 包装器与各种不同的语言模型进行集成，使应用程序能够与聊天模型及文本嵌入模型进行交互。支持的提供商包括 OpenAI、HuggingFace、Azure 和 Anthropic。提供标准化接口意味着能够毫不费力地更换模型，以节省资金和能源，以及获得更好的性能。LangChain 的核心构建块是 prompt 类，它允许用户通过简洁的指令或示例来与 LLM 进行交互。提示工程有助于优化提示，以实现最佳模型性能。模板在输入方面提供了灵活性，并且可用的提示集合在一系列应用程序中经过了实战测试。在处理大型文档时会出现向量存储，其中需要对文档进行分块才能传递给 LLM。文档的这些部分将存储为嵌入，这意味着它们是信息的向量表示。所有这些工具都增强了 LLM 的知识，并提高了它们在问答和总结等应用中的表现。向量存储方面有许多集成，其中包括阿里云、OpenSearch、AnalyticDB for PostgreSQL、Meta AI 的 Annoy 库、Cassandra、Chroma、Elasticsearch、Facebook AI 相似性搜索（Faiss）、MongoDB Atlas 向量搜索、PGVector 作为 Postgres 的向量相似性搜索、Pinecone、Scikit-learn（用于 k 最近邻搜索的 SKLearnVectorStore）等。

（6）数据连接器和加载程序：这些组件提供用于连接到外部数据源的接口。

（7）回调：回调用于记录和流式传输任何链的中间步骤。

数据连接器包括用于存储数据的模块和用于与外部系统（如 Web 搜索或数据库）交互的实用程序，其中最重要的是数据检索。一些例子包括 Microsoft 文档（DOCX）、超文本标记语言（HTML）和其他常见格式，如 PDF、文本文件、JSON 和 CSV，其他的工具可以实现向潜在客户发送电子邮件，向用户的关注者发送有趣的双关语，或向用户的同事发送消息。接下来更加深入地介绍 Agent 可以做什么及它们是如何做出决策的。

2.4.1 初识 Agent

在 LangChain 中，Agent 用于控制应用程序的执行流程，以便与用户、环境和其他 Agent 进行交互。Agent 可用于决定要执行哪些操作、与外部数据源交互及随时间推移存储和重用信息。Agent商可以转账、预订航班或与客户交谈。

Agent是一个软件实体,可以在世界中执行操作和任务,并与其环境进行交互。在LangChain中,Agent使用工具和链将它们组合在一起,以决定使用哪种方式。

Agent可以与外界建立联系。例如,搜索引擎或向量数据库可用于查找最新的相关信息,然后可以将此信息提供给模型,这称为检索增强。通过整合外部信息源,LLM可以从当前信息和扩展知识中汲取灵感。这是智能体如何克服LLM固有的弱点并通过将工具与模型相结合来增强它们的一个例子。在关于LLM的局限性的部分中已经看到,对于计算,一个简单的计算器优于由数十亿个参数组成的模型。在这种情况下,Agent可以决定将计算传递给计算器或Python解释器。

LangChain中的Agent可用于执行各种任务,例如搜索信息、调用API、访问数据库、代码执行。每个Agent都可以决定使用哪个工具及何时使用。由于这对于理解LangChain的工作原理至关重要,因此下面将详细阐述细节问题。

每个Agent都配备了子组件:工具(功能组件)、工具包(工具的集合)、Agent执行器。

Agent执行器是允许在工具之间进行选择的执行机制。Agent执行器可以看作Agent和执行环境之间的中介。它接收来自Agent的请求或命令,并将它们转换为可由底层系统或软件执行的操作。它管理这些操作的执行,并向Agent提供反馈或结果。LangChain中有不同类型的执行或决策模式。ReAct模式[8]是Reason and Act的缩写,其中Agent主动将任务分配给适当的工具,为其自定义输入,并解析其输出以解决问题。在本书中,使用了一个文档存储,在其中搜索答案——这是作为ReAct文档存储模式实现的。在LangChain中,在默认情况下,Agent遵循零样本ReAct模式(Zero Shot),其中决策仅基于工具(Tool)的描述。这种机制可以通过记忆进行扩展,以考虑完整的对话历史记录。使用ReAct,开发者可以提示它在思想、行为、观察循环中作出响应,而不是要求LLM自动完成文本。LLM的提示是逐步响应的并能够将操作与这些步骤相关联,然后这些步骤的结果(例如搜索结果)被传递回LLM,以便在其迭代实现目标时进行下一次审议。对于Zero Shot模式,提示非常重要,它是通过连接前缀、描述工具及其用途的字符串、格式说明和后缀创建的。Zero Shot模式的代码如下:

```
//第2章/ZeroShot.py
PREFIX = """Answer the following questions as best you can. You have access to the following tools:"""
FORMAT_INSTRUCTIONS = """Use the following format:
Question: the input question you must answer
Thought: you should always think about what to do
Action: the action to take, should be one of [{tool_names}]
Action Input: the input to the action
Observation: the result of the action
... (this Thought/Action/Action Input/Observation can repeat N times)
Thought: I now know the final answer
```

```
Final Answer: the final answer to the original input question"""
SUFFIX = """Begin!
Question: {input}
Thought:{agent_scratchpad}"""
```

例如,在下面的例子中,询问LLM"LangChain Agent 执行器和 LangChain 执行计划之间的区别",代码如下:

```
//第 2 章/example.py
I'm not familiar with these terms, so I should search for information about them.
Action: Search
Action Input: " difference between langchain agent executor and langchain execution plan"
Observation: The Concept of Agents in LangChain Action Agents decide an action to take and execute that action one step at a time. They are more conventional and suitable for small tasks. On the other hand, Plan-and-Execute Agents first decide a plan of actions to take, and then execute those actions one at a time.
Thought:Based on the observation, a langchain agent executor is an agent that decides and executes actions one step at a time, while a langchain execution plan is an agent that first decides a plan of actions and then executes them one at a time.
Final Answer: A langchain agent executor executes actions one step at a time, while a langchain execution plan first decides a plan of actions and then executes them.
```

此外,还有一些已经被实施的机制。华盛顿大学、Mosaic AI、Meta AI Research 和 Allen Institute 的研究人员[9]发现,对于需要组合推理的问题,LLM 可能通常无法给出正确和完整的答案,其中必须将多条信息放在一起。具有搜索模式的自问将问题分解为成分,并调用搜索引擎方法以检索必要的信息,从而回答问题。用户 Nkov 在 LangChain 的 GitHub 上讨论了这种强大机制的一个例子。这个例子是穆罕默德·阿里(Muhammad Ali)或艾伦·图灵(Alan Turing)的寿命有多长,对话是这样展开的:

```
//第 2 章/age.py
Question: Who lived longer, Muhammad Ali or Alan Turing?
Are follow up questions needed here: Yes.
Follow up: How old was Muhammad Ali when he died?
Intermediate answer: Muhammad Ali was 74 years old when he died.
Follow up: How old was Alan Turing when he died?
Intermediate answer: Alan Turing was 41 years old when he died.

So the final answer is: Muhammad Ali
```

在每个步骤中,LLM 决定是否需要后续搜索,并将此信息反馈给 LLM。截至本书完稿前,OpenAI 模型(gpt-3.5-turbo-0613、gpt-4-0613)已经过微调,以检测何时应该执行函数调用及应该将哪些输入函数中。为此,还可以在对这些语言模型的 API 调用中描述函数。这也在 LangChain 中实现。在 LangChain 中,有以下一些策略尚未作为执行机制实现:

(1) 递归批评和改进其输出(Recursively Criticizes and Improves,RCI)方法使用 LLM 作为构建智能体的计划者,其中前者使用 LLM 在执行行动之前产生想法,而后者则促使 LLM 思考经验教训以改进后续情节。

(2) Tree of Thought 算法[10]通过遍历搜索树来推进模型推理。基本策略可以是深度优先或广度优先的树遍历,但许多其他策略也可以实现这个功能并且已经过测试,例如 Best First、Monte Carlo 和 A*。这些策略已被发现可以显著提高解决问题的成功率。

这些决定可以提前计划,也可以在每步都做出。创建 Agent 可以采取的一系列行动以实现目标的过程称为行动计划生成。通过行动计划生成,有以下两种不同类型的 Agent 可以根据任务所需的动态性进行选择:

(1) 行动 Agent(Action Agent)在每次迭代时根据所有先前操作的输出决定下一个操作。

(2) 计划和执行 Agent(Plan-and-execute Agent)在开始时决定完整的行动计划,然后它们在不更新序列的情况下执行所有操作。

2.4.2 初识链

LangChain 的核心思想是 LLM 和其他组件的组合进行协同工作。例如,用户和开发人员可以按顺序将多个 LLM 调用和其他组件组合在一起,以创建复杂的应用程序,例如类似聊天机器人的社交交互、数据提取和数据分析。

提示链接是一种可用于提高 LangChain 应用程序性能的技术。提示链接涉及将多个提示链接在一起以自动完成更复杂的响应。简单地讲,链和 Agent 都是组件的包装器。两者都可以通过使 LLM 与外部系统交互并收集最新信息来扩展 LLM 的功能。将应用程序模块化为 Chain 和 Agent 等构建块可以使调试和维护它们变得更加容易。Chain 最简单的例子可能是 PromptTemplate,它将格式化的响应传递给语言模型。更有趣的 Chain 示例包括用于数学相关查询的 LLMMath 和用于查询数据库的 SQLDatabaseChain。这些被称为实用链,因为它们将语言模型与特定工具相结合。一些 Chain 可以做出自主决策。与代理类似,Router Chain 可以根据其描述决定使用哪个工具。RouterChain 可以动态地选择要使用的检索系统,例如提示或索引。LangChain 通过链确保输出的内容没有违反法律道德或以其他方式违反 OpenAI 的审核规则(OpenAIModerationChain)或符合道德、法律或习俗原则(ConstitutionalChain)。LLMCheckerChain 可以通过验证所提供的陈述和问题背后的假设来防止出现幻觉现象并减少不准确的响应。在卡内基-梅隆大学、艾伦研究所、华盛顿大学、英伟达公司、加州大学圣地亚哥分校和谷歌研究院的研究人员于 2023 年 5 月发表的一篇论文[11]中,研究人员发现这种策略在包括对话响应、数学推理和代码推理在内的基准测试中平均将任务性能绝对提高约 20%。接下来介绍记忆策略。

2.4.3 初识记忆

LLM 和工具是无状态的,因为它们不保留有关先前响应和对话的任何信息。记忆是 LangChain 中的一个关键概念,它可以通过存储以前对语言模型的调用结果、用户、Agent 运行环境的状态及 Agent 的目标来提高 LangChain 应用程序的性能。这有助于减少需要调用语言模型的次数,并有助于确保即使环境发生变化 Agent 也可以继续运行。

记忆有助于为应用程序提供上下文,并可以使 LLM 的输出更加连贯,并且使输出和上下文相关。例如,用户可以使用 ConversationBufferMemory 存储所有会话或使用缓冲区 ConversationBufferWindowMemor 来保留使用的对话中的最后一条消息。在每次调用期间,记录的消息都包含在模型的历史记录参数中,但是应该注意,这将增加 Token 使用量(从而增加 API 费用)和响应的延迟。它还可能影响模型的 Token 限制。还有一个会话摘要记忆策略,其中 LLM 用于汇总会话历史记录,这可能会为额外的 API 调用产生额外的费用。这些记忆选项有一些很有意思的细微差别。例如,一个有趣的功能是与 LLM 的对话可以编码为知识图谱(ConversationKGMemory),它可以被整合回提示中,或者用于预测响应,而不必再回到 LLM 去查询信息。

知识图谱是一种数据表示形式,它使用图结构化数据模型来集成数据,通常以三元组、主语、谓语和宾语的形式出现,例如 subject＝Sam、predicate＝loves、object＝apples。此图存储有关人物、地点或事件等实体的信息,以及它们之间的联系。接下来将讲解可以使用的工具。

2.4.4 LangChain 中的工具

Tool 是 LangChain 中的组件,可以与模型结合使用以扩展其功能。LangChain 提供了文档加载器、索引和向量存储等工具,这些工具有助于检索和存储数据,以增强 LLM 中的数据检索。LangChain 中有许多可用的工具,以下是可以使用工具执行的几个示例。

(1) 机器翻译器:语言模型可以使用机器翻译器来更好地理解和处理多种语言的文本。此工具使非翻译专用语言模型能够理解和回答不同语言的问题。

(2) 计算器:语言模型可以利用简单的计算器工具来解决数学计算问题。该计算器支持基本的算术运算,使模型能够准确地求解专为解决数学问题而设计的数据集中的数学查询。

(3) 地图:通过与 Bing 地图 API 或类似服务连接,语言模型可以检索位置信息、协助路线规划、提供行驶距离计算及提供有关附近兴趣点的详细信息。

(4) 天气:天气 API 为语言模型提供全球城市的实时天气信息。模型可以回答有关当前天气状况的查询,或预测不同时间范围内特定位置的天气。

(5) 股票:与股票市场 API 连接,允许语言模型查询特定的股票市场信息,例如开盘价和收盘价、最高价和最低价等。

(6) 幻灯片:配备幻灯片制作工具的语言模型可以使用 API 提供的高级语义(如 python-pptx 库)或基于给定主题从 Internet 检索图像来创建幻灯片。这些工具有助于创建各个专业领域所需的幻灯片。

（7）表处理：使用 Pandas DataFrame 构建的 API 使语言模型能够对表执行数据分析和可视化任务。通过连接到这些工具，模型可以为用户提供更简化、更自然的表格数据处理功能。

（8）知识图谱：语言模型可以使用模拟人类查询过程的 API 查询知识图谱，例如查找候选实体或关系、发送 SPARQL 查询和检索结果。这些工具有助于根据存储在知识图谱中的事实知识回答问题。

（9）搜索引擎：通过利用必应搜索等搜索引擎 API，语言模型可以与搜索引擎交互以提取信息并为实时查询提供答案。这些工具增强了模型从 Web 收集信息并提供准确响应的能力。

（10）维基百科：配备维基百科搜索工具的语言模型可以在维基百科页面上搜索特定实体，在页面中查找关键字，或消除具有相似名称的实体的歧义。这些工具有助于使用从维基百科检索到的内容执行问答任务。

（11）在线购物：将语言模型与在线购物工具连接，使它们能够执行搜索商品、加载有关产品的详细信息、选择商品、浏览购物页面及根据特定用户说明做出购买决策等操作。

其他工具包括 AI Painting，它允许语言模型使用 AI 图像生成模型生成图像；三维模型构建，使语言模型能够使用复杂的三维渲染引擎创建三维模型；化学性质，协助解决有关化学性质的科学问题，可以使用 PubChem 等 API；数据库工具有助于以自然语言方式访问数据库数据，以执行 SQL 查询和检索结果。这些不同的工具为语言模型提供了额外的功能和能力，以执行文本处理以外的任务。通过 API 与这些工具连接，语言模型可以增强它们在翻译、数学问题解决、基于位置的查询、天气预报、股票市场分析、幻灯片创建、表格处理和分析、图像生成、文本到语音转换等领域的能力。所有这些工具都可以为开发者提供高级 AI 功能，而且工具几乎没有限制。开发者可以轻松地构建自定义工具来扩展 LLM 的功能。不同工具的使用扩大了语言模型的应用范围，使它们能够更高效地处理各种实际问题。

2.5 总结

在当今世界，准确理解和处理语言对于开发智能应用程序及创建个性化应用和提高用户体验至关重要，因此，LLM 非常适合将该功能提供给应用程序。然而，正如在本章中所讨论的那样，独立的 LLM 有其局限性。如果使用外部工具补充 LLM，则可克服其中一些限制，并大大提高它们在创建 LLM 应用程序时的性能。这就是 LangChain 的用武之地，它是一个针对 AI 开发人员的框架，是一个用于设置代理的应用程序。这些代理由计算实体组成，如 LLM 和其他可以自主执行某些任务的工具。本章中已经讨论了它的重要概念，包括代理和链等概念。总之，LangChain 是一个有价值的开源框架，用于使用来自 OpenAI 和 HuggingFace 等模型提供商提供的 LLM 来简化应用程序的开发。该框架在释放生成式 AI 的力量方面提供了巨大的价值。在接下来的章节中将通过构建 LLM 应用程序来深刻体现 LangChain 的这些核心原则。通过利用 LangChain 的功能，开发人员可以释放 LLM 的全部潜力。在第 3 章中将使用 LangChain 实现一个应用程序。

第 3 章 LangChain 入门

3.1 安装 LangChain 方法

可以通过简单地从终端输入 pip install langchain 命令来安装 LangChain，但是，本书还将在几个不同的用例中使用各种其他工具和集成。为了确保所有示例和代码片段都能按预期工作，本书提供了不同的设置环境的方法。读者有多种方法可以设置 Python 环境。这里提供了 4 种流行的安装相关依赖的方法：Docker、Conda、Pip 和 Poetry。如果读者在安装过程中遇到问题，则可查阅相应的文档。截至本书撰写时，已经对不同的安装进行了测试，但是，情况可能会发生一些变化。接下来将从 Python 开始介绍如何设置环境。

3.1.1 安装 Python

在设置 Python 环境和安装相关依赖项之前，通常应该安装 Python。读者可以从 python.org 下载适用于自己的操作系统的最新版本的 Python，也可以使用平台的包管理器。先来看 macOS 的 Homebrew 和 Ubuntu 的 apt-get。在 macOS 上，使用 Homebrew，命令如下：

```
brew install python
```

对于 Ubuntu，使用 apt-get，命令如下：

```
sudo apt-get updatesudo apt-get install python 3.10
```

提示：如果用户是编程或 Python 的新手，则建议在继续使用 LangChain 和本书中的应用程序之前，先学习一些初级教程。

除了基本的 Python 命令之外，以交互方式尝试数据处理的一个重要工具是 Jupyter 系列的工具。接下来介绍 Jupyter Notebook 和 JupyterLab。

3.1.2 Jupyter Notebook 和 JupyterLab

Jupyter Notebook 和 JupyterLab 是基于 Web 的开源交互式环境，用于创建、共享和协作处理文档。它们使用户能够编写代码、显示可视化效果,并在称为笔记本的单个文档中包含解释性文本。两者之间的主要区别在于它们的界面和功能。

Jupyter Notebook 旨在支持各种编程语言,如 Julia、Python 和 R——事实上,项目名称是对这 3 种语言的引用。Jupyter Notebook 提供了一个简单的用户界面,允许用户创建、编辑和运行具有线性布局的笔记本。它还支持附加功能和自定义的扩展。

另外,JupyterLab 是 Jupyter Notebook 的增强版本。JupyterLab 于 2018 年推出,为处理笔记本和其他文件类型提供了更强大、更灵活的环境。它提供了一个模块化、可扩展和可自定义的界面,用户可以在其中并排排列多个窗口(例如,笔记本、文本编辑器、终端),从而使工作流程更加高效。

用户可以从终端启动计算机上的笔记本服务器,命令如下:

```
jupyter notebook
```

用户应该会看到浏览器打开了一个包含 Jupyter 笔记本的新选项卡,如图 3-1 所示。

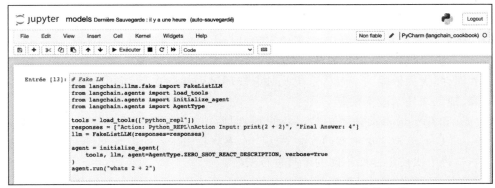

图 3-1　LangChain Agent 代码在 Jupyter Notebook 中的实现

或者,用户也可以使用 JupyterLab。JupyterLab 是下一代 Notebook 服务器,可显著提高可用性。可以从终端启动 JupyterLab Notebook 服务器,如图 3-2 所示。

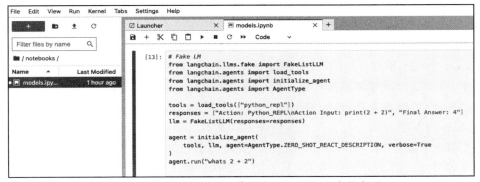

图 3-2　LangChain Agent 代码在 JupyterLab 中的实现

Jupyter Notebook 或 JupyterLab 中的任何一个都将为用户提供一个集成开发环境（IDE），用于处理本书中介绍的一些代码。在安装了 Python 和 Jupyter Notebook 或 JupyterLab 之后，接下来本书将带领读者快速探索依赖管理工具（Docker、Conda、Pip 和 Poetry）之间的区别，并使用它们来为 LangChain 项目设置完整的环境。

3.1.3 环境管理

在探索在 LangChain 中建立 Python 环境以使用生成模型的各种方法之前，必须了解主要依赖管理工具（Docker、Conda、Pip 和 Poetry）之间的区别。这 4 种工具都是广泛用于软件开发和部署领域的工具。

Docker 是一个开源平台，通过容器化提供操作系统级虚拟化。它可以在轻量级、可移植的容器中自动部署应用程序，这些容器可在任何安装了 Docker 的系统上一致地运行。

Conda 是一个跨平台的包管理器，擅长从多个渠道安装和管理包，不限于 Python。它主要面向数据科学和机器学习项目，可以稳健地处理复杂的依赖关系树，满足具有大量依赖关系的复杂项目的需求。

Pip 是 Python 最常用的包管理器，允许用户轻松安装和管理第三方库，但是，Pip 在处理复杂的依赖项时存在局限性，增加了包安装过程中出现依赖项冲突的风险。

Poetry 是一个较新的包管理器，它结合了 Pip 和 Conda 的最佳功能。Poetry 拥有现代直观的界面、强大的依赖关系解析系统及对虚拟环境创建的支持，提供了额外的功能，例如依赖隔离、锁定文件和版本控制。

Poetry 和 Conda 都简化了虚拟环境管理，而使用 Pip 通常需要使用单独的工具，例如 Virtualenv。Conda 是这里推荐的安装方法。本书将提供 Pip 的需求文件和 Poetry 的说明文档。本书将依次使用这些不同的工具进行安装。

如果读者不熟悉 Git，则可以下载 ZIP 格式的文档，然后使用工具解压缩存档。或者，如果需要使用 Git 克隆存储库并更改为项目目录，则命令如下：

```
git clone https://github.com/Roonielu/langchain.git
cd langchain
```

现在读者的本地计算机上有了本书的存储库，接下来介绍 Docker 方式。

1. Docker

Docker 是一个平台，使开发人员能够自动部署、打包和管理应用程序。Docker 使用容器化技术，有助于标准化和隔离环境。使用容器的优点是，它可以保护本地环境免受在容器中运行的任何（可能不安全）代码的影响。缺点是镜像可能需要时间来构建，并且可能需要大约 10 GB 的存储空间。与其他环境管理工具类似，Docker 非常有用，因为它可以为项目创建可重现的环境。用户可以使用 Docker 创建一个包含项目所需的所有库和环境，并与他人共享该环境。若要开始使用 Docker，则需要执行以下步骤：

用户可以访问 Docker 的官方网站 https://docs.docker.com/get-docker/，并按照该网

址上的安装指南进行操作。

在终端中，运行以下命令以构建 Docker 镜像（需要注意：这需要在项目根目录中才能正常工作），命令如下：

```
docker build -t langchain
```

这将从 Docker Hub 拉取 continuumio/miniconda3 镜像，并生成该镜像，然后使用创建的镜像以交互方式启动 Docker 容器，命令如下：

```
docker run -it langchain
```

在执行完以上命令后，Notebook 将在容器中启动。用户应该能够从浏览器导航到 Jupyter Notebook。可以在 http://localhost:8080/ 这个地址找到它。接下来介绍 Conda 方式。

2. Conda

Conda 允许用户为不同的项目管理多个环境。它适用于 Python、R 和其他语言，并通过维护与 Python 库关联的库列表来帮助安装系统库。开始使用 Conda 的最佳方法是按照以下链接中的说明安装 Anaconda 或 Miniconda：https://docs.continuum.io/anaconda/install/。虽然 Conda 环境占用的磁盘空间比 Docker 少，但从 Anaconda 开始，整个环境仍占用大约 2.5GB。使用 Miniconda 可能会节省一些磁盘空间。Conda 还有一个图形界面叫作 Anaconda Navigator，它可以安装在 macOS 和 Windows 系统上，并且可以从终端安装任何依赖项及 Conda 工具。接下来继续使用 Conda 工具并安装本书的依赖。用户如果要创建一个新环境，则命令如下：

```
conda env create --file langchain_ai.yml
```

Conda 允许创建具有许多不同库的环境，以及不同版本的 Python。在本书中使用了 Python 3.10。可以通过运行以下命令激活环境，命令如下：

```
conda activate langchain_ai
```

现在可以在环境中启动 Jupyter Notebook 或 JupyterLab，命令如下：

```
jupyter notebook
```

3. Pip

Pip 是 Python 的默认包管理器。它能让用户轻松安装和管理第三方库。用户可以安装单个库，但也可以维护 Python 库的完整列表。使用 pip 安装库中的说明安装 pip，然后使用 pip 命令安装库。例如，要安装 NumPy 库，命令如下：

```
pip install numpy==1.0
```

设置一个完整的环境一般从一个依赖需求列表开始。按照惯例,这个列表位于一个名为 requirements.txt 的文件中。此文件已经被包含在项目的根目录中,该目录列出了所有基本库。用户可以使用以下命令安装所有库,命令如下:

```
pip install -r requirements.txt
```

但是,需要注意,如前所述,Pip 和所属的环境并没有关联。Virtualenv 是一个可以帮助用户维护环境的工具,例如维护不同版本的库。用户可以创建自定义的 Virtualenv,命令如下:

```
//第 3 章/env.py
#创建一个新环境 myenv
virtualenv myenv
#激活 myenv 环境
source myenv/bin/activate
#安装依赖项或运行 Python
python
#再次离开环境
deactivate
Please note that in Windows, the activation command is slightly different - you'd run a shell script:
#激活 myenv 环境
myenv\Scripts\activate.bat
```

4. Poetry

Poetry 是 Python 的依赖管理工具,可简化库安装和版本控制。

用户可以按照 https://python-poetry.org/ 上的说明安装 Poetry。在终端中运行 poetry install 命令(如前所述从项目根目录)。该命令将自动创建一个新环境(如果尚未创建环境)并安装所有依赖项。接下来本书将介绍模型的集成。

3.2 导入模型

在正确使用生成式模型之前,需要设置对 LLM 或其他模型的访问,以便将它们集成到应用程序中。正如本书开头所讨论的,科技巨头公司有各种语言模型,例如 OpenAI 的 GPT-4、谷歌的 BERT 和 PaLM-2、MetaAI 的 LLaMA 等。通过 LangChain 开发者可以与所有这些模型交互,例如通过应用程序接口(API)调用或者调用用户在本地计算机上下载的开源模型,其中一些模型支持文本生成和嵌入。本章将重点讨论文本生成类模型,后续章节将讨论嵌入、向量数据库和神经搜索等高级功能。当前 LangChain 支持的语言模型供应商包括 OpenAI、HuggingFace、Cohere、Anthropic、Azure、Google Cloud Platform Vertex AI(PaLM-2)和 Jina AI 等，但这个列表仍在不断增长。

用户可以访问 https://integrations.langchain.com/llms 了解 LangChain 当前支持的所有模型。在图像模型方面，主要的供应商包括 OpenAI（DALL-E）、Midjourney Inc.（Midjourney）和 Stability AI（Stable Diffusion）。LangChain 目前没有开箱即用的组件来处理非文本模型，但是其文档详细描述了如何通过 Replicate 工具调用 Stable Diffusion 模型接口。对于每个模型供应商提供的 API，用户需要先创建账号并获取 API 密钥，在大多数情况下是免费的。对一些模型供应商而言，用户甚至不需要提供信用卡信息。要在 Python 环境中设置 API 密钥，可以执行的命令如下：

```
import os
os.environ["OPENAI_API_KEY"] = "<your token>"
```

这里 OPENAI_API_KEY 是适用于 OpenAI 的环境密钥。在环境中设置密钥的优点是不会在代码中暴露它们。用户可以从终端公开这些变量，命令如下：

```
export OPENAI_API_KEY=<your token>
```

接下来依次介绍一些著名的模型提供商。本书将为每个示例提供用法示例。首先从用于测试的 Fake LLM 开始介绍。

3.2.1　虚拟 LLM（Fake LLM）

Fake LLM 主要用于测试。LangChain 文档中有一个工具与 LLM 一起使用的示例。用户可以直接在 Python 中执行此示例，命令如下：

```
//第 3 章/FakeLLM.py
from langchain.llms.fake import FakeListLLM
from langchain.agents import load_tools
from langchain.agents import initialize_agent
from langchain.agents import AgentType
tools = load_tools(["python_repl"])
responses = ["Action: Python_REPL\nAction Input: print(2 + 2)", "Final Answer: 4"]
llm = FakeListLLM(responses=responses)
agent = initialize_agent(
    tools, llm, agent=AgentType.ZERO_SHOT_REACT_DESCRIPTION, verbose=True
)
agent.run("what's 2 + 2")
```

这里演示的是连接一个工具，一个 Python 读取-求值-打印循环（REPL），它将根据 LLM 的输出进行调用。Fake List LLM 将给出两个响应 responses 并且不会根据输入而改变，然后建立一个 Agent，该 Agent 根据第 2 章中介绍过的 ReAct 策略做出决策。用户输入一个文本运行这个 Agent,这个输入是"什么是 2 + 2"。用户可以观察到 Fake LLM 的输出如何触发对 Python Interprete 的调用,最终该解释器返回 4。读者注意,该操作必须与 Tool 的名称属性 PythonREPLTool 匹配,该属性的开头代码如下:

```
//第 3 章/PythonREPLTool.py
class PythonREPLTool(BaseTool):
    """用于在 REPL 中运行 Python 代码的工具。"""
    name = "Python_REPL"
    description = (
        "一个 Python Shell。使用它来执行 Python 命令。"
        "输入应该是有效的 Python 命令。"
        "如果您想查看某个值的输出,则应该将其打印出来"
        "使用 `print(...)`."
    )
```

Tool 的名称和描述将传递给 LLM,然后 LLM 根据提供的信息做出决定。Python 解释器的输出被传递给 Fake LLM,后者忽略观察结果并返回 4。很显然,如果用户将第 2 个响应更改为"最终答案:5",则 Agent 的输出将与问题不对应。在接下来的章节中将通过实际的 LLM 而不是虚假的 LLM 来使其更有实际意义。首先介绍 OpenAI。

3.2.2 OpenAI

OpenAI 是一家美国 AI 研究实验室,目前是生成式模型(尤其是 LLM)的市场领导者。它们提供了一系列具有不同能力水平的模型,适用于不同的任务。在本章中将看到如何使用 LangChain 和 OpenAIPython 客户端库与 OpenAI 模型进行交互。OpenAI 还为文本嵌入模型提供了一个 Embedding 类。本书主要将 OpenAI 用于接下来的应用程序开发。OpenAI 提供了几种模型供用户选择,每种模型都有自己的优点、Token 使用计数和用例,其中主要的 LLM 是 GPT-3.5 和 GPT-4,它们具有不同的 Token 长度。如果要创建 API 密钥,用户则需要按照下列步骤操作:

(1) 在 https://platform.openai.com/创建一个登录名。
(2) 设置账户信息。
(3) 用户可以在"个人"→"查看 API 密钥"下查看 API 密钥。
(4) 单击"创建新密钥"按钮并为其指定名称。

创建密钥,用户应该会看到网页消息"API 密钥已生成"。用户需要将密钥复制到剪贴板并保存它。在这里可以将密钥设置为环境变量(OPENAI_API_KEY),或者在每次为 OpenAI 调用构造类时将其作为参数传递。用户可以使用 OpenAI 语言模型类设置要与之交互的 LLM。现在创建一个使用此模型进行计算的 Agent,代码如下:

```
//第 3 章/agent.py
from langchain.llms import OpenAI
```

```
llm = OpenAI(temperature=0., model="text-davinci-003")
agent = initialize_agent(
    tools, llm, agent=AgentType.ZERO_SHOT_REACT_DESCRIPTION, verbose=True
)
agent.run("whats 4 + 4")
```

输出的结果如下：

```
//第 3 章/output.py
> Entering new   chain...
   I need to add two numbers
Action: Python_REPL
Action Input: print(4 + 4)
Observation: 8
Thought: I now know the final answer
Final Answer: 4 + 4 = 8
> Finished chain.
'4 + 4 = 8'
```

3.2.3　HuggingFace

　　HuggingFace 是 NLP 领域非常突出的参与者，在开源和托管解决方案方面具有相当大的吸引力。该公司是一家美国公司，开发用于构建机器学习应用程序的工具，其员工开发和维护 TransformersPython 库，该库用于自然语言处理任务，包括 BERT 和 GPT-2 等最先进和流行模型的实现，并与 PyTorch、TensorFlow 和 JAX 兼容。HuggingFace 还提供了 HuggingFaceHub，这是一个用于托管基于 Git 的代码存储库、机器学习模型、数据集和 Web 应用程序的平台，它为机器学习提供了超过 12 万个模型、2 万个数据集和 5 万个演示应用程序（Spaces）。它是一个在线平台，人们可以在其中协作并共同构建 ML。这些工具允许用户加载和使用来自 HuggingFace 的模型、嵌入和数据集。例如，HuggingFaceHub 集成为文本生成和文本分类等任务提供了对不同模型的访问。HuggingFaceEmbeddings 集成允许用户使用句子转换器模型，并在生态系统中提供了各种其他库，包括用于数据集处理的 Datasets、用于模型评估的 Assess、用于模拟的 Simulate 和用于机器学习演示的 Gradio。除了这些产品之外，HuggingFace 还参与了 BigScience 研究研讨会等计划，在那里发布了一个名为 BLOOM 的开放 LLM，该模型具有 1760 亿个参数。HuggingFace 获得了大量资金，包括 4000 万美元的 B 轮融资和最近由 Coatue 和 Sequoia 领投的 C 轮融资，估值为 20 亿美元。HuggingFace 还与 Graphcore 和 AmazonWebServices 等公司建立了合作伙伴关系，以优化其产品并将其提供给更广泛的客户群。

　　用户可以登录 https://huggingface.co/settings/profile 创建一个账号和 API 密钥，然后可以把 API 存储在环境变量 HUGGINGFACEHUB_API_TOKEN 中。接下来使用谷歌公司开发的开源模型作为一个例子，即 Flan-T5-XXL 模型，代码如下：

```
//第 3 章/T5.py
from langchain.llms import HuggingFaceHub
llm = HuggingFaceHub(
    model_kwargs={"temperature": 0.5, "max_length": 64},
    repo_id="google/flan-t5-xxl"
)
prompt = "In which country is Tokyo?"
completion = llm(prompt)
print(completion)
```

我们得到的回答是"日本"。LLM 接受文本输入,在本例中为一个问题,并返回一个完成的值。该模型拥有大量知识,可以给出知识问题的答案。

3.2.4 微软云

微软云(Azure)是由微软公司运营的云计算平台,它与 OpenAI 集成,以提供强大的语言模型,如 GPT-3、Codex 和嵌入。它通过其全球数据中心提供应用程序和服务的访问、管理和开发,用于编写帮助、摘要、代码生成和语义搜索等用例。Azure 提供软件即服务(SaaS)、平台即服务(PaaS)和基础架构即服务(IaaS)等功能。通过 GitHub 或微软凭据进行身份验证,用户可以访问 https://azure.microsoft.com/在 Azure 上创建一个账号,然后可以在"认知服务"→"AzureOpenAI"下创建新的 API 密钥。设置完成后,模型应该可以通过 LangChain 中的 AzureOpenAI()LLM 类访问。

3.2.5 谷歌云

谷歌云平台(Google Cloud Platform,GCP)及其机器学习平台 Vertex 提供了许多模型和功能。GCP 提供对 Lambda、T5 和 PaLM 等 LLM 的访问。谷歌还更新了谷歌云 NaturalLanguage(NL)API,其中包含新的基于 LLM 的内容分类模型。此更新版本提供了广泛的预训练分类法,以帮助进行广告定位和基于内容的过滤。NLAPI 改进的 v2 分类模型得到了增强,增加了 1000 多个标签,并支持 11 种语言,准确性更高。如果需要在 GCP 上运行模型,用户则需要安装 gcloud 命令行界面(CLI),可以在访问 https://cloud.google.com/sdk/docs/install 获得详细的说明。

可以从终端进行身份验证并打印密钥 Token,代码如下:

```
gcloud auth application-default login
```

通常用户还需要为项目启用 Vertex。如果尚未启用,用户则应该会收到一条错误指示消息以引导指向正确的网站,然后用户必须在其中单击"启用"。接下来演示在 GCP 中运行一个模型,代码如下:

```
//第 3 章/GCP.py
```

```
from langchain.llms import VertexAI
from langchain import PromptTemplate, LLMChain
template = """Question: {question}
Answer: Let's think step by step."""
prompt = PromptTemplate(template=template, input_variables=["question"])
llm = VertexAI()
llm_chain = LLMChain(prompt=prompt, llm=llm, verbose=True)
question = "What NFL team won the Super Bowl in the year Justin Beiber was born? "
llm_chain.run(question)
```

用户应该能看到这样的输出:

```
//第 3 章/GCPresponse.py
Entering new chain...
Prompt after formatting:
Question: What NFL team won the Super Bowl in the year Justin Beiber was born?
Answer: Let's think step by step.
Finished chain.
Justin Beiber was born on March 1, 1994. The Super Bowl in 1994 was won by the San
Francisco 49ers.
```

上面已将 verbose 设置为 True,以便查看模型的推理过程。令人感叹的是,即使考虑到名称的拼写错误,它也能做出正确的响应。循序渐进的提示指令是获得正确答案的关键。Vertex 给开发者提供了各种模型,如表 3-1 所示。

表 3-1　Vertex 中可用的模型

模　　型	描　　述	属　　性
text-bison	遵循自然语言指令进行了微调	最大输入 Token:8192 最大输出 Token:1024 训练数据:截至 2023 年 2 月
chat-bison	针对多轮对话进行了微调	最大输入 Token:4096 最大输出 Token:1024 训练数据:截至 2023 年 2 月
code-bison	经过微调可根据自然语言描述生成代码	最大输入 Token:4096 最大输出 Token:2048
codechat-bison	针对聊天机器人对话进行了微调,有助于解决与代码相关的问题	最大输入 Token:4096 最大输出 Token:2048
code-gecko	针对代码补全进行了微调	最大输入 Token:2048 最大输出 Token:64

还可以使用模型生成代码。接下来看 Code-Bison 模型是否能解决 FizzBuzz 问题,这是一个入门级和中级软件开发人员职位的常见面试问题,代码如下:

```
//第 3 章/FizzBuzz.py
question = """
Given an integer n, return a string array answer (1-indexed) where:
answer[i] == "FizzBuzz" if i is divisible by 3 and 5.
answer[i] == "Fizz" if i is divisible by 3.
answer[i] == "Buzz" if i is divisible by 5.
answer[i] == i (as a string) if none of the above conditions are true.
"""
llm = VertexAI(model_name="code-bison")
llm_chain = LLMChain(prompt=prompt, llm=llm)
print(llm_chain.run(question))
```

可以从模型中得到回答,代码如下:

```
//第 3 章/FizzBuzzAnswer.py
```python
answer = []
for i in range(1, n + 1):
    if i % 3 == 0 and i % 5 == 0:
        answer.append("FizzBuzz")
    elif i % 3 == 0:
        answer.append("Fizz")
    elif i % 5 == 0:
        answer.append("Buzz")
    else:
        answer.append(str(i))
return answer
```
```

3.2.6 Jina AI

Jina AI 由 Han Xiao 和 Xuanbin He 于 2020 年 2 月创立,是一家总部位于柏林的德国 AI 公司,专门提供具有文本、图像、音频和视频模型的云原生神经搜索解决方案。他们的开源神经搜索生态系统使企业和开发人员能够轻松地构建可扩展且高度可用的神经搜索解决方案,从而实现高效的信息检索。最近,Jina AI 推出了 Finetuner,该工具可以根据特定用例和要求对任何深度神经网络进行微调。该公司已通过三轮融资共筹集了 3750 万美元,最近一次融资来自 2021 年 11 月的 A 轮融资。Jina AI 著名投资者包括 GGV 纪源资本和嘉楠耘智。用户可以在 https://cloud.jina.ai 设置登录。在平台上,可以针对不同的用例设置 API,例如图像标题、文本嵌入、图像嵌入、视觉问答、视觉推理、图像放大或中文文本嵌入。不幸的是,Jina AI API 截至本书撰写的时候还没有通过 LangChain 集成。不过用户可以通过将 LangChain 中的 LLM 类子类化为自定义 LLM 接口实现此类调用。接下来演示设置一个聊天机器人,由 Jina AI 提供支持。用户可以在 https://chat.jina.ai/api 生成 APIToken,这里可以将其设置为 JINA_AUTH_TOKEN,代码如下:

```
//第 3 章/Jina.py
from langchain.chat_models import JinaChat
from langchain.schema import HumanMessage
chat = JinaChat(temperature=0.)
messages = [
    HumanMessage(
        content="Translate this sentence from English to French: I love
generative AI!"
    )
]
chat(messages)

#用户可以看到
AIMessage(content="J'adore l'IA générative !", additional_kwargs={}, example=
False).
```

用户可以设置不同的温度,其中低温度使响应更可预测。在这种情况下,它几乎没有区别。首先这里以一条系统消息开始对话,阐明聊天机器人的用途。本例中演示询问一些食物建议,代码如下:

```
//第 3 章/JinaFood.py
chat = JinaChat(temperature=0.)
chat(
    [
        SystemMessage(
            content="您可以用一个词帮助用户找到一种营养丰富且美味的食物。"
        ),
        HumanMessage(
            content="I like pasta with cheese, but I need to eat more vegetables,
what should I eat?"
        )
    ]
)
```

在 Jupyter Notebook 中可以看到回答,代码如下:

```
//第 3 章/JinaFoodResponse.py
AIMessage(content = 'A tasty and nutritious option could be a vegetable pasta
dish. Depending on your taste, you can choose a sauce that complements the
vegetables. Try adding broccoli, spinach, bell peppers, and zucchini to your
pasta with some grated parmesan cheese on top. This way, you get to enjoy your
pasta with cheese while incorporating some veggies into your meal.', additional_
kwargs={}, example=False)
```

了解 LangChain 中 LLM 和聊天模型之间的区别非常重要。LLM 是文本模型,它以字符串提示作为输入并输出字符串。聊天模型类似于 LLM,但专为对话而设计。它们将聊天消息列表作为输入,并标记有说话者,并返回聊天消息作为输出。LLM 和聊天模型都实现了基本语言模型接口，其中包括 predict() 和 predict_messages() 等方法。这种共享接口促成了应用程序中不同类型的模型之间及聊天和 LLM 模型之间的互换性。

3.2.7 Replicate

Replicate Inc.成立于 2019 年，是一家总部位于旧金山的初创公司，该公司为 AI 开发人员提供了一个简化的流程，他们可以利用云技术以最少的代码输入实现和发布 AI 模型。该平台适用于私有和公共模型，并支持模型推理和微调。该公司最近的一次融资来自 A 轮融资，投资总额为 1250 万美元，由 Andreessen Horowitz 牵头，Y Combinator、Sequoia 和各种独立投资者参与其中。Ben Firshman 在 Docker 推动了开源产品的工作，Andreas Jansson 是 Spotify 的前机器学习工程师，他们共同创立了 Replicate Inc.，共同希望消除阻碍人工智能大规模接受的技术障碍，因此他们创建了 Cog。这是一个开源工具，可将机器学习模型打包到一个标准的生产就绪的容器中，该容器可以在任何当前的操作系统上运行并自动生成 API。这些容器还可以通过复制平台部署在 GPU 集群上，因此，开发人员可以专注于其他基本任务，从而提高他们的生产力。

用户可以在 https://replicate.com/ 上使用 GitHub 凭据进行身份验证，然后单击左上角的用户图标，可以找到 APIToken。此时只需复制 API 密钥并将其作为 REPLICATE_API_TOKEN 在自己的环境中提供。

使用 replicate 创建一个镜像的方法很简单，代码如下：

```
//第 3 章/replicate.py
from langchain.llms import Replicate
text2image = Replicate(
    model="stability-ai/stable-diffusion:db21e45d3f7023abc2a46ee38a23973f6-dce16bb082a930b0c49861f96d1e5bf",
    input={"image_dimensions": "512x512"},
)
image_url = text2image("关于用 Python 创建生成式人工智能应用程序的书的封面")
```

运行过后可以得到一张图片，如图 3-3 所示。

接下来快速了解如何在 HuggingFace 或 Llama.cpp 文件中运行本地模型。

3.2.8 本地模型

用户也可以从 LangChain 运行本地模型。不过有一点需要注意：LLM 很大，这意味着它会占用很大存储空间。如果用户只有一台旧计算机，则可以尝试托管服务，例如 Google Colab。这些将允许用户在具有大量内存和不同硬件(包括张量处理单元(TPU)或 GPU)的机器上运行。由

图 3-3 一本关于使用 Python 的生成式模型的书的封面

于接下来的两个用例都可能需要很长时间才能运行完或使 Jupyter Notebook 崩溃,因此本书没有将此代码包含进去,也没有将依赖项包含在设置说明中,但是本地模型仍然值得讨论。本地运行模型能够让用户完全控制模型,并且这个过程不会通过互联网共享任何数据。接下来先介绍 HuggingFace 的 Transformers 库。

1. HuggingFace Transformers

这里将快速展示设置和运行管道(Pipeline)的一般方法,代码如下:

```
//第 3 章/transformers.py
from transformers import pipeline
import torch
generate_text = pipeline(
    model="aisquared/dlite-v1-355m",
    torch_dtype=torch.bfloat16,
    trust_remote_code=True,
    device_map="auto",
    framework="pt"
)
generate_text("In this chapter, we'll discuss first steps with generative AI in Python.")
```

该模型非常小(3.55 亿个参数),但性能相对较高,并且模型指令针对对话进行了调整。需要注意,本地模型不需要 API。这将下载模型所需的所有内容,例如分词器和模型权重,然后用户可以运行文本补全。为了将这个 Pipeline 插入 LangChain Agent 或 Chain 中,代码如下:

```
//第 3 章/pipeline.py
from langchain import PromptTemplate, LLMChain
template = """Question: {question}
Answer: Let's think step by step."""
prompt = PromptTemplate(template=template, input_variables=["question"])
llm_chain = LLMChain(prompt=prompt, llm=generate_text)
question = "What is electroencephalography?"
print(llm_chain.run(question))
```

在此示例中包括了 PromptTemplate 的使用,该模板为任务提供了特定说明。接下来介绍 Llama.cpp。

2. Llama.cpp

Llama.cpp 是一个 C++ 程序,它执行基于 Llama 架构的模型,Llama 是 MetaAI 发布的首批大型开源模型之一,它反过来催生了许多其他模型的开发。需要注意,用户需要安装 MD5 校验和工具。在默认情况下,这包含在多个 Linux 发行版(如 Ubuntu)中。在 macOS 上可以使用 brew 安装它,代码如下:

```
brew install md5sha1sum
```

用户需要从 GitHub 下载 llama.cpp 存储库。可以在线执行此操作，选择 GitHub 上的下载选项之一，或者可以从终端使用 Git 命令，命令如下：

```
git clone https://github.com/ggerganov/llama.cpp.git
```

然后用户需要安装 Python 依赖，可以使用 pip 包安装程序来完成。为了方便起见，这里切换到 llama.cpp 项目的根目录，命令如下：

```
cd llama.cpp
pip install -r requirements.txt
```

根据自己的需求，用户可能希望在安装依赖之前创建一个 Python 环境。接下来需要编译 llama.cpp，命令如下：

```
make -C . -j4 #runs make in subdir with 4 processes
```

这里可以使用 4 个进程并行构建。为了获得 Llama 模型的权重，用户需要同意注册条款和条件，并等待 Meta 的注册电子邮件。Pyllama 项目中有一些工具，例如 Llama 模型下载器，但需要注意，它们可能不符合 Meta 的许可规定。开发者可以从 HuggingFace 下载模型，这些模型应该与 llama.cpp 兼容，例如 Vicuna 或 Alpaca。假设用户已将 7B Llama 模型的模型权重下载到 models/7B 目录中，这时可以下载更大尺寸的模型，例如 13B、30B、65B，但是，这里需要注意的是，这些模型占用很大的内存和磁盘空间。用户必须使用转换脚本将模型转换为 llama.cpp 格式，称为 ggml。接着可以选择性地量化模型，以便在进行推理时节省内存。这里所讲的量化是指减少用于存储权重的位数，命令如下：

```
python3 convert.py models/7B/
./quantize ./models/ggml-model-f16.bin ./models/7B/ggml-model-q4_0.bin q4_0
```

最后一个文件比以前的文件小得多，占用的内存空间也小得多，这意味着用户可以在配置较低的计算机上运行它。一旦选择了想要运行的模型，用户就可以将其集成到 Agent 或 Chain 中，命令如下：

```
llm = LlamaCpp(
    model_path="./ggml-model-q4_0.bin",
    verbose=True
```

到这里就完成了所有对 LangChain 模型集成的介绍。

3.3 模型输出解析

通常 LLM 输出的是文本,但很多时候,用户希望获得更多的结构化信息,而不仅是文本。这就是输出解析器的用武之地。输出分析器是帮助构建语言模型响应的类。输出解析器必须实现两种主要方法。

(1) 获取格式说明:此方法返回一个字符串,其中包含有关如何格式化语言模型输出的说明方法。

(2) 解析:此方法接受一个字符串(假定来自语言模型的响应)并将其解析为某种结构的方法。

(3) 用提示进行解析:此方法接受字符串(假定来自语言模型的响应)和提示(假定是生成此类响应的提示)并将其解析为某种结构。当输出解析器想要以某种方式重试或修复输出并需要根据提示中的信息来执行此操作时会提供提示。

下面的例子用于输出解析器的主要类型,即 PydanticOutputParser,代码如下:

```python
//第 3 章/ PydanticOutputParser.py
from typing import List

from langchain.llms import OpenAI
from langchain.output_parsers import PydanticOutputParser
from langchain.prompts import PromptTemplate
from langchain.pydantic_v1 import BaseModel, Field, validator


model = OpenAI(model_name="text-davinci-003", temperature=0.0)


#定义用户想要的数据结构
class Joke(BaseModel):
    setup: str = Field(description="question to set up a joke")
    punchline: str = Field(description="answer to resolve the joke")

    #用户可以使用 Pydantic 轻松地添加自定义验证逻辑
    @validator("setup")
    def question_ends_with_question_mark(cls, field):
        if field[-1] != "?":
            raise ValueError("Badly formed question!")
        return field


#设置解析器+将指令注入提示模板中
parser = PydanticOutputParser(pydantic_object=Joke)
```

```
prompt = PromptTemplate(
    template="Answer the user query.\n{format_instructions}\n{query}\n",
    input_variables=["query"],
    partial_variables={"format_instructions": parser.get_format_instructions()},
)

#旨在提示语言模型填充数据结构的查询
prompt_and_model = prompt | model
output = prompt_and_model.invoke({"query": "Tell me a joke."})
parser.invoke(output)
```

此外,输出解析器还可以和 LangChain 表达式语言(LangChain Expression Language, LCEL)结合,LCEL 将在后面详细介绍。输出解析器实现了 Runnable 接口,这是 LCEL 的基本构建块。这意味着它们支持 invoke、ainvoke、stream、astream、batch、abatchastream_log 调用。输出解析器接受字符串或 BaseMessage 作为输入,并且可以返回任意类型,命令如下:

```
parser.invoke(output)
```

输出的代码如下:

```
Joke(setup='Why did the chicken cross the road?', punchline='To get to the other side!')
```

与其手动调用解析器,用户不如将其添加到 Runnable 序列中,代码如下:

```
chain = prompt | model | parser
chain.invoke({"query": "Tell me a joke."})
```

输出的代码如下:

```
Joke(setup='Why did the chicken cross the road?', punchline='To get to the other side!')
```

虽然所有分析器都支持流式处理接口,但只有某些分析器可以通过部分分析的对象进行流式处理,因为这高度依赖于输出类型。无法构造部分对象的解析器将只生成完全解析的输出。

例如,SimpleJsonOutputParser 可以通过部分输出进行流式处理,代码如下:

```
//第 3 章 / SimpleJsonOutputParser.py
from langchain.output_parsers.json import SimpleJsonOutputParser
```

```
json_prompt = PromptTemplate.from_template(
    "Return a JSON object with an `answer` key that answers the following question: {question}"
)
json_parser = SimpleJsonOutputParser()
json_chain = json_prompt | model | json_parser

list(json_chain.stream({"question": "Who invented the microscope?"}))
```

输出的代码如下：

```
[{},
 {'answer': ''},
 {'answer': 'Ant'},
 {'answer': 'Anton'},
 {'answer': 'Antonie'},
 {'answer': 'Antonie van'},
 {'answer': 'Antonie van Lee'},
 {'answer': 'Antonie van Leeu'},
 {'answer': 'Antonie van Leeuwen'},
 {'answer': 'Antonie van Leeuwenho'},
 {'answer': 'Antonie van Leeuwenhoek'}]
```

下面介绍一些具体的具有其他功能的输出解析器。

3.3.1 列表解析器

当需要返回逗号分隔项的列表时，可以使用列表解析器（List Parser），代码如下：

```
//第 3 章/ ListParser.py
from langchain.output_parsers import CommaSeparatedListOutputParser
from langchain.prompts import PromptTemplate
from langchain.llms import OpenAI

output_parser = CommaSeparatedListOutputParser()

format_instructions = output_parser.get_format_instructions()
prompt = PromptTemplate(
    template="List five {subject}.\n{format_instructions}",
    input_variables=["subject"],
    partial_variables={"format_instructions": format_instructions}
)

model = OpenAI(temperature=0)

_input = prompt.format(subject="ice cream flavors")
```

```
output = model(_input)

output_parser.parse(output)
```

输出的结果如下：

```
['Vanilla',
 'Chocolate',
 'Strawberry',
 'Mint Chocolate Chip',
 'Cookies and Cream']
```

3.3.2 日期解析器

日期解析器(Datetime Parser)可用于将 LLM 输出解析为日期时间格式，代码如下：

```
//第 3 章/ DatetimeParser.py
from langchain.prompts import PromptTemplate
from langchain.output_parsers import DatetimeOutputParser
from langchain.chains import LLMChain
from langchain.llms import OpenAI

output_parser = DatetimeOutputParser()
template = """Answer the users question:

{question}

{format_instructions}"""
prompt = PromptTemplate.from_template(
    template,
    partial_variables={"format_instructions": output_parser.get_format_instructions()},
)

chain = LLMChain(prompt=prompt, llm=OpenAI())
output = chain.run("around when was bitcoin founded?")
```

输出的结果如下：

```
Output
  '\n\n2008-01-03T18:15:05.000000Z'
output_parser.parse(output)
  datetime.datetime(2008, 1, 3, 18, 15, 5)
```

3.3.3 自动修复解析器

自动修复解析器(Auto-fixing Parser)嵌套在另一个输出解析器之上,如果第1个输出解析器失败,则会调用另一个 LLM 来修复任何错误,但是除了可以抛出错误之外,开发者使用它还可以做其他事情。具体来讲可以将格式错误的输出及格式化的指令传递给模型,并要求它修复它。在以下示例中将使用上面的 Pydantic 输出解析器。查看向它传递一个不符合架构的结果会发生什么情况,代码如下:

```
//第 3 章/Autofixingparser.py
from langchain.chat_models import ChatOpenAI
from langchain.output_parsers import PydanticOutputParser
from langchain.pydantic_v1 import BaseModel, Field
from typing import List
class Actor(BaseModel):
    name: str = Field(description="name of an actor")
    film_names: List[str] = Field(description="list of names of films they starred in")

actor_query = "Generate the filmography for a random actor."

parser = PydanticOutputParser(pydantic_object=Actor)

misformatted = "{'name': 'Tom Hanks', 'film_names': ['Forrest Gump']}"
parser.parse(misformatted)
```

输出的结果如下:

```
---------------------------------------------------------------------------
JSONDecodeError                           Traceback (most recent call last)
File ~/workplace/langchain/langchain/output_parsers/pydantic.py:23, in PydanticOutputParser.parse(self, text)
     22     json_str = match.group()
---> 23 json_object = json.loads(json_str)
     24 return self.pydantic_object.parse_obj(json_object)

File ~/.pyenv/versions/3.9.1/lib/python3.9/json/__init__.py:346, in loads(s, cls, object_hook, parse_float, parse_int, parse_constant, object_pairs_hook, **kw)
    343 if (cls is None and object_hook is None and
    344         parse_int is None and parse_float is None and
    345         parse_constant is None and object_pairs_hook is None and not kw):
```

```
--> 346     return _default_decoder.decode(s)
    347 if cls is None:


File ~/.pyenv/versions/3.9.1/lib/python3.9/json/decoder.py:337, in JSONDecoder.decode(self, s, _w)
    333 """Return the Python representation of ``s`` (a ``str`` instance
    334 containing a JSON document).
    335
    336 """
--> 337 obj, end = self.raw_decode(s, idx=_w(s, 0).end())
    338 end = _w(s, end).end()


File ~/.pyenv/versions/3.9.1/lib/python3.9/json/decoder.py:353, in JSONDecoder.raw_decode(self, s, idx)
    352 try:
--> 353     obj, end = self.scan_once(s, idx)
    354 except StopIteration as err:


JSONDecodeError: Expecting property name enclosed in double quotes: line 1 column 2 (char 1)


During handling of the above exception, another exception occurred:


OutputParserException                     Traceback (most recent call last)

Cell In[6], line 1
----> 1 parser.parse(misformatted)


File ~/workplace/langchain/langchain/output_parsers/pydantic.py:29, in PydanticOutputParser.parse(self, text)
     27 name = self.pydantic_object.__name__
     28 msg = f"Failed to parse {name} from completion {text}. Got: {e}"
---> 29 raise OutputParserException(msg)


OutputParserException: Failed to parse Actor from completion {'name': 'Tom Hanks', 'film_names': ['Forrest Gump']}. Got: Expecting property name enclosed in double quotes: line 1 column 2 (char 1)
```

现在用户可以构造和使用OutputFixingParser了。此输出解析器将另一个输出解析

器作为参数,但也将 LLM 作为参数,以尝试使用该解析器来纠正任何格式错误,代码如下:

```
from langchain.output_parsers import OutputFixingParser
new_parser = OutputFixingParser.from_llm(parser=parser, llm=ChatOpenAI())
new_parser.parse(misformatted)
Actor(name='Tom Hanks', film_names=['Forrest Gump'])
```

3.3.4　Pydantic(JSON)解析器

此输出解析器允许用户指定任意 JSON 架构,并在 LLM 中查询符合该架构的 JSON 输出。本书已经在前面的章节中阐述,LLM 很多时候存在幻觉现象,导致输出的格式不符合用户的要求。用户必须使用具有足够强大的 LLM 来生成格式正确的 JSON 文件。OpenAI 的模型目前拥有这个能力,但是其他的一些模型还没有这样的能力。这时 JSON 解析器就非常重要了。

用户可以使用 Pydantic 声明数据模型。Pydantic 的 BaseModel 类似于 Python 数据类,但具有实际的类型检查等功能,代码如下:

```
//第 3 章/Pydantic.py
from typing import List

from langchain.llms import OpenAI
from langchain.output_parsers import PydanticOutputParser
from langchain.prompts import PromptTemplate
from langchain.pydantic_v1 import BaseModel, Field, validator

model_name = "text-davinci-003"
temperature = 0.0
model = OpenAI(model_name=model_name, temperature=temperature)

#定义用户想要的数据结构
class Joke(BaseModel):
    setup: str = Field(description="question to set up a joke")
    punchline: str = Field(description="answer to resolve the joke")

    #可以使用 Pydantic 轻松地添加自定义验证逻辑
    @validator("setup")
    def question_ends_with_question_mark(cls, field):
        if field[-1] != "?":
            raise ValueError("Badly formed question!")
        return field


#旨在提示语言模型填充数据结构的查询
```

```python
joke_query = "Tell me a joke."

#设置解析器+将指令注入提示模板中
parser = PydanticOutputParser(pydantic_object=Joke)

prompt = PromptTemplate(
    template="Answer the user query.\n{format_instructions}\n{query}\n",
    input_variables=["query"],
    partial_variables={"format_instructions": parser.get_format_instructions()},
)

_input = prompt.format_prompt(query=joke_query)

output = model(_input.to_string())

parser.parse(output)
```

输出的结果如下：

```
Joke(setup='Why did the chicken cross the road? ', punchline='To get to the other side!')
```

下面是另一个示例，但使用复合类型字段，代码如下：

```
//第 3 章/PydanticCompound.py
class Actor(BaseModel):
    name: str = Field(description="name of an actor")
    film_names: List[str] = Field(description="list of names of films they starred in")


actor_query = "Generate the filmography for a random actor."

parser = PydanticOutputParser(pydantic_object=Actor)

prompt = PromptTemplate(
    template="Answer the user query.\n{format_instructions}\n{query}\n",
    input_variables=["query"],
    partial_variables={"format_instructions": parser.get_format_instructions()},
)

_input = prompt.format_prompt(query=actor_query)

output = model(_input.to_string())

parser.parse(output)
```

输出的结果如下:

```
Actor(name='Tom Hanks', film_names=['Forrest Gump', 'Saving Private Ryan', 'The Green Mile', 'Cast Away', 'Toy Story'])
```

3.3.5 重试解析器

虽然在某些情况下可以通过仅查看 LLM 输出来修复任何解析错误,但在其他情况下则没有那么简单。例如,输出不仅格式不正确,而且不完整。下面是一个例子,代码如下:

```
//第 3 章/RetryParser.py
from langchain.prompts import (
    PromptTemplate,
    ChatPromptTemplate,
    HumanMessagePromptTemplate,
)
from langchain.llms import OpenAI
from langchain.chat_models import ChatOpenAI
from langchain.output_parsers import (
    PydanticOutputParser,
    OutputFixingParser,
    RetryOutputParser,
)
from pydantic import BaseModel, Field, validator
from typing import List

template = """Based on the user question, provide an Action and Action Input for what step should be taken.
{format_instructions}
Question: {query}
Response:"""


class Action(BaseModel):
    action: str = Field(description="action to take")
    action_input: str = Field(description="input to the action")


parser = PydanticOutputParser(pydantic_object=Action)

prompt = PromptTemplate(
    template="Answer the user query.\n{format_instructions}\n{query}\n",
    input_variables=["query"],
    partial_variables={"format_instructions": parser.get_format_instructions()},
)
```

```
prompt_value = prompt.format_prompt(query="who is leo di caprios gf?")

bad_response = '{"action": "search"}'
```

如果用户尝试按原样解析此响应,则将得到一个错误,代码如下:

```
parser.parse(bad_response)
//第 3 章/RetryParserError.py
---------------------------------------------------------------
ValidationError                           Traceback (most recent call last)
    File ~/workplace/langchain/langchain/output_parsers/pydantic.py:24, in
PydanticOutputParser.parse(self, text)
     23      json_object = json.loads(json_str)
---> 24      return self.pydantic_object.parse_obj(json_object)
     26 except (json.jsonDecodeError, ValidationError) as e:

    File ~/.pyenv/versions/3.9.1/envs/langchain/lib/python3.9/site-packages/
pydantic/main.py:527, in pydantic.main.BaseModel.parse_obj()

    File ~/.pyenv/versions/3.9.1/envs/langchain/lib/python3.9/site-packages/
pydantic/main.py:342, in pydantic.main.BaseModel.__init__()

ValidationError: 1 validation error for Action
action_input
  field required (type=value_error.missing)

During handling of the above exception, another exception occurred:

OutputParserException                     Traceback (most recent call last)
Cell In[6], line 1
----> 1 parser.parse(bad_response)

    File ~/workplace/langchain/langchain/output_parsers/pydantic.py:29, in
PydanticOutputParser.parse(self, text)
```

```
        27 name = self.pydantic_object.__name__
        28 msg = f"Failed to parse {name} from completion {text}. Got: {e}"
---> 29 raise OutputParserException(msg)


OutputParserException: Failed to parse Action from completion {"action":
"search"}. Got: 1 validation error for Action
    action_input
        field required (type=value_error.missing)
```

如果用户尝试使用 OutputFixingParser 来改正此错误，程序则会感到困惑。也就是说，它不知道实际要为操作输入什么值，代码如下：

```
fix_parser = OutputFixingParser.from_llm(parser=parser, llm=ChatOpenAI())
fix_parser.parse(bad_response)
Action(action='search', action_input='')
```

相反用户可以使用重试解析器（Retry Output Parser），它通过传入提示（及原始输出）再次尝试以获得更好的响应，代码如下：

```
from langchain.output_parsers import RetryWithErrorOutputParser
retry_parser = RetryWithErrorOutputParser.from_llm(
    parser=parser, llm=OpenAI(temperature=0)
)
retry_parser.parse_with_prompt(bad_response, prompt_value)
Action(action='search', action_input='who is leo di caprios gf?')
```

3.3.6　结构化输出解析器

当用户想要返回多个字段时，可以使用结构化输出解析器（Structured Output Parser），代码如下：

```
from langchain.output_parsers import StructuredOutputParser, ResponseSchema
from langchain.prompts import PromptTemplate, ChatPromptTemplate,
    HumanMessagePromptTemplate
from langchain.llms import OpenAI
from langchain.chat_models import ChatOpenAI
```

同时定义用户期望接收的响应模式，代码如下：

```
response_schemas = [
    ResponseSchema(name="answer", description="answer to the user's question"),
```

```
        ResponseSchema(name="source", description="source used to answer the user's 
question, should be a website.")
]
output_parser = StructuredOutputParser.from_response_schemas(response_
schemas)
```

现在得到一个字符串,其中包含有关如何格式化响应的说明,然后将其插入提示符中,代码如下:

```
format_instructions = output_parser.get_format_instructions()
prompt = PromptTemplate(
    template="answer the users question as best as possible.\n{format_
instructions}\n{question}",
    input_variables=["question"],
    partial_variables={"format_instructions": format_instructions}
)
```

现在用户可以使用它设置要发送到 LLM 的提示的格式,然后解析返回的结果,代码如下:

```
model = OpenAI(temperature=0)
_input = prompt.format_prompt(question="what's the capital of france?")
output = model(_input.to_string())
output_parser.parse(output)

{'answer': 'Paris',
    'source': 'https://www.worldatlas.com/articles/what-is-the-capital-of-
france.html'}
```

下面是在聊天模型中使用它的示例,代码如下:

```
//第 3 章/StructuredOutputParser.py
chat_model = ChatOpenAI(temperature=0)

prompt = ChatPromptTemplate(
    messages=[
        HumanMessagePromptTemplate.from_template("answer the users question as 
best as possible.\n{format_instructions}\n{question}")
    ],
    input_variables=["question"],
    partial_variables={"format_instructions": format_instructions}
)
_input = prompt.format_prompt(question="what's the capital of france?")
output = chat_model(_input.to_messages())
output_parser.parse(output.content)
```

得到的最终结果如下：

```
{'answer': 'Paris', 'source': 'https://en.wikipedia.org/wiki/Paris'}
```

3.3.7　XML 解析器

XML 解析器允许用户以流行的 XML 格式从 LLM 获取结果。如果没有 XML 解析器，则用户通常必须使用具有足够能力的 LLM 来生成格式正确的 XML。在下面的示例中使用了本书在前面介绍过的 Claude 模型，该模型非常适合做 XML 标记，代码如下：

```
//第 3 章/XmlParser.py

from langchain.prompts import PromptTemplate
from langchain.llms import Anthropic
from langchain.output_parsers import XMLOutputParser

model = Anthropic(model="claude-2", max_tokens_to_sample=512, temperature=0.1)

/Users/harrisonchase/workplace/langchain/libs/langchain/langchain/llms/
anthropic.py:171: UserWarning: This Anthropic LLM is deprecated. Please use
`from langchain.chat_models import ChatAnthropic` instead
    warnings.warn(
```

接下来从对模型的简单请求开始，代码如下：

```
//第 3 章/simpleRequest.py
actor_query = "Generate the shortened filmography for Tom Hanks."
output = model(
    f"""
```

```
<movie>A League of Their Own (1992)</movie>
<movie>Sleepless in Seattle (1993)</movie>
<movie>Forrest Gump (1994)</movie>
<movie>Apollo 13 (1995)</movie>
<movie>Toy Story (1995)</movie>
<movie>Saving Private Ryan (1998)</movie>
<movie>Cast Away (2000)</movie>
<movie>The Da Vinci Code (2006)</movie>
<movie>Toy Story 3 (2010)</movie>
<movie>Captain Phillips (2013)</movie>
<movie>Bridge of Spies (2015)</movie>
<movie>Toy Story 4 (2019)</movie>
```

现在将使用 XMLOutputParser 来获得结构化输出，代码如下：

```
//第 3 章/xmlParser.py
parser = XMLOutputParser(tags=["movies", "actor", "film", "name", "genre"])
prompt = PromptTemplate(
    template="""

    Human:
    {query}
    {format_instructions}
    Assistant:""",
    input_variables=["query"],
    partial_variables={"format_instructions": parser.get_format_instructions()},
)


chain = prompt | model | parser

output = chain.invoke({"query": actor_query})

print(output)
```

输出的结果如下：

```
{'filmography': [{'movie': [{'title': 'Splash'}, {'year': '1984'}]}, {'movie': [{'title': 'Big'}, {'year': '1988'}]}, {'movie': [{'title': 'A League of Their Own'}, {'year': '1992'}]}, {'movie': [{'title': 'Sleepless in Seattle'}, {'year': '1993'}]}, {'movie': [{'title': 'Forrest Gump'}, {'year': '1994'}]}, {'movie': [{'title': 'Toy Story'}, {'year': '1995'}]}, {'movie': [{'title': 'Apollo 13'}, {'year': '1995'}]}, {'movie': [{'title': 'Saving Private Ryan'}, {'year': '1998'}]}, {'movie': [{'title': 'Cast Away'}, {'year': '2000'}]}, {'movie': [{'title': 'Catch Me If You Can'}, {'year': '2002'}]}, {'movie': [{'title': 'The Polar Express'}, {'year': '2004'}]}, {'movie': [{'title': 'Bridge of Spies'}, {'year': '2015'}]}]}
```

最后添加一些标签来根据用户的需求定制输出,代码如下:

```python
//第 3 章/customParser.py
parser = XMLOutputParser(tags=["movies", "actor", "film", "name", "genre"])
prompt = PromptTemplate(
    template="""

    Human:
    {query}
    {format_instructions}
    Assistant:""",
    input_variables=["query"],
    partial_variables={"format_instructions": parser.get_format_instructions()},
)


chain = prompt | model | parser

output = chain.invoke({"query": actor_query})

print(output)
```

输出的结果如下:

```
{'movies': [{'actor': [{'name': 'Tom Hanks'}, {'film': [{'name': 'Splash'},
{'genre': 'Comedy'}]}, {'film': [{'name': 'Big'}, {'genre': 'Comedy'}]}, {'film':
[{'name': 'A League of Their Own'}, {'genre': 'Comedy'}]}, {'film': [{'name':
'Sleepless in Seattle'}, {'genre': 'Romance'}]}, {'film': [{'name': 'Forrest
Gump'}, {'genre': 'Drama'}]}, {'film': [{'name': 'Toy Story'}, {'genre':
'Animation'}]}, {'film': [{'name': 'Apollo 13'}, {'genre': 'Drama'}]}, {'film':
[{'name': 'Saving Private Ryan'}, {'genre': 'War'}]}, {'film': [{'name': 'Cast
Away'}, {'genre': 'Adventure'}]}, {'film': [{'name': 'The Green Mile'}, {'genre':
'Drama'}]}]}]}
```

3.4　LangChain 表达式语言

　　LangChain 表达式语言(LCEL)是一种声明式语言,可以轻松地将链组合在一起。使用 LCEL 的优点如下。

　　(1) 流媒体支持:当用户使用 LCEL 构建链时将获得最佳的首次 Token 时间(直到第 1 个输出块出来之前经过的时间)。对于某些链,这意味着用户将 Token 直接从 LLM 流式传输到流式输出解析器,然后以与 LLM 提供程序输出原始 Token 相同的速率返回解析的增量输出块。

(2) 异步支持：任何使用 LCEL 构建的链都可以使用同步 API(例如，在原型设计时在 Jupyter Notebook 中)和异步 API(例如，在 LangServe 服务器中)调用。这样就可以在原型和生产中使用相同的代码，并且能够在同一服务器中处理许多并发请求。

(3) 优化的并行执行：每当用户的 LCEL 链具有可以并行执行的步骤时(例如，如果从多个检索器获取文档)都会在同步和异步接口中自动以尽可能小的延迟执行此操作。

(4) 重试和回退：为 LCEL 链的任何部分配置重试和回退操作。这是使用户的链在规模上更加可靠的好方法。LangChain 目前正在努力添加对重试/回退的流式处理支持，因此用户可以在不产生任何延迟成本的情况下获得更高的可靠性。

(5) 访问中间结果：对于更复杂的链，甚至在产生最终输出之前访问中间步骤的结果通常非常有用。这可以用来让最终用户知道发生了什么，甚至只是用来调试链。用户可以流式传输中间结果，并且它在每个 LangServe 服务器上都可用。

(6) 输入和输出模式：输入和输出模式为每个 LCEL 链提供从链结构推断的 Pydantic 和 JSONSchema 模式。这可用于验证输入和输出，并且是 LangServe 的一个组成部分。

(7) 无缝 LangSmith 跟踪集成：随着用户的链变得越来越复杂，了解每步到底发生了什么变得越来越重要。使用 LCEL，所有步骤都会被自动地记录到 LangSmith 中，以实现最大的可观察性和可调试性。

3.4.1 LCEL 接口简介

为了尽可能轻松地创建自定义链，LangChain 实现了一个 Runnable 协议。Runnable 协议是为大多数组件实现的。这是一个标准接口，可以很容易地定义自定义链，并以标准方式调用它们。标准接口包括以下几个同步方法。

(1) stream：回流响应的块。

(2) invoke：在输入上调用链。

(3) batch：在输入列表上调用链。

它们还具有相应的异步方法。

(1) astream：流回响应异步的块。

(2) ainvoke：在输入异步上调用链。

(3) abatch：在输入异步列表中调用链。

(4) astream_log：除了最终响应之外，还会在中间步骤发生时流回它们。

为了深入地查看这些方法，这里将创建一个简单的 PromptTemplate + ChatModel 链，代码如下：

```
//第 3 章/PromptTemplateChatModel.py
from langchain.prompts import ChatPromptTemplate
from langchain.chat_models import ChatOpenAI

model = ChatOpenAI()
```

```
prompt = ChatPromptTemplate.from_template("tell me a joke about {topic}")
chain = prompt | model
```

1. 输入架构

输入架构是 Runnable 接受的输入的描述。这是一个 Pydantic 模型，由任何 Runnable 的结构动态地生成。用户可以对其调用.schema()获取 JSONSchema 表示形式，代码如下：

```
//第 3 章/runnableSchema.py

#链的输入模式是其第一部分(提示)的输入模式
chain.input_schema.schema()

    {'title': 'PromptInput',
     'type': 'object',
     'properties': {'topic': {'title': 'Topic', 'type': 'string'}}}

prompt.input_schema.schema()
    {'title': 'PromptInput',
     'type': 'object',
     'properties': {'topic': {'title': 'Topic', 'type': 'string'}}}
```

模型输出的代码如下：

```
//第 3 章/InputSchemaOutput.py

model.input_schema.schema()

{'title': 'ChatOpenAIInput',
    'anyOf': [{'type': 'string'},
     {'$ref': '#/definitions/StringPromptValue'},
     {'$ref': '#/definitions/ChatPromptValueConcrete'},
     {'type': 'array',
      'items': {'anyOf': [{'$ref': '#/definitions/AIMessage'},
        {'$ref': '#/definitions/HumanMessage'},
        {'$ref': '#/definitions/ChatMessage'},
        {'$ref': '#/definitions/SystemMessage'},
        {'$ref': '#/definitions/FunctionMessage'}]}}],
    'definitions': {'StringPromptValue': {'title': 'StringPromptValue',
      'description': 'String prompt value.',
      'type': 'object',
      'properties': {'text': {'title': 'Text', 'type': 'string'},
       'type': {'title': 'Type',
        'default': 'StringPromptValue',
        'enum': ['StringPromptValue'],
        'type': 'string'}},
```

```
          'required': ['text']},
         'AIMessage': {'title': 'AIMessage',
          'description': 'A Message from an AI.',
          'type': 'object',
          'properties': {'content': {'title': 'Content', 'type': 'string'},
           'additional_kwargs': {'title': 'Additional Kwargs', 'type': 'object'},
           'type': {'title': 'Type',
            'default': 'ai',
            'enum': ['ai'],
            'type': 'string'},
           'example': {'title': 'Example', 'default': False, 'type': 'boolean'}},
          'required': ['content']},
         'HumanMessage': {'title': 'HumanMessage',
          'description': 'A Message from a human.',
          'type': 'object',
          'properties': {'content': {'title': 'Content', 'type': 'string'},
           'additional_kwargs': {'title': 'Additional Kwargs', 'type': 'object'},
           'type': {'title': 'Type',
            'default': 'human',
            'enum': ['human'],
            'type': 'string'},
           'example': {'title': 'Example', 'default': False, 'type': 'boolean'}},
          'required': ['content']},
         'ChatMessage': {'title': 'ChatMessage',
          'description': 'A Message that can be assigned an arbitrary speaker (i.e. role).',
          'type': 'object',
          'properties': {'content': {'title': 'Content', 'type': 'string'},
           'additional_kwargs': {'title': 'Additional Kwargs', 'type': 'object'},
           'type': {'title': 'Type',
            'default': 'chat',
            'enum': ['chat'],
            'type': 'string'},
           'role': {'title': 'Role', 'type': 'string'}},
          'required': ['content', 'role']},
         'SystemMessage': {'title': 'SystemMessage',
          'description': 'A Message for priming AI behavior, usually passed in as the first of a sequence\nof input messages.',
          'type': 'object',
          'properties': {'content': {'title': 'Content', 'type': 'string'},
           'additional_kwargs': {'title': 'Additional Kwargs', 'type': 'object'},
           'type': {'title': 'Type',
            'default': 'system',
            'enum': ['system'],
            'type': 'string'}},
          'required': ['content']},
```

```
            'FunctionMessage': {'title': 'FunctionMessage',
             'description': 'A Message for passing the result of executing a function
    back to a model.',
             'type': 'object',
             'properties': {'content': {'title': 'Content', 'type': 'string'},
              'additional_kwargs': {'title': 'Additional Kwargs', 'type': 'object'},
              'type': {'title': 'Type',
               'default': 'function',
               'enum': ['function'],
               'type': 'string'},
              'name': {'title': 'Name', 'type': 'string'}},
             'required': ['content', 'name']},
            'ChatPromptValueConcrete': {'title': 'ChatPromptValueConcrete',
             'description': 'Chat prompt value which explicitly lists out the message
    types it accepts.\nFor use in external schemas.',
             'type': 'object',
             'properties': {'messages': {'title': 'Messages',
               'type': 'array',
               'items': {'anyOf': [{'$ref': '#/definitions/AIMessage'},
                 {'$ref': '#/definitions/HumanMessage'},
                 {'$ref': '#/definitions/ChatMessage'},
                 {'$ref': '#/definitions/SystemMessage'},
                 {'$ref': '#/definitions/FunctionMessage'}]}},
              'type': {'title': 'Type',
               'default': 'ChatPromptValueConcrete',
               'enum': ['ChatPromptValueConcrete'],
               'type': 'string'}},
             'required': ['messages']}}}
```

2．输出架构

输出架构是对 Runnable 生成的输出的描述。这是一个 Pydantic 模型，由任何 Runnable 的结构动态地生成。用户可以对其调用.schema()获取 JSONSchema 表示形式，代码如下：

```
#链的输出架构是其最后一部分的输出架构,在本例中为 ChatModel,它输出 ChatMessage
chain.output_schema.schema()
```

输出的代码如下：

```
//第 3 章/OuputSchemaOutput.py
{'title': 'ChatOpenAIOutput',
    'anyOf': [{'$ref': '#/definitions/HumanMessage'},
     {'$ref': '#/definitions/AIMessage'},
     {'$ref': '#/definitions/ChatMessage'},
```

```
            {'$ref': '#/definitions/FunctionMessage'},
            {'$ref': '#/definitions/SystemMessage'}],
 'definitions': {'HumanMessage': {'title': 'HumanMessage',
   'description': 'A Message from a human.',
   'type': 'object',
   'properties': {'content': {'title': 'Content', 'type': 'string'},
    'additional_kwargs': {'title': 'Additional Kwargs', 'type': 'object'},
    'type': {'title': 'Type',
     'default': 'human',
     'enum': ['human'],
     'type': 'string'},
    'example': {'title': 'Example', 'default': False, 'type': 'boolean'}},
   'required': ['content']},
  'AIMessage': {'title': 'AIMessage',
   'description': 'A Message from an AI.',
   'type': 'object',
   'properties': {'content': {'title': 'Content', 'type': 'string'},
    'additional_kwargs': {'title': 'Additional Kwargs', 'type': 'object'},
    'type': {'title': 'Type',
     'default': 'ai',
     'enum': ['ai'],
     'type': 'string'},
    'example': {'title': 'Example', 'default': False, 'type': 'boolean'}},
   'required': ['content']},
  'ChatMessage': {'title': 'ChatMessage',
   'description': 'A Message that can be assigned an arbitrary speaker (i.e. role).',
   'type': 'object',
   'properties': {'content': {'title': 'Content', 'type': 'string'},
    'additional_kwargs': {'title': 'Additional Kwargs', 'type': 'object'},
    'type': {'title': 'Type',
     'default': 'chat',
     'enum': ['chat'],
     'type': 'string'},
    'role': {'title': 'Role', 'type': 'string'}},
   'required': ['content', 'role']},
  'FunctionMessage': {'title': 'FunctionMessage',
   'description': 'A Message for passing the result of executing a function back to a model.',
   'type': 'object',
   'properties': {'content': {'title': 'Content', 'type': 'string'},
    'additional_kwargs': {'title': 'Additional Kwargs', 'type': 'object'},
    'type': {'title': 'Type',
     'default': 'function',
     'enum': ['function'],
     'type': 'string'},
```

```
            'name': {'title': 'Name', 'type': 'string'}},
           'required': ['content', 'name']},
          'SystemMessage': {'title': 'SystemMessage',
           'description': 'A Message for priming AI behavior, usually passed in as the
first of a sequence\nof input messages.',
           'type': 'object',
           'properties': {'content': {'title': 'Content', 'type': 'string'},
            'additional_kwargs': {'title': 'Additional Kwargs', 'type': 'object'},
            'type': {'title': 'Type',
             'default': 'system',
             'enum': ['system'],
             'type': 'string'}},
           'required': ['content']}}}
```

Stream 方法的实现代码如下：

```
for s in chain.stream({"topic": "bears"}):
    print(s.content, end="", flush=True)

Why don't bears wear shoes?

    Because they already have bear feet!
```

Invoke 方法的实现代码如下：

```
chain.invoke({"topic": "bears"})
    AIMessage(content="Why don't bears wear shoes? \n\nBecause they already have
bear feet!")
```

Batch 方法的实现代码如下：

```
chain.batch([{"topic": "bears"}, {"topic": "cats"}])
    [AIMessage(content="Why don't bears wear shoes? \n\nBecause they have bear
feet!"),
     AIMessage(content="Why don't cats play poker in the wild? \n\nToo many
cheetahs!")]
```

用户可以使用 max_concurrency 参数设置并发请求数，代码如下：

```
chain.batch([{"topic": "bears"}, {"topic": "cats"}], config={"max_
concurrency": 5})
    [AIMessage(content="Why don't bears wear shoes? \n\nBecause they have bear
feet!"),
     AIMessage(content="Why don't cats play poker in the wild? \n\nToo many
cheetahs!")]
```

3. 异步流中间步骤

所有的 Runnable 对象还有一种方法.astream_log()，用于流式传输（当它们发生时）链/序列的全部或部分中间步骤。这对于向用户显示进度、使用中间结果或调试链非常有用。用户可以流式传输所有步骤（默认），也可以按名称、标记或元数据包含/排除步骤。此方法生成 JSONPatch 操作，当接收相同的顺序应用时会生成 RunState，代码如下：

```
//第 3 章/Asyncinter.py
class LogEntry(TypedDict):
    id: str
    """子运行的 ID"""
    name: str
    """正在运行的对象的名称"""
    type: str
    """正在运行的对象的类型,例如提示、连锁、LLM 等"""
    tags: List[str]
    """运行的标签列表"""
    metadata: Dict[str, Any]
    """运行的元数据的键-值对"""
    start_time: str
    """运行开始时的 ISO-8601 时间戳"""

    streamed_output_str: List[str]
    """本次运行流式传输的 LLM Token 列表(如果适用)"""
    final_output: Optional[Any]
    """本次运行的最终输出
    仅在运行成功完成后可用"""
    end_time: Optional[str]
    """运行结束时的 ISO-8601 时间戳
    仅在运行完成后可用"""


class RunState(TypedDict):
    id: str
    """运行的 ID"""
    streamed_output: List[Any]
    """由 Runnable.stream() 流式传输的输出块列表"""
    final_output: Optional[Any]
    """运行的最终输出,通常是聚合(`+`)streamed_output 的结果
    仅在运行成功完成后可用"""

    logs: Dict[str, LogEntry]
    """运行名称到子运行的映射 如果提供了过滤器,则此列表将仅包含与过滤器匹配的运行"""
```

如果用户需要流式处理 JSONPatch 块，例如，在 HTTP 服务器中流式传输 JSONPatch，则需要在客户端上应用操作以在其中重建运行状态。用户可以参阅 LangServe 的工具，以便更

轻松地从任何 Runnable 线程构建 Web 服务器,代码如下:

```python
//第 3 章/Jsonpatch.py
from langchain.embeddings import OpenAIEmbeddings
from langchain.schema.output_parser import StrOutputParser
from langchain.schema.runnable import RunnablePassthrough
from langchain.vectorstores import FAISS

template = """Answer the question based only on the following context:
{context}

Question: {question}
"""
prompt = ChatPromptTemplate.from_template(template)

vectorstore = FAISS.from_texts(
    ["harrison worked at kensho"], embedding=OpenAIEmbeddings()
)
retriever = vectorstore.as_retriever()

retrieval_chain = (
    {
        "context": retriever.with_config(run_name="Docs"),
        "question": RunnablePassthrough(),
    }
    | prompt
    | model
    | StrOutputParser()
)

async for chunk in retrieval_chain.astream_log(
    "where did harrison work?", include_names=["Docs"]
):
    print("-" * 40)
    print(chunk)
```

```
//第 3 章/Jsonpatchout.py
----------------------------------------
    RunLogPatch({'op': 'replace',
      'path': '',
      'value': {'final_output': None,
              'id': 'e2f2cc72-eb63-4d20-8326-237367482efb',
              'logs': {},
              'streamed_output': []}})
```

```
----------------------------------------
RunLogPatch({'op': 'add',
  'path': '/logs/Docs',
  'value': {'end_time': None,
            'final_output': None,
            'id': '8da492cc-4492-4e74-b8b0-9e60e8693390',
            'metadata': {},
            'name': 'Docs',
            'start_time': '2023-10-19T17:50:13.526',
            'streamed_output_str': [],
            'tags': ['map:key:context', 'FAISS'],
            'type': 'retriever'}})
----------------------------------------
RunLogPatch({'op': 'add',
  'path': '/logs/Docs/final_output',
  'value': {'documents': [Document(page_content='harrison worked at kensho')]}},
 {'op': 'add',
  'path': '/logs/Docs/end_time',
  'value': '2023-10-19T17:50:13.713'})
----------------------------------------
RunLogPatch({'op': 'add', 'path': '/streamed_output/-', 'value': ''})
----------------------------------------
RunLogPatch({'op': 'add', 'path': '/streamed_output/-', 'value': 'H'})
----------------------------------------
RunLogPatch({'op': 'add', 'path': '/streamed_output/-', 'value': 'arrison'})
----------------------------------------
RunLogPatch({'op': 'add', 'path': '/streamed_output/-', 'value': ' worked'})
----------------------------------------
RunLogPatch({'op': 'add', 'path': '/streamed_output/-', 'value': ' at'})
----------------------------------------
RunLogPatch({'op': 'add', 'path': '/streamed_output/-', 'value': ' Kens'})
----------------------------------------
RunLogPatch({'op': 'add', 'path': '/streamed_output/-', 'value': 'ho'})
----------------------------------------
RunLogPatch({'op': 'add', 'path': '/streamed_output/-', 'value': '.'})
----------------------------------------
RunLogPatch({'op': 'add', 'path': '/streamed_output/-', 'value': ''})
----------------------------------------
RunLogPatch({'op': 'replace',
  'path': '/final_output',
  'value': {'output': 'Harrison worked at Kensho.'}})
```

如果用户需要流式处理增量 RunState，则只需传递 diff＝False 便可获取 RunState 的增量值。同时也可以获得更详细的输出和更多重复的部分，代码如下：

```
async for chunk in retrieval_chain.astream_log(
    "where did harrison work?", include_names=["Docs"], diff=False
):
    print("-" * 70)
    print(chunk)
```

输出的代码如下:

```
//第 3 章/Jsonpatchincrement.py
----------------------------------------------------------------------
RunLog({'final_output': None,
 'id': 'afe66178-d75f-4c2d-b348-b1d144239cd6',
 'logs': {},
 'streamed_output': []})
----------------------------------------------------------------------
RunLog({'final_output': None,
 'id': 'afe66178-d75f-4c2d-b348-b1d144239cd6',
 'logs': {'Docs': {'end_time': None,
                   'final_output': None,
                   'id': '88d51118-5756-4891-89c5-2f6a5e90cc26',
                   'metadata': {},
                   'name': 'Docs',
                   'start_time': '2023-10-19T17:52:15.438',
                   'streamed_output_str': [],
                   'tags': ['map:key:context', 'FAISS'],
                   'type': 'retriever'}},
 'streamed_output': []})
----------------------------------------------------------------------
RunLog({'final_output': None,
 'id': 'afe66178-d75f-4c2d-b348-b1d144239cd6',
 'logs': {'Docs': {'end_time': '2023-10-19T17:52:15.738',
                   'final_output': {'documents': [Document(page_content='harrison worked at kensho')]},
                   'id': '88d51118-5756-4891-89c5-2f6a5e90cc26',
                   'metadata': {},
                   'name': 'Docs',
                   'start_time': '2023-10-19T17:52:15.438',
                   'streamed_output_str': [],
                   'tags': ['map:key:context', 'FAISS'],
                   'type': 'retriever'}},
 'streamed_output': []})
----------------------------------------------------------------------
RunLog({'final_output': None,
 'id': 'afe66178-d75f-4c2d-b348-b1d144239cd6',
 'logs': {'Docs': {'end_time': '2023-10-19T17:52:15.738',
```

```
                              'final_output': {'documents': [Document(page_content=
'harrison worked at kensho')]},
                              'id': '88d51118-5756-4891-89c5-2f6a5e90cc26',
                              'metadata': {},
                              'name': 'Docs',
                              'start_time': '2023-10-19T17:52:15.438',
                              'streamed_output_str': [],
                              'tags': ['map:key:context', 'FAISS'],
                              'type': 'retriever'}},
 'streamed_output': ['']})
-------------------------------------------------------------
RunLog({'final_output': None,
 'id': 'afe66178-d75f-4c2d-b348-b1d144239cd6',
 'logs': {'Docs': {'end_time': '2023-10-19T17:52:15.738',
                              'final_output': {'documents': [Document(page_content=
'harrison worked at kensho')]},
                              'id': '88d51118-5756-4891-89c5-2f6a5e90cc26',
                              'metadata': {},
                              'name': 'Docs',
                              'start_time': '2023-10-19T17:52:15.438',
                              'streamed_output_str': [],
                              'tags': ['map:key:context', 'FAISS'],
                              'type': 'retriever'}},
 'streamed_output': ['', 'H']})
-------------------------------------------------------------
RunLog({'final_output': None,
 'id': 'afe66178-d75f-4c2d-b348-b1d144239cd6',
 'logs': {'Docs': {'end_time': '2023-10-19T17:52:15.738',
                              'final_output': {'documents': [Document(page_content=
'harrison worked at kensho')]},
                              'id': '88d51118-5756-4891-89c5-2f6a5e90cc26',
                              'metadata': {},
                              'name': 'Docs',
                              'start_time': '2023-10-19T17:52:15.438',
                              'streamed_output_str': [],
                              'tags': ['map:key:context', 'FAISS'],
                              'type': 'retriever'}},
 'streamed_output': ['', 'H', 'arrison']})
-------------------------------------------------------------
RunLog({'final_output': None,
 'id': 'afe66178-d75f-4c2d-b348-b1d144239cd6',
 'logs': {'Docs': {'end_time': '2023-10-19T17:52:15.738',
                              'final_output': {'documents': [Document(page_content=
'harrison worked at kensho')]},
                              'id': '88d51118-5756-4891-89c5-2f6a5e90cc26',
                              'metadata': {},
```

```
                        'name': 'Docs',
                        'start_time': '2023-10-19T17:52:15.438',
                        'streamed_output_str': [],
                        'tags': ['map:key:context', 'FAISS'],
                        'type': 'retriever'}},
    'streamed_output': ['', 'H', 'arrison', ' worked']})
---------------------------------------------------------------
    RunLog({'final_output': None,
    'id': 'afe66178-d75f-4c2d-b348-b1d144239cd6',
    'logs': {'Docs': {'end_time': '2023-10-19T17:52:15.738',
                        'final_output': {'documents': [Document(page_content=
'harrison worked at kensho')]},
                        'id': '88d51118-5756-4891-89c5-2f6a5e90cc26',
                        'metadata': {},
                        'name': 'Docs',
                        'start_time': '2023-10-19T17:52:15.438',
                        'streamed_output_str': [],
                        'tags': ['map:key:context', 'FAISS'],
                        'type': 'retriever'}},
    'streamed_output': ['', 'H', 'arrison', ' worked', ' at']})
---------------------------------------------------------------
    RunLog({'final_output': None,
    'id': 'afe66178-d75f-4c2d-b348-b1d144239cd6',
    'logs': {'Docs': {'end_time': '2023-10-19T17:52:15.738',
                        'final_output': {'documents': [Document(page_content=
'harrison worked at kensho')]},
                        'id': '88d51118-5756-4891-89c5-2f6a5e90cc26',
                        'metadata': {},
                        'name': 'Docs',
                        'start_time': '2023-10-19T17:52:15.438',
                        'streamed_output_str': [],
                        'tags': ['map:key:context', 'FAISS'],
                        'type': 'retriever'}},
    'streamed_output': ['', 'H', 'arrison', ' worked', ' at', ' Kens']})
---------------------------------------------------------------
    RunLog({'final_output': None,
    'id': 'afe66178-d75f-4c2d-b348-b1d144239cd6',
    'logs': {'Docs': {'end_time': '2023-10-19T17:52:15.738',
                        'final_output': {'documents': [Document(page_content=
'harrison worked at kensho')]},
                        'id': '88d51118-5756-4891-89c5-2f6a5e90cc26',
                        'metadata': {},
                        'name': 'Docs',
                        'start_time': '2023-10-19T17:52:15.438',
                        'streamed_output_str': [],
                        'tags': ['map:key:context', 'FAISS'],
```

```
                                'type': 'retriever'}},
       'streamed_output': ['', 'H', 'arrison', ' worked', ' at', ' Kens', 'ho']})
----------------------------------------------------------------
    RunLog({'final_output': None,
     'id': 'afe66178-d75f-4c2d-b348-b1d144239cd6',
     'logs': {'Docs': {'end_time': '2023-10-19T17:52:15.738',
                       'final_output': {'documents': [Document(page_content=
'harrison worked at kensho')]},
                       'id': '88d51118-5756-4891-89c5-2f6a5e90cc26',
                       'metadata': {},
                       'name': 'Docs',
                       'start_time': '2023-10-19T17:52:15.438',
                       'streamed_output_str': [],
                       'tags': ['map:key:context', 'FAISS'],
                       'type': 'retriever'}},
      'streamed_output': ['', 'H', 'arrison', ' worked', ' at', ' Kens', 'ho', '.']})
----------------------------------------------------------------
    RunLog({'final_output': None,
     'id': 'afe66178-d75f-4c2d-b348-b1d144239cd6',
     'logs': {'Docs': {'end_time': '2023-10-19T17:52:15.738',
                       'final_output': {'documents': [Document(page_content=
'harrison worked at kensho')]},
                       'id': '88d51118-5756-4891-89c5-2f6a5e90cc26',
                       'metadata': {},
                       'name': 'Docs',
                       'start_time': '2023-10-19T17:52:15.438',
                       'streamed_output_str': [],
                       'tags': ['map:key:context', 'FAISS'],
                       'type': 'retriever'}},
      'streamed_output': ['',
                          'H',
                          'arrison',
                          ' worked',
                          ' at',
                          ' Kens',
                          'ho',
                          '.',
                          '']})
----------------------------------------------------------------
    RunLog({'final_output': {'output': 'Harrison worked at Kensho.'},
     'id': 'afe66178-d75f-4c2d-b348-b1d144239cd6',
     'logs': {'Docs': {'end_time': '2023-10-19T17:52:15.738',
                       'final_output': {'documents': [Document(page_content=
'harrison worked at kensho')]},
                       'id': '88d51118-5756-4891-89c5-2f6a5e90cc26',
                       'metadata': {},
```

```
                    'name': 'Docs',
                    'start_time': '2023-10-19T17:52:15.438',
                    'streamed_output_str': [],
                    'tags': ['map:key:context', 'FAISS'],
                    'type': 'retriever'}},
   'streamed_output': ['',
                      'H',
                      'arrison',
                      ' worked',
                      ' at',
                      ' Kens',
                      'ho',
                      '.',
                      '']})
```

4. 并行

LCEL 支持并行请求。例如，当使用 RunnableParallel（通常编写为字典）时，它会并行执行每个元素，代码如下：

```
//第 3 章/RunnableParallel.py

from langchain.schema.runnable import RunnableParallel

chain1 = ChatPromptTemplate.from_template("tell me a joke about {topic}") | model
chain2 = (
    ChatPromptTemplate.from_template("write a short (2 line) poem about {topic}")
    | model
)
combined = RunnableParallel(joke=chain1, poem=chain2)
chain1.invoke({"topic": "bears"})

    CPU times: user 54.3 ms, sys: 0 ns, total: 54.3 ms
    Wall time: 2.29 s




    AIMessage(content="Why don't bears wear shoes? \n\nBecause they already have bear feet!")



chain2.invoke({"topic": "bears"})
```

```
CPU times: user 7.8 ms, sys: 0 ns, total: 7.8 ms
Wall time: 1.43 s
```

```
AIMessage(content="In wild embrace,\nNature's strength roams with grace.")
```

```
combined.invoke({"topic": "bears"})
```

```
CPU times: user 167 ms, sys: 921 µs, total: 168 ms
Wall time: 1.56 s
```

```
{'joke': AIMessage(content="Why don't bears wear shoes? \n\nBecause they already have bear feet!"),
 'poem': AIMessage(content="Fierce and wild, nature's might,\nBears roam the woods, shadows of the night.")}
```

并行可以与其他 Runnable 接口/线程结合使用,并行可以用于批处理,代码如下:

```
//第 3 章/RunnableParallel2.py

chain1.batch([{"topic": "bears"}, {"topic": "cats"}])

CPU times: user 159 ms, sys: 3.66 ms, total: 163 ms
Wall time: 1.34 s
```

```
[AIMessage(content="Why don't bears wear shoes? \n\nBecause they already have bear feet!"),
 AIMessage(content="Sure, here's a cat joke for you: \n\nWhy don't cats play poker in the wild? \n\nBecause there are too many cheetahs!")]

chain2.batch([{"topic": "bears"}, {"topic": "cats"}])
```

```
    CPU times: user 165 ms, sys: 0 ns, total: 165 ms
    Wall time: 1.73 s




    [AIMessage(content="Silent giants roam, \nNature's strength, love's emblem
shown."),
     AIMessage(content='Whiskers aglow, paws tiptoe, \nGraceful hunters, hearts
aglow.')]

combined.batch([{"topic": "bears"}, {"topic": "cats"}])

    CPU times: user 507 ms, sys: 125 ms, total: 632 ms
    Wall time: 1.49 s




    [{'joke': AIMessage(content="Why don't bears wear shoes? \n\nBecause they
already have bear feet!"),
      'poem': AIMessage(content="Majestic bears roam, \nNature's wild guardians
of home.")},
     {'joke': AIMessage(content="Sure, here's a cat joke for you:\n\nWhy did the
cat sit on the computer? \n\nBecause it wanted to keep an eye on the mouse!"),
      'poem': AIMessage(content='Whiskers twitch, eyes gleam, \nGraceful
creatures, feline dream.')}]
```

接下来介绍 LCEL 的几个实际应用案例。

3.4.2　绑定运行时参数

有时，用户希望使用常量参数在 Runnable 序列中调用 Runnable，这些常量参数不属于序列中前一个 Runnable 的输出，也不属于用户输入。这里可以使用 Runnable.bind() 轻松地传递这些参数。假设有一个简单的提示＋模型序列，代码如下：

```
//第 3 章/bindRunningTimes.py
from langchain.chat_models import ChatOpenAI
from langchain.prompts import ChatPromptTemplate
from langchain.schema import StrOutputParser
from langchain.schema.runnable import RunnablePassthrough

prompt = ChatPromptTemplate.from_messages(
```

```
    [
        (
            "system",
            "Write out the following equation using algebraic symbols then solve it. Use the format\n\nEQUATION:...\nSOLUTION:...\n\n",
        ),
        ("human", "{equation_statement}"),
    ]
)
model = ChatOpenAI(temperature=0)
runnable = (
    {"equation_statement": RunnablePassthrough()} | prompt | model | StrOutputParser()
)
print(runnable.invoke("x raised to the third plus seven equals 12"))
```

输出的结果如下:

```
//第 3 章/bindRunningTimesOutput.py
EQUATION: x^3 + 7 = 12

    SOLUTION:
    Subtracting 7 from both sides of the equation, we get:
    x^3 = 12 - 7
    x^3 = 5

    Taking the cube root of both sides, we get:
    x = ∛5

    Therefore, the solution to the equation x^3 + 7 = 12 is x = ∛5.
```

如果希望使用某些停用词调用模型,则代码如下:

```
runnable = (
    {"equation_statement": RunnablePassthrough()}
    | prompt
    | model.bind(stop="SOLUTION")
    | StrOutputParser()
)
print(runnable.invoke("x raised to the third plus seven equals 12"))

EQUATION: x^3 + 7 = 12
```

3.4.3 运行自定义函数

用户可以在管道中使用任意函数。需要注意，这些函数的所有输入都只能是单个参数。如果有一个接受多个参数的函数，就应该编写一个接受单个输入并将其解压缩为多个参数的包装器，代码如下：

```python
//第 3 章/customFunction.py
from operator import itemgetter

from langchain.chat_models import ChatOpenAI
from langchain.prompts import ChatPromptTemplate
from langchain.schema.runnable import RunnableLambda


def length_function(text):
    return len(text)


def _multiple_length_function(text1, text2):
    return len(text1) * len(text2)


def multiple_length_function(_dict):
    return _multiple_length_function(_dict["text1"], _dict["text2"])


prompt = ChatPromptTemplate.from_template("what is {a} + {b}")
model = ChatOpenAI()

chain1 = prompt | model

chain = (
    {
        "a": itemgetter("foo") | RunnableLambda(length_function),
        "b": {"text1": itemgetter("foo"), "text2": itemgetter("bar")}
        | RunnableLambda(multiple_length_function),
    }
    | prompt
    | model
)

chain.invoke({"foo": "bar", "bar": "gah"})

    AIMessage(content='3 + 9 equals 12.', additional_kwargs={}, example=False)
```

此外，可运行的 lambda 可以选择接受 RunnableConfig，它们可以使用该配置将回调、标签和其他配置信息传递给嵌套的进程，代码如下：

```python
//第 3 章/acceptRunnableConfig.py
from langchain.schema.output_parser import StrOutputParser
from langchain.schema.runnable import RunnableConfig

import json

def parse_or_fix(text: str, config: RunnableConfig):
    fixing_chain = (
        ChatPromptTemplate.from_template(
            "Fix the following text:\n\n```text\n{input}\n```\nError: {error}"
            " Don't narrate, just respond with the fixed data."
        )
        | ChatOpenAI()
        | StrOutputParser()
    )
    for _ in range(3):
        try:
            return json.loads(text)
        except Exception as e:
            text = fixing_chain.invoke({"input": text, "error": e}, config)
    return "Failed to parse"

from langchain.callbacks import get_openai_callback

with get_openai_callback() as cb:
    RunnableLambda(parse_or_fix).invoke(
        "{foo: bar}", {"tags": ["my-tag"], "callbacks": [cb]}
    )
    print(cb)
```

输出的代码如下：

```
Tokens Used: 65
    Prompt Tokens: 56
    Completion Tokens: 9
Successful Requests: 1
Total Cost (USD): $0.00010200000000000001
```

3.4.4 流式传输自定义生成器函数

用户可以在 LCEL 管道中使用生成器函数（使用 yield 关键字且行为类似于迭代器的

函数)。这些生成器的签名应为 Iterator[Input]->Iterator[Output]。或者对于异步生成器其签名应为 AsyncIterator[Input]->AsyncIterator[Output]。这些对于实现自定义输出解析器、修改上一步的输出同时保留流式处理功能非常有用。

下面是使用逗号分隔列表实现一个自定义输出解析器的例子,原始的输出代码如下:

```python
//第 3 章/customGenerator.py
from typing import Iterator, List

from langchain.chat_models import ChatOpenAI
from langchain.prompts.chat import ChatPromptTemplate
from langchain.schema.output_parser import StrOutputParser

prompt = ChatPromptTemplate.from_template(
    "Write a comma-separated list of 5 animals similar to: {animal}"
)
model = ChatOpenAI(temperature=0.0)

str_chain = prompt | model | StrOutputParser()

for chunk in str_chain.stream({"animal": "bear"}):
    print(chunk, end="", flush=True)

=========================================
    lion, tiger, wolf, gorilla, panda

str_chain.invoke({"animal": "bear"})

=========================================
'lion, tiger, wolf, gorilla, panda'
```

接下来自定义输出解析器,代码如下:

```python
//第 3 章/customParser.py
#这是一个自定义解析器,用于拆分 LLM 标记的迭代器
#放入以逗号分隔的字符串列表中
def split_into_list(input: Iterator[str]) -> Iterator[List[str]]:
    #保留部分输入,直到得到逗号
    buffer = ""
    for chunk in input:
        #将当前块添加到缓冲区
        buffer += chunk
        #当缓冲区中有逗号时
        while "," in buffer:
            #以逗号分隔缓冲区
            comma_index = buffer.index(",")
```

```
        #产生逗号之前的所有内容
        yield [buffer[:comma_index].strip()]
        #保存其余部分以供下一次迭代使用
        buffer = buffer[comma_index + 1 :]
    #产生最后一个块
    yield [buffer.strip()]
```

自定义输出解析后输出的结果如下：

```
list_chain = str_chain | split_into_list

for chunk in list_chain.stream({"animal": "bear"}):
    print(chunk, flush=True)


===============
    ['lion']
    ['tiger']
    ['wolf']
    ['gorilla']
    ['panda']


list_chain.invoke({"animal": "bear"})

    ['lion', 'tiger', 'wolf', 'gorilla', 'panda']
```

3.4.5 并行化步骤

RunnableParallel（又名 RunnableMap）可以很容易地并行执行多个 Runnable，并将这些 Runnable 的输出作为映射返回，代码如下：

```
//第 3 章/parallelSteps.py
from langchain.chat_models import ChatOpenAI
from langchain.prompts import ChatPromptTemplate
from langchain.schema.runnable import RunnableParallel

model = ChatOpenAI()
joke_chain = ChatPromptTemplate.from_template("tell me a joke about {topic}") 
| model
poem_chain = (
    ChatPromptTemplate.from_template("write a 2-line poem about {topic}") 
| model
)

map_chain = RunnableParallel(joke=joke_chain, poem=poem_chain)
```

```
map_chain.invoke({"topic": "bear"})

    {'joke': AIMessage(content="Why don't bears wear shoes? \n\nBecause they
have bear feet!", additional_kwargs={}, example=False),
     'poem': AIMessage(content="In woodland depths, bear prowls with might, \
nSilent strength, nature's sovereign, day and night.", additional_kwargs={},
example=False)}
```

映射可用于操作一个 Runnable 的输出,以匹配序列中下一个 Runnable 的输入格式,代码如下:

```
//第 3 章/manipulateOutputs.py
from langchain.embeddings import OpenAIEmbeddings
from langchain.schema.output_parser import StrOutputParser
from langchain.schema.runnable import RunnablePassthrough
from langchain.vectorstores import FAISS

vectorstore = FAISS.from_texts(
    ["harrison worked at kensho"], embedding=OpenAIEmbeddings()
)
retriever = vectorstore.as_retriever()
template = """Answer the question based only on the following context:
{context}

Question: {question}
"""
prompt = ChatPromptTemplate.from_template(template)

retrieval_chain = (
    {"context": retriever, "question": RunnablePassthrough()}
    | prompt
    | model
    | StrOutputParser()
)
retrieval_chain.invoke("where did harrison work?")

    'Harrison worked at Kensho.'
```

这里提示的输入应该是带有键 context 和 question 的映射。用户输入只是一个问题,因此,这里需要使用检索器获取上下文,并在 question 键下传递用户输入。

RunnableMaps 对于并行运行独立进程也很有用,因为映射中的每个 Runnable 都是并行执行的。例如,早期的 joke_chain、poem_chain 和 map_chain 都具有大致相同的运行时,也就是使 map_chain 同时执行其他两个运行时。

```
joke_chain.invoke({"topic": "bear"})

========================================
    958 ms ± 402 ms per loop (mean ± std. dev. of 7 runs, 1 loop each)

poem_chain.invoke({"topic": "bear"})

========================================
    1.22 s ± 508 ms per loop (mean ± std. dev. of 7 runs, 1 loop each)

map_chain.invoke({"topic": "bear"})

========================================
    1.15 s ± 119 ms per loop (mean ± std. dev. of 7 runs, 1 loop each)
```

3.4.6 根据输入的动态路由逻辑

此示例介绍了如何在 LangChain 表达式语言中执行路由。路由允许用户创建非确定性链,其中上一步的输出定义了下一步。路由有助于提供与 LLM 交互的结构和一致性。有两种方法可以执行路由:使用 RunnableBranch 或编写自定义工厂函数(该函数接受上一步的输入并返回 runnable)。

接下来将使用两步序列来说明这两种方法,其中第 1 步将输入问题分类为 LangChain、Anthropic 或 Other,然后路由到相应的提示链。

RunnableBranch 使用(condition,runnable)对列表和默认的 runnable 进行初始化。它通过传递调用它的输入的每个条件来选择哪个分支。它选择要计算为 True 的第 1 个条件,并使用输入运行与该条件对应的可运行条件。如果提供的条件不匹配,则将运行默认的 runnable。下面是它的实际效果示例。

首先创建一个链,将传入的问题识别为与 LangChain、Anthropic 或其他有关的问题,代码如下:

```
//第 3 章/RunnableBranch.py
from langchain.chat_models import ChatAnthropic
from langchain.prompts import PromptTemplate
from langchain.schema.output_parser import StrOutputParser

chain = (
    PromptTemplate.from_template(
        """Given the user question below, classify it as either being about 
`LangChain`, `Anthropic`, or `Other`.

Do not respond with more than one word.
```

```
<question>
{question}
</question>

Classification:"""
    )
    | ChatAnthropic()
    | StrOutputParser()
)

chain.invoke({"question": "how do I call Anthropic?"})

    'Anthropic'
```

接下来创建 3 个子链，代码如下：

```
//第 3 章/subChain.py
langchain_chain = (
    PromptTemplate.from_template(
        """You are an expert in langchain. \
Always answer questions starting with "As Harrison Chase told me". \
Respond to the following question:

Question: {question}
Answer:"""
    )
    | ChatAnthropic()
)
anthropic_chain = (
    PromptTemplate.from_template(
        """You are an expert in anthropic. \
Always answer questions starting with "As Dario Amodei told me". \
Respond to the following question:

Question: {question}
Answer:"""
    )
    | ChatAnthropic()
)
general_chain = (
    PromptTemplate.from_template(
        """Respond to the following question:

Question: {question}
Answer:"""
```

```
    )
    | ChatAnthropic()
)

from langchain.schema.runnable import RunnableBranch

branch = RunnableBranch(
    (lambda x: "anthropic" in x["topic"].lower(), anthropic_chain),
    (lambda x: "langchain" in x["topic"].lower(), langchain_chain),
    general_chain,
)
full_chain = {"topic": chain, "question": lambda x: x["question"]} | branch

full_chain.invoke({"question": "how do I use Anthropic?"})
```

输出的代码如下:

```
    AIMessage(content=" As Dario Amodei told me, here are some ways to use
Anthropic:\n\n- Sign up for an account on Anthropic's website to access tools
like Claude, Constitutional AI, and Writer. \n\n- Use Claude for tasks like email
generation, customer service chat, and QA. Claude can understand natural
language prompts and provide helpful responses.\n\n- Use Constitutional AI if
you need an AI assistant that is harmless, honest, and helpful. It is designed to
be safe and aligned with human values. \n\n- Use Writer to generate natural
language content for things like marketing copy, stories, reports, and more. Give
it a topic and prompt and it will create high-quality written content.\n\n- Check
out Anthropic's documentation and blog for tips, tutorials, examples, and
announcements about new capabilities as they continue to develop their AI
technology.\n\n- Follow Anthropic on social media or subscribe to their
newsletter to stay up to date on new features and releases.\n\n- For most people,
the easiest way to leverage Anthropic's technology is through their website -
just create an account to get started!", additional_kwargs={}, example=False)

full_chain.invoke({"question": "whats 2 + 2"})

    AIMessage(content=' 2 + 2 = 4', additional_kwargs={}, example=False)
```

用户还可以使用自定义函数在不同输出之间路由,代码如下:

```
//第 3 章/routebetween.py

def route(info):
    if "anthropic" in info["topic"].lower():
        return anthropic_chain
    elif "langchain" in info["topic"].lower():
```

```
        return langchain_chain
    else:
        return general_chain

from langchain.schema.runnable import RunnableLambda

full_chain = {"topic": chain, "question": lambda x: x["question"]} | RunnableLambda(
    route
)

full_chain.invoke({"question": "how do I use Anthropic?"})
```

输出的代码如下：

```
AIMessage(content=' As Dario Amodei told me, to use Anthropic IPC you first need to import it:\n\n```python\nfrom anthropic import ic\n```\n\nThen you can create a client and connect to the server:\n\n```python \nclient = ic.connect() \n```\n\nAfter that, you can call methods on the client and get responses:\n\n```python\nresponse = client.ask("What is the meaning of life?")\nprint(response)\n```\n\nYou can also register callbacks to handle events:\n\n```python\ndef on_poke(event):\n    print("Got poked!")\n\nclient.on(\'poke\', on_poke)\n```\n\nAnd that\'s the basics of using the Anthropic IPC client library for Python! Let me know if you have any other questions!', additional_kwargs={}, example=False)
```

尝试询问另外两个问题，输出的代码如下：

```
full_chain.invoke({"question": "how do I use LangChain?"})

    AIMessage(content=' As Harrison Chase told me, to use LangChain you first need to sign up for an API key at platform.langchain.com. Once you have your API key, you can install the Python library and write a simple Python script to call the LangChain API. Here is some sample code to get started:\n\n```python\nimport langchain\n\napi_key = "YOUR_API_KEY"\n\nlangchain.set_key(api_key)\n\nresponse = langchain.ask("What is the capital of France?")\nprint(response.response)\n```\n\nThis will send the question "What is the capital of France?" to the LangChain API and print the response. You can customize the request by providing parameters like max_tokens, temperature, etc. The LangChain Python library documentation has more details on the available options. The key things are getting an API key and calling langchain.ask() with your question text. Let me know if you have any other questions!', additional_kwargs={}, example=False)

full_chain.invoke({"question": "whats 2 + 2"})

    AIMessage(content=' 4', additional_kwargs={}, example=False)
```

3.5 链

在前面的章节中简单介绍了链的概念，这里深入讲解。对于简单的应用程序来讲，单独使用 LLM 是没问题的，但更复杂的应用程序，则需要链接 LLM——要么相互链接，要么与其他组件链接。LangChain 提供了两个用于"链接"组件的高级框架。传统方法是使用 Chain 接口，更新后的方法是使用 LCEL，关于 LCEL 将在后面详细介绍。在构建新应用程序时建议使用 LCEL 进行链组合，但是 LangChain 本身也支持许多有用的内置链，因此本书在这里同时介绍这两个框架。Chain 本身也可以在 LCEL 中使用，因此两者并不相互排斥。

1. LCEL 版本

LCEL 最大的优点是它提供了直观且可读的语法。例如，用户可以将提示、模型和输出解析器组合在一起，代码如下：

```python
//第 3 章/lcelChain.py
from langchain.chat_models import ChatAnthropic
from langchain.prompts import ChatPromptTemplate
from langchain.schema import StrOutputParser

model = ChatAnthropic()
prompt = ChatPromptTemplate.from_messages(
    [
        (
            "system",
            "You're a very knowledgeable historian who provides accurate and eloquent answers to historical questions.",
        ),
        ("human", "{question}"),
    ]
)
runnable = prompt | model | StrOutputParser()

for chunk in runnable.stream({"question": "How did Mansa Musa accumulate his wealth?"}):
    print(chunk, end="", flush=True)
```

输出的代码如下：

```
//第 3 章/lcelChainOutput.py
Mansa Musa was the emperor of the Mali Empire in West Africa during the 14th century. He accumulated immense wealth through several means:
```

        - Gold mining - Mali contained very rich gold deposits, especially in the region of Bambuk. Gold mining and gold trade was a major source of wealth for the empire.

        - Control of trade routes - Mali dominated the trans-Saharan trade routes connecting West Africa to North Africa and beyond. By taxing the goods that passed through its territory, Mali profited greatly.

        - Tributary states - Many lands surrounding Mali paid tribute to the empire. This came in the form of gold, slaves, and other valuable resources.

        - Agriculture - Mali also had extensive agricultural lands irrigated by the Niger River. Surplus food produced could be sold or traded.

        - Royal monopolies - The emperor claimed monopoly rights over the production and sale of certain goods like salt from the Taghaza mines. This added to his personal wealth.

        - Inheritance - As an emperor, Mansa Musa inherited a wealthy state. His predecessors had already consolidated lands and accumulated riches which fell to Musa.

        So in summary, mining, trade, taxes,

2. 传统的 Chain 接口版本

在 LangChain 中可以笼统地将 Chain 定义为对组件的一系列调用，其中可以包括其他链。基本的界面很简单，代码如下：

```python
//第 3 章/legacyChain.py
class Chain(BaseModel, ABC):
    """Base interface that all chains should implement."""

    memory: BaseMemory
    callbacks: Callbacks

    def __call__(
        self,
        inputs: Any,
        return_only_outputs: bool = False,
        callbacks: Callbacks = None,
    ) -> Dict[str, Any]:
        ...
```

用户可以使用内置的 LLMChain 重新创建上面的 LCEL 可运行对象，代码如下：

```
from langchain.chains import LLMChain


chain = LLMChain(llm=model, prompt=prompt, output_parser=StrOutputParser())
chain.run(question="How did Mansa Musa accumulate his wealth?")
```

输出的结果如下：

```
" Mansa Musa was the emperor of the Mali Empire in West Africa in the early 14th
century. He accumulated his vast wealth through several means:\n\n- Gold mining -
Mali contained very rich gold deposits, especially in the southern part of the
empire. Gold mining and trade was a major source of wealth.\n\n- Control of trade
routes - Mali dominated the trans-Saharan trade routes connecting West Africa to
North Africa and beyond. By taxing and controlling this lucrative trade, Mansa
Musa reaped great riches.\n\n- Tributes from conquered lands - The Mali Empire
expanded significantly under Mansa Musa's rule. As new lands were conquered, they
paid tribute to the mansa in the form of gold, salt, and slaves.\n\n- Inheritance -
Mansa Musa inherited a wealthy empire from his predecessor. He continued to build
the wealth of Mali through the factors above.\n\n- Sound fiscal management - Musa
is considered to have managed the empire and its finances very effectively,
including keeping taxes reasonable and promoting a robust economy. This allowed
him to accumulate and maintain wealth.\n\nSo in summary, conquest, trade, taxes,
mining, and inheritance all contributed to Mansa Musa growing the M"
```

3.5.1 链接口中的方法调用

从 Chain 继承的所有类都提供了几种运行链逻辑的方法。最直接的方法是使用 __call__，代码如下：

```
chat = ChatOpenAI(temperature=0)
prompt_template = "Tell me a {adjective} joke"
llm_chain = LLMChain(llm=chat, prompt=PromptTemplate.from_template(prompt_
template))

llm_chain(inputs={"adjective": "corny"})

    {'adjective': 'corny',
    'text': 'Why did the tomato turn red? Because it saw the salad dressing!'}
```

在默认情况下，__call__ 会同时返回输入键值和输出键值。用户可以通过将 return_only_outputs 设置为 True 来将其配置为仅返回输出键值，代码如下：

```
llm_chain("corny", return_only_outputs=True)

    {'text': 'Why did the tomato turn red? Because it saw the salad dressing!'}
```

如果 Chain 只输出一个输出键值（其 output_keys 中只有一个元素），则可以使用 run 方法。需要注意，run 输出的是字符串而不是字典，代码如下：

```
#因为 llm_chain 只输出一个输出键值，所以可以使用 run 方法
llm_chain.output_keys
    ['text']
llm_chain.run({"adjective": "corny"})
    'Why did the tomato turn red? Because it saw the salad dressing!'
```

对于一个输入键可以直接输入字符串，而无须指定输入映射，代码如下：

```
#这两者是等价的
llm_chain.run({"adjective": "corny"})
llm_chain.run("corny")
#这两者也是等价的
llm_chain("corny")
llm_chain({"adjective": "corny"})

    {'adjective': 'corny',
     'text': 'Why did the tomato turn red? Because it saw the salad dressing!'}
```

3.5.2 自定义链的创建

要实现自己的自定义链可以对 Chain 进行子类化并实现以下方法，代码如下：

```
//第 3 章/customChain.py
from __future__ import annotations

from typing import Any, Dict, List, Optional

from langchain.callbacks.manager import (
    AsyncCallbackManagerForChainRun,
    CallbackManagerForChainRun,
)
from langchain.chains.base import Chain
from langchain.prompts.base import BasePromptTemplate
from langchain.schema.language_model import BaseLanguageModel
from pydantic import Extra


class MyCustomChain(Chain):
    """
    An example of a custom chain.
    """
```

```python
        prompt: BasePromptTemplate
        """提示要使用的对象"""
        llm: BaseLanguageModel
        output_key: str = "text"  #: :meta private:

        class Config:
            """Configuration for this pydantic object."""

            extra = Extra.forbid
            arbitrary_types_allowed = True

        @property
        def input_keys(self) -> List[str]:
            """将是提示所需的任何键

          :元私有
            """
            return self.prompt.input_variables

        @property
        def output_keys(self) -> List[str]:
            """将始终返回文本键

          :元私有:
            """
            return [self.output_key]

        def _call(
            self,
            inputs: Dict[str, Any],
            run_manager: Optional[CallbackManagerForChainRun] = None,
        ) -> Dict[str, str]:
            #自定义链的逻辑在这里
            #这只是一个模仿 LLMChain 的例子
            prompt_value = self.prompt.format_prompt(**inputs)

            #每当调用一种语言模型或其他链时,都应该传递
            #它的回调管理器。这允许通过以下方式跟踪内部运行
            #在外部运行中注册的任何回调
            #总可以调用
            #'run_manager.get_child()'
            response = self.llm.generate_prompt(
                [prompt_value], callbacks=run_manager.get_child() if run_manager else None
            )
```

```python
        # 如果想记录一些关于这次运行的信息,则可以通过调用
        # run_manager 方法实现。这将触发任何为该事件注册的回调
        if run_manager:
            run_manager.on_text("Log something about this run")

        return {self.output_key: response.generations[0][0].text}

    async def _acall(
        self,
        inputs: Dict[str, Any],
        run_manager: Optional[AsyncCallbackManagerForChainRun] = None,
    ) -> Dict[str, str]:
        # 自定义链的逻辑在这里
        # 这只是一个模仿 LLMChain 的例子
        prompt_value = self.prompt.format_prompt(**inputs)

        # 每当调用一种语言模型或其他链时,都应该传递
        # 它的回调管理器。这允许通过以下方式跟踪内部运行
        # 在外部运行中注册的任何回调
        # 总可以调用
        # 'run_manager.get_child()'
        response = await self.llm.agenerate_prompt(
            [prompt_value], callbacks=run_manager.get_child() if run_manager else None
        )

        # 如果想记录一些关于这次运行的信息,则可以通过调用
        # run_manager 方法实现。这将触发任何为该事件注册的回调
        if run_manager:
            await run_manager.on_text("Log something about this run")

        return {self.output_key: response.generations[0][0].text}

    @property
    def _chain_type(self) -> str:
        return "my_custom_chain"
```

然后可以使用刚才自定义好的链,代码如下:

```python
//第 3 章/customChainCreation.py
from langchain.callbacks.stdout import StdOutCallbackHandler
from langchain.chat_models.openai import ChatOpenAI
from langchain.prompts.prompt import PromptTemplate
```

```
chain = MyCustomChain(
    prompt=PromptTemplate.from_template("tell us a joke about {topic}"),
    llm=ChatOpenAI(),
)

chain.run({"topic": "callbacks"}, callbacks=[StdOutCallbackHandler()])
```

输出的结果如下：

```
> Entering new MyCustomChain chain...
Log something about this run
> Finished chain.




'Why did the callback function feel lonely? Because it was always waiting for someone to call it back!'
```

3.5.3 几种常见的链

1. LLM 链

任何 LLM 应用程序中最常见的链接类型是将提示模板与 LLM 和可选的输出解析器组合在一起。推荐的方法是使用 LangChain 表达式语言，其中 BasePromptTemplate、BaseLanguageModel 和 BaseOutputParser 都实现了 Runnable 接口，并被设计为通过管道相互连接，使 LCEL 组合变得非常容易，代码如下：

```
//第 3 章/LLMChain.py
from langchain.chat_models import ChatOpenAI
from langchain.prompts import PromptTemplate
from langchain.schema import StrOutputParser

prompt = PromptTemplate.from_template(
    "What is a good name for a company that makes {product}?"
)
runnable = prompt | ChatOpenAI() | StrOutputParser()
runnable.invoke({"product": "colorful socks"})

    'VibrantSocks'
```

如果不选择使用 LCEL 用户，则可以选用传统的 LLMchain 接口。LLMChain 是一个简单的链，它围绕语言模型添加了一些功能。它在整个 LangChain 中被广泛使用，包括其他链和代理。LLMChain 由 PromptTemplate 和 LLM 组成。它使用提供的输入键值（内存键值，如果可用）设置提示模板的格式，将格式化的字符串传递给 LLM 并返回 LLM 输出，一个完整的实现流程的代码如下：

```python
//第 3 章/LLMChainLegacy.py
from langchain.chains import LLMChain
from langchain.llms import OpenAI
from langchain.prompts import PromptTemplate

prompt_template = "What is a good name for a company that makes {product}?"

llm = OpenAI(temperature=0)
llm_chain = LLMChain(llm=llm, prompt=PromptTemplate.from_template(prompt_template))
llm_chain("colorful socks")

    {'product': 'colorful socks', 'text': '\n\nSocktastic!'}
```

通常用户需要解析输出。在默认情况下，LLMChain 不会解析输出，即使基础提示对象具有输出解析器也是如此。如果要在 LLM 输出上应用该输出解析器，用户则需要使用 predict_and_parse 而不是 predict，使用 apply_and_parse 而不是 apply。如果使用 predict，则代码如下：

```python
//第 3 章/predict.py
from langchain.output_parsers import CommaSeparatedListOutputParser

output_parser = CommaSeparatedListOutputParser()
template = """List all the colors in a rainbow"""
prompt = PromptTemplate(
    template=template, input_variables=[], output_parser=output_parser
)
llm_chain = LLMChain(prompt=prompt, llm=llm)

llm_chain.predict()

    '\n\nRed, orange, yellow, green, blue, indigo, violet'
```

如果使用 predict_and_parse，则代码如下：

```
llm_chain.predict_and_parse()

    /Users/bagatur/langchain/libs/langchain/langchain/chains/llm.py: 280:
UserWarning: The predict_and_parse method is deprecated, instead pass an output
parser directly to LLMChain.
```

```
warnings.warn()
```

```
['Red', 'orange', 'yellow', 'green', 'blue', 'indigo', 'violet']
```

用户也可以直接从字符串模板构造 LLMChain，代码如下：

```
template = """Tell me a {adjective} joke about {subject}."""
llm_chain = LLMChain.from_string(llm=llm, template=template)

llm_chain.predict(adjective="sad", subject="ducks")

'\n\nQ: What did the duck say when his friend died? \nA: Quack, quack, goodbye.'
```

2. 路由链

路由允许用户创建非确定性链，其中上一步的输出定义下一步。路由有助于提供与 LLM 交互的结构和一致性。举个非常简单的例子，假设有两个针对不同类型的优化问题的模板，希望根据用户输入来选择模板，代码如下：

```
//第 3 章/twotemplate.py
from langchain.prompts import PromptTemplate

physics_template = """You are a very smart physics professor. \
You are great at answering questions about physics in a concise and easy to understand manner. \
When you don't know the answer to a question you admit that you don't know.

Here is a question:
{input}"""
physics_prompt = PromptTemplate.from_template(physics_template)

math_template = """You are a very good mathematician. You are great at answering math questions. \
You are so good because you are able to break down hard problems into their component parts, \
answer the component parts, and then put them together to answer the broader question.

Here is a question:
{input}"""
math_prompt = PromptTemplate.from_template(math_template)
```

如果使用LCEL，用户则可以使用RunnableBranch轻松做到这一点。RunnableBranch使用(condition, runnable)对列表和默认的runnable进行初始化。它通过传递调用它输入的每个条件来选择哪个分支。它选择计算结果为True的第1个条件，并使用输入运行与该条件对应的可运行条件。如果提供的条件不匹配，则将运行默认的runnable，代码如下：

```python
//第 3 章/lcelRouter.py
from langchain.chat_models import ChatOpenAI
from langchain.schema.output_parser import StrOutputParser
from langchain.schema.runnable import RunnableBranch

general_prompt = PromptTemplate.from_template(
    "You are a helpful assistant. Answer the question as accurately as you can.\n\n{input}"
)
prompt_branch = RunnableBranch(
    (lambda x: x["topic"] == "math", math_prompt),
    (lambda x: x["topic"] == "physics", physics_prompt),
    general_prompt,
)

from typing import Literal

from langchain.output_parsers.openai_functions import PydanticAttrOutputFunctionsParser
from langchain.pydantic_v1 import BaseModel
from langchain.utils.openai_functions import convert_pydantic_to_openai_function


class TopicClassifier(BaseModel):
    "Classify the topic of the user question"

    topic: Literal["math", "physics", "general"]
    "The topic of the user question. One of 'math', 'physics' or 'general'."


classifier_function = convert_pydantic_to_openai_function(TopicClassifier)
llm = ChatOpenAI().bind(
    functions=[classifier_function], function_call={"name": "TopicClassifier"}
)
parser = PydanticAttrOutputFunctionsParser(
    pydantic_schema=TopicClassifier, attr_name="topic"
)
classifier_chain = llm | parser
```

```python
from operator import itemgetter

from langchain.schema.output_parser import StrOutputParser
from langchain.schema.runnable import RunnablePassthrough

final_chain = (
    RunnablePassthrough.assign(topic=itemgetter("input") | classifier_chain)
    | prompt_branch
    | ChatOpenAI()
    | StrOutputParser()
)

final_chain.invoke(
    {
        "input": "What is the first prime number greater than 40 such that one plus the prime number is divisible by 3?"
    }
)
```

输出的结果如下：

```
 "Thank you for your kind words! I'll be happy to help you with this math question.\n\nTo find the first prime number greater than 40 that satisfies the given condition, we need to follow a step-by-step approach. \n\nFirstly, let's list the prime numbers greater than 40:\n41, 43, 47, 53, 59, 61, 67, 71, ...\n\nNow, we need to check if one plus each of these prime numbers is divisible by 3. We can do this by calculating the remainder when dividing each number by 3.\n\nFor 41, (41 + 1) % 3 = 42 % 3 = 0. It is divisible by 3.\n\nFor 43, (43 + 1) % 3 = 44 % 3 = 2. It is not divisible by 3.\n\nFor 47, (47 + 1) % 3 = 48 % 3 = 0. It is divisible by 3.\n\nSince 41 and 47 are both greater than 40 and satisfy the condition, the first prime number greater than 40 such that one plus the prime number is divisible by 3 is 41.\n\nTherefore, the answer to the question is 41."
```

当然用户也可以使用传统的 LLMchain 接口。接下来将展示如何使用 RouterChain 范式创建一个链，该链动态地选择下一个链用于给定输入。RouterChain 由两个组件组成：RouterChain 本身（负责选择要调用的下一个链）和 destination_chains（路由器链可以路由到的链）。在这个示例中将展示 MultiPromptChain 中使用的这些路由链，以创建一个问答链，该链选择与给定问题最相关的提示，然后使用该提示回答问题。

LLMRouterChain 使用 LLM 确定如何路由，代码如下：

```
//第 3 章/LLMRouterChain.py
from langchain.chains import ConversationChain
from langchain.chains.llm import LLMChain
from langchain.chains.router import MultiPromptChain
```

```python
from langchain.llms import OpenAI

prompt_infos = [
    {
        "name": "physics",
        "description": "Good for answering questions about physics",
        "prompt_template": physics_template,
    },
    {
        "name": "math",
        "description": "Good for answering math questions",
        "prompt_template": math_template,
    },
]
llm = OpenAI()
destination_chains = {}
for p_info in prompt_infos:
    name = p_info["name"]
    prompt_template = p_info["prompt_template"]
    prompt = PromptTemplate(template=prompt_template, input_variables=["input"])
    chain = LLMChain(llm=llm, prompt=prompt)
    destination_chains[name] = chain
default_chain = ConversationChain(llm=llm, output_key="text")
from langchain.chains.router.llm_router import LLMRouterChain, RouterOutputParser
from langchain.chains.router.multi_prompt_prompt import MULTI_PROMPT_ROUTER_TEMPLATE
destinations = [f"{p['name']}: {p['description']}" for p in prompt_infos]
destinations_str = "\n".join(destinations)
router_template = MULTI_PROMPT_ROUTER_TEMPLATE.format(destinations=destinations_str)
router_prompt = PromptTemplate(
    template=router_template,
    input_variables=["input"],
    output_parser=RouterOutputParser(),
)
router_chain = LLMRouterChain.from_llm(llm, router_prompt)
chain = MultiPromptChain(
    router_chain=router_chain,
    destination_chains=destination_chains,
    default_chain=default_chain,
    verbose=True,
)
```

接下来问几个问题,输出的代码如下:

```
print(chain.run("What is black body radiation? "))


    > Entering new MultiPromptChain chain...


    /Users/bagatur/langchain/libs/langchain/langchain/chains/llm. py: 280:
UserWarning: The predict_and_parse method is deprecated, instead pass an output
parser directly to LLMChain.
      warnings.warn(


    physics: {'input': 'What is black body radiation? '}
    > Finished chain.
=================================================
print(
    chain.run(
        "What is the first prime number greater than 40 such that one plus the
prime number is divisible by 3? "
    )
)


    > Entering new MultiPromptChain chain...


    /Users/bagatur/langchain/libs/langchain/langchain/chains/llm. py: 280:
UserWarning: The predict_and_parse method is deprecated, instead pass an output
parser directly to LLMChain.
      warnings.warn(


    math: {'input': 'What is the first prime number greater than 40 such that one
plus the prime number is divisible by 3? '}
    > Finished chain.


    The first prime number greater than 40 such that one plus the prime number is
divisible by 3 is 43. This can be seen by breaking down the problem:

    1) We know that a prime number is a number that is only divisible by itself
and one.
    2) We also know that if a number is divisible by 3, the sum of its digits must be
divisible by 3.

    So, if we want to find the first prime number greater than 40 such that one plus the
prime number is divisible by 3, we can start counting up from 40, testing each number to
see if it is prime and if the sum of the number and one is divisible by three.
```

```
    The first number we come to that satisfies these conditions is 43.


    Black body radiation is the thermal electromagnetic radiation within or
surrounding a body in thermodynamic equilibrium with its environment, or emitted
by a black body (an idealized physical body which absorbs all incident
electromagnetic radiation). It is a characteristic of the temperature of the
body; if the body has a uniform temperature, the radiation is also uniform across
the spectrum of frequencies. The spectral characteristics of the radiation are
determined by the temperature of the body, which implies that a black body at a
given temperature will emit the same amount of radiation at every frequency.
=================================================
print(chain.run("What is the name of the type of cloud that rains?"))

    > Entering new MultiPromptChain chain...


    /Users/bagatur/langchain/libs/langchain/langchain/chains/llm.py: 280:
UserWarning: The predict_and_parse method is deprecated, instead pass an output
parser directly to LLMChain.
    warnings.warn(


    physics: {'input': 'What is the name of the type of cloud that rains?'}
    > Finished chain.


    The type of cloud that rains is called a cumulonimbus cloud.
```

EmbeddingRouterChain 使用嵌入和相似性在目标链之间路由,代码如下:

```
//第 3 章/EmbeddingRouterChain.py
from langchain.chains.router.embedding_router import EmbeddingRouterChain
from langchain.embeddings import CohereEmbeddings
from langchain.vectorstores import Chroma

names_and_descriptions = [
    ("physics", ["for questions about physics"]),
    ("math", ["for questions about math"]),
]
router_chain = EmbeddingRouterChain.from_names_and_descriptions(
    names_and_descriptions, Chroma, CohereEmbeddings(), routing_keys=["input"]
)
chain = MultiPromptChain(
```

```
        router_chain=router_chain,
        destination_chains=destination_chains,
        default_chain=default_chain,
        verbose=True,
)
```

同样询问几个问题,输出的代码如下:

```
print(chain.run("What is black body radiation?"))

> Entering new MultiPromptChain chain...
physics: {'input': 'What is black body radiation? '}
> Finished chain.


    Black body radiation is the electromagnetic radiation emitted by a black
body, which is an idealized physical body that absorbs all incident electromagnetic
radiation. This radiation is related to the temperature of the body, with higher
temperatures leading to higher radiation levels. The spectrum of the radiation is
continuous, and is described by the Planck's law of black body radiation.
=============================================================================
print(
    chain.run(
        "What is the first prime number greater than 40 such that one plus the
prime number is divisible by 3?"
    )
)

> Entering new MultiPromptChain chain...
math: {'input': 'What is the first prime number greater than 40 such that one
plus the prime number is divisible by 3? '}
> Finished chain.


The first prime number greater than 40 such that one plus the prime number is
divisible by 3 is 43. This is because 43 is a prime number, and 1 + 43 = 44, which is
divisible by 3.
```

3. Sequential 链

调用 LLM 后的下一步是对 LLM 进行一系列调用。当用户想要从一个调用获取输出并将其用作另一个调用的输入时,这特别有用。推荐的方法是使用 LCEL。传统方法是使用 SequentialChain,为了向后兼容,这里继续在此处记录它。这里举个玩具的例子,假设我们想要创建一个链,首先创建一个游戏概要,然后根据概要生成一个游戏评论,代码如下:

```
//第 3 章/sequentialChain.py
from langchain.prompts import PromptTemplate

synopsis_prompt = PromptTemplate.from_template(
    """You are a playwright. Given the title of play, it is your job to write a
synopsis for that title.

Title: {title}
Playwright: This is a synopsis for the above play:"""
)

review_prompt = PromptTemplate.from_template(
    """You are a play critic from the New York Times. Given the synopsis of play,
it is your job to write a review for that play.

Play Synopsis:
{synopsis}
Review from a New York Times play critic of the above play:"""
)
```

首先使用 LCEL 的方式,创建一系列调用(对 LLM 或任何其他组件/任意函数)正是 LangChain 表达式语言的设计目的,代码如下:

```
//第 3 章/sequentialChainLCEL.py
from langchain.chat_models import ChatOpenAI
from langchain.schema import StrOutputParser

llm = ChatOpenAI()
chain = (
    {"synopsis": synopsis_prompt | llm | StrOutputParser()}
    | review_prompt
    | llm
    | StrOutputParser()
)
chain.invoke({"title": "Tragedy at sunset on the beach"})
```

输出的结果如下:

```
' In "Tragedy at Sunset on the Beach," playwright has crafted a deeply
affecting drama that delves into the complexities of human relationships and the
consequences that arise from one fateful evening. Set against the breathtaking
backdrop of a serene beach at sunset, the play takes audiences on an emotional
journey as it explores the lives of four individuals whose paths intertwine in
unexpected and tragic ways.\n\nAt the center of the story is Sarah, a young woman
grappling with the recent loss of her husband. Seeking solace and a fresh start,
```

she embarks on a solitary trip to the beach, hoping to find peace and clarity. It is here that she encounters James, a charismatic but troubled artist, lost in his own world of anguish and self-doubt. The unlikely connection they form becomes the catalyst for a series of heart-wrenching events, as their emotional baggage and personal demons collide. \n \nThe play skillfully weaves together the narratives of Sarah, James, and Rachel, Sarah\'s best friend. As Rachel arrives on the beach with the intention of helping Sarah heal, she unknowingly carries a secret that threatens to shatter their friendship forever. Against the backdrop of crashing waves and vibrant sunsets, the characters\' lives unravel, exposing hidden desires, betrayals, and deeply buried secrets. The boundaries of love, friendship, and loyalty blur, forcing each character to confront their own vulnerabilities and face the consequences of their choices. \n \nWhat sets "Tragedy at Sunset on the Beach" apart is its ability to evoke genuine emotion from its audience. The playwright\'s poignant exploration of the human condition touches upon universal themes of loss, forgiveness, and the lengths we go to protect the ones we love. The richly drawn characters come alive on stage, their struggles and triumphs resonating deeply with the audience. Moments of intense emotion are skillfully crafted, leaving spectators captivated and moved.\n\nThe play\'s evocative setting adds another layer of depth to the storytelling. The picturesque beach at sunset becomes a metaphor for the fragility of life and the fleeting nature of happiness. The crashing waves and vibrant colors serve as a backdrop to the characters\' unraveling lives, heightening the emotional impact of their stories.\n\nWhile "Tragedy at Sunset on the Beach" is undeniably a heavy and somber play, it ultimately leaves audiences questioning the power of redemption. The characters\' journeys, though tragic, offer glimpses of hope and the potential for healing. It reminds us that even amidst the darkest moments, there is still a chance for redemption and forgiveness.\n\nOverall, "Tragedy at Sunset on the Beach" is a thought-provoking and emotionally charged play that will captivate audiences from start to finish. The playwright \ ' s skillful storytelling, evocative setting, and richly drawn characters make for a truly memorable theatrical experience. This is a play that will leave spectators questioning their own lives and the choices they make, long after the curtain falls.'

如果用户也想得到概要,则可以做如下操作,代码如下:

```
from langchain.schema.runnable import RunnablePassthrough

synopsis_chain = synopsis_prompt | llm | StrOutputParser()
review_chain = review_prompt | llm | StrOutputParser()
chain = {"synopsis": synopsis_chain} | RunnablePassthrough.assign(review=review_chain)
chain.invoke({"title": "Tragedy at sunset on the beach"})
```

{'synopsis': 'Tragedy at Sunset on the Beach is a gripping and emotionally charged drama that delves into the complexities of human relationships and the fragility of life. Set against the backdrop of a picturesque beach at sunset, the play follows a group of friends who gather to celebrate a joyous occasion.\n\nAs the sun begins its descent, tensions simmer beneath the surface, and long-held secrets and resentments come to light. The characters find themselves entangled in a web of love, betrayal, and loss, as they confront their deepest fears and desires.\n\nThe main focus revolves around Sarah, a vibrant and free-spirited woman who becomes the center of a tragic event. Through a series of flashback scenes, we witness the unraveling of her life, exploring her complicated relationships with her closest friends and romantic partners.\n\nThe play explores themes of regret, redemption, and the consequences of our choices. It delves into the human condition, questioning the nature of happiness and the value of time. The audience is taken on an emotional rollercoaster, experiencing moments of laughter, heartache, and profound reflection.\n\nTragedy at Sunset on the Beach challenges conventional notions of tragedy, evoking a sense of empathy and understanding for the flawed and vulnerable characters. It serves as a reminder that life is unpredictable and fragile, urging us to cherish every moment and embrace the beauty that exists even amidst tragedy.',

'review': "In Tragedy at Sunset on the Beach, playwright John Smithson delivers a powerful and thought-provoking exploration of the human experience. Set against the stunning backdrop of a beach at sunset, this emotionally charged drama takes the audience on a journey through the complexities of relationships, the fragility of life, and the profound impact of our choices.\n\nSmithson skillfully weaves together a tale of love, betrayal, and loss, as a group of friends gather to celebrate a joyous occasion. As the sun sets, tensions rise, and long-held secrets and resentments are exposed, leaving the characters entangled in a web of emotions. Through a series of poignant flashback scenes, we witness the unraveling of Sarah's life, a vibrant and free-spirited woman who becomes the center of a tragic event.\n\nWhat sets Tragedy at Sunset on the Beach apart is its ability to challenge conventional notions of tragedy. Smithson masterfully portrays flawed and vulnerable characters with such empathy and understanding that the audience can't help but empathize with their struggles. This play serves as a reminder that life is unpredictable and fragile, urging us to cherish every moment and embrace the beauty that exists even amidst tragedy.\n\nThe performances in this production are nothing short of extraordinary. The actors effortlessly navigate the emotional rollercoaster of the script, eliciting moments of laughter, heartache, and profound reflection from the audience. Their ability to convey the complexities of their characters' relationships and inner turmoil is truly commendable.\n\nThe direction by Jane Anderson is impeccable, capturing the essence of the beach at sunset and utilizing the space to create an immersive experience for the audience. The use of flashbacks adds depth and nuance to the narrative, allowing for a deeper understanding of the characters and their motivations.\n\nTragedy at Sunset on the Beach is not a play for the faint of heart. It tackles heavy themes of regret, redemption, and the consequences of our choices. However, it is precisely this raw and unflinching exploration of the human condition that makes it such a compelling piece of theater.

```
Smithson's writing, combined with the exceptional performances and direction,
make this play a must-see for theatergoers looking for a thought-provoking and
emotionally resonant experience.\n\nIn a city renowned for its theater scene,
Tragedy at Sunset on the Beach stands out as a shining example of the power of live
performance to evoke empathy, provoke contemplation, and remind us of the fragile
beauty of life. It is a production that will linger in the minds and hearts of its
audience long after the final curtain falls."}
```

同理，如果使用传统的接口，则 SequentialChain 允许连接多个链，并将它们组合到执行某些特定方案的管道中。有两种类型的顺序链：SimpleSequentialChain 和 SequentialChain。

SimpleSequentialChain 是 SequentialChain 的最简单形式，其中每个步骤都有一个单一的输入/输出，一个步骤的输出是下一个步骤的输入，代码如下：

```python
//第 3 章/ SimpleSequentialChain.py
from langchain.chains import LLMChain
from langchain.llms import OpenAI
from langchain.prompts import PromptTemplate

#这是一个 LLMChain,用于编写一个剧本标题的概要
llm = OpenAI(temperature=0.7)
synopsis_chain = LLMChain(llm=llm, prompt=synopsis_prompt)

#这是一个 LLMChain,用于根据概要为戏剧撰写评论
llm = OpenAI(temperature=0.7)
review_chain = LLMChain(llm=llm, prompt=review_prompt)

#这是我们按顺序运行这两条链的整体链
from langchain.chains import SimpleSequentialChain

overall_chain = SimpleSequentialChain(
    chains=[synopsis_chain, review_chain], verbose=True
)

review = overall_chain.run("Tragedy at sunset on the beach")
```

输出的结果如下：

```
> Entering new SimpleSequentialChain chain...


Tragedy at Sunset on the Beach is a modern tragedy about a young couple in
love. The couple, Jack and Jill, are deeply in love and plan to spend the day
together on the beach at sunset. However, when they arrive, they are shocked to
discover that the beach is an abandoned, dilapidated wasteland. With no one else
around, they explore the beach and start to reminisce about their relationship
```

and the good times they've shared.

But then, out of the blue, a mysterious figure emerges from the shadows and reveals a dark secret. The figure tells the couple that the beach is no ordinary beach, but is in fact the site of a terrible tragedy that took place many years ago. As the figure explains what happened, Jack and Jill become overwhelmed with grief.

In the end, Jack and Jill are forced to confront the truth about the tragedy and its consequences. The play is ultimately a reflection on the power of tragedy and the human capacity to confront and overcome it.

Tragedy at Sunset on the Beach is a powerful, thought-provoking modern tragedy that is sure to leave a lasting impression on its audience. The play follows the story of Jack and Jill, a young couple deeply in love, as they explore an abandoned beach and discover a dark secret from the past.

The play brilliantly captures the raw emotions of Jack and Jill as they learn of the tragedy that has occurred on the beach. The writing is masterful, and the actors do a wonderful job of conveying the couple's grief and pain. The play is ultimately a reflection on the power of tragedy and the human capacity to confront and overcome it.

Overall, Tragedy at Sunset on the Beach is a must-see for anyone looking for a thought-provoking and emotionally moving play. This play is sure to stay with its audience long after the curtain closes. Highly recommended.

> Finished chain.

```
print(review)
```

Tragedy at Sunset on the Beach is a powerful, thought-provoking modern tragedy that is sure to leave a lasting impression on its audience. The play follows the story of Jack and Jill, a young couple deeply in love, as they explore an abandoned beach and discover a dark secret from the past.

The play brilliantly captures the raw emotions of Jack and Jill as they learn of the tragedy that has occurred on the beach. The writing is masterful, and the actors do a wonderful job of conveying the couple's grief and pain. The play is ultimately a reflection on the power of tragedy and the human capacity to confront and overcome it.

Overall, Tragedy at Sunset on the Beach is a must-see for anyone looking for a thought-provoking and emotionally moving play. This play is sure to stay with its audience long after the curtain closes.

当然，并非所有的 SequentialChain 都像将单个字符串作为参数传递并获取单个字符串作为链中所有步骤的输出一样简单。在下一个示例中将尝试涉及多个输入的更复杂的链，并且还有多个最终输出。特别重要的是如何命名输入/输出变量。在上面的例子中不必考虑这一点，因为只是将一条链的输出作为输入直接传递给下一条链，但在这里需要注意这一点，因为有多个输入，代码如下：

```
//第 3 章/ SequentialChain.py
#这是一个 LLMChain,用剧本的标题和它所处的时代来写一个概要
llm = OpenAI(temperature=0.7)
synopsis_template = """You are a playwright. Given the title of play and the era
it is set in, it is your job to write a synopsis for that title.

Title: {title}
Era: {era}
Playwright: This is a synopsis for the above play:"""
synopsis_prompt_template = PromptTemplate(
    input_variables=["title", "era"], template=synopsis_template
)
synopsis_chain = LLMChain(
    llm=llm, prompt=synopsis_prompt_template, output_key="synopsis"
)

#这是一个 LLMChain,用于根据概要撰写戏剧评论
llm = OpenAI(temperature=0.7)
template = """You are a play critic from the New York Times. Given the synopsis of
play, it is your job to write a review for that play.

Play Synopsis:
{synopsis}
Review from a New York Times play critic of the above play:"""
prompt_template = PromptTemplate(input_variables=["synopsis"], template=
template)
review_chain = LLMChain(llm=llm, prompt=prompt_template, output_key="review")

#这是按顺序运行这两个链的整体链
from langchain.chains import SequentialChain

overall_chain = SequentialChain(
    chains=[synopsis_chain, review_chain],
    input_variables=["era", "title"],
    #这里返回多个变量
    output_variables=["synopsis", "review"],
    verbose=True,
)
```

```
overall_chain({"title": "Tragedy at sunset on the beach", "era": "Victorian 
England"})
```

输出的结果如下:

```
> Entering new SequentialChain chain...

> Finished chain.




{'title': 'Tragedy at sunset on the beach',
 'era': 'Victorian England',
 'synopsis': "\n\nThe play is set in Victorian England and follows the story of 
a young couple, Mary and John, who were deeply in love and had just gotten engaged. 
On the night of their engagement, they decided to take a romantic walk along the 
beach at sunset. Unexpectedly, John is shot by a stranger and killed right in 
front of Mary. In a state of shock and anguish, Mary is left alone, struggling to 
comprehend what has just occurred. \n\nThe play follows Mary as she searches for 
answers to John's death. As Mary's investigation begins, she discovers that John 
was actually involved in a dark and dangerous plot to overthrow the government. 
Unbeknownst to Mary, John had been working as a spy in a secret mission to uncover 
the truth behind a political scandal. \n\nNow, Mary must face the consequences of 
her beloved's actions and find a way to save the future of England. As the story 
unfolds, Mary must confront her own beliefs as well as the powerful people who are 
determined to end her mission. \n\nAt the end of the play, all of Mary's questions 
are answered and she is able to make a choice that will ultimately decide the fate 
of the nation. Tragedy at Sunset on the Beach is a",
 'review': "\n\nSet against the backdrop of Victorian England, Tragedy at 
Sunset on the Beach tells a heart-wrenching story of love, loss, and tragedy. The 
play follows Mary and John, a young couple deeply in love, who experience an 
unexpected tragedy on the night of their engagement. When John is shot and killed 
by a stranger, Mary is left alone to uncover the truth behind her beloved's death.\
n\nWhat follows is an intense and gripping journey as Mary discovers that John was 
a spy in a secret mission to uncover a powerful political scandal. As Mary faces 
off against those determined to end her mission, she must confront her own beliefs 
and ultimately decide the fate of the nation.\n\nThe play is skillfully crafted 
and brilliantly performed. The actors portray a range of emotions from joy to 
sorrow that will leave the audience moved and captivated. The production is a 
beautiful testament to the power of love and the strength of the human spirit, and 
it is sure to leave a lasting impression. Highly recommended."}
```

4. Transformation 链

通常用户希望在输入从一个组件传递到另一个组件时对其进行转换,这时就要用到 Transformation 链。接下来将创建一个虚拟转换,该转换接收超长文本,将文本过滤为仅有前 3 段,然后将其传递到链中以总结这些段落,代码如下:

```python
from langchain.prompts import PromptTemplate

prompt = PromptTemplate.from_template(
    """Summarize this text:

{output_text}

Summary:"""
)

with open("../../state_of_the_union.txt") as f:
    state_of_the_union = f.read()
```

按照惯例如果使用 LCEL,用户则可以在任何 RunnableSequence 中添加函数,代码如下:

```python
//第 3 章/ LCELTransformationChain.py
from langchain.chat_models import ChatOpenAI
from langchain.schema import StrOutputParser

runnable = (
    {"output_text": lambda text: "\n\n".join(text.split("\n\n")[:3])}
    | prompt
    | ChatOpenAI()
    | StrOutputParser()
)
runnable.invoke(state_of_the_union)

'The speaker acknowledges the presence of important figures in the government 
and addresses the audience as fellow Americans. They highlight the impact of 
COVID-19 on keeping people apart in the previous year but express joy in being 
able to come together again. The speaker emphasizes the unity of Democrats, 
Republicans, and Independents as Americans.'
```

如果使用传统的链,则代码如下:

```python
//第 3 章/ TransformationChain.py
from langchain.chains import LLMChain, SimpleSequentialChain, TransformChain
from langchain.llms import OpenAI
```

```python
def transform_func(inputs: dict) -> dict:
    text = inputs["text"]
    shortened_text = "\n\n".join(text.split("\n\n")[:3])
    return {"output_text": shortened_text}


transform_chain = TransformChain(
    input_variables=["text"], output_variables=["output_text"], transform=
transform_func
)

template = """Summarize this text:

{output_text}

Summary:"""
prompt = PromptTemplate(input_variables=["output_text"], template=template)
llm_chain = LLMChain(llm=OpenAI(), prompt=prompt)

sequential_chain = SimpleSequentialChain(chains=[transform_chain, llm_chain])
sequential_chain.run(state_of_the_union)

' In an address to the nation, the speaker acknowledges the hardships of the past
year due to the COVID-19 pandemic, but emphasizes that regardless of political
affiliation, all Americans can come together.'
```

3.6 实战案例：客户服务助手应用程序开发

本节将使用 LangChain 为客户服务代理构建一个文本分类应用程序。背景是当给定一个文档时，如电子邮件，希望将其分类为与意图相关的不同类别，提取文本的情绪，并提供摘要。客户服务代理负责回答客户查询、解决问题和处理投诉。他们的工作对于保持客户满意度和忠诚度至关重要，这将直接影响公司的声誉。生成式模型可以通过多种方式为客户服务代理提供帮助。

（1）情绪分类：这有助于识别客户情绪，并允许客户服务代理个性化他们的响应。

（2）摘要：这使客户服务代理能够了解冗长的客户消息的关键点并节省时间。

（3）意图分类：类似于摘要，这有助于预测客户的目的，并允许更快地解决问题。

（4）回答建议：这为客户服务代理提供了对常见查询的建议响应，确保提供准确和一致的消息传递。

将这些方法结合起来可以帮助客户服务代理更准确、更及时地作出响应，最终提高客户满意度。在这里将集中讨论前三点。LangChain 是一个非常灵活的库，具有许多集成，能够解决各种文本问题。开发者可以在许多不同的集成之间进行选择，以此来执行这些任务。

开发者可以要求任何 LLM 给出一个开放领域（任何类别）的分类，或者在多个类别之间进行选择。由于训练规模很大，所以 LLM 是非常强大的模型，特别是在给定少量提示时 LLM 能很好地被用于不需要任何额外训练的情感分析。对此，2023 年 4 月研究人员在《ChatGPT 是一个好的情感分析器吗？》的论文中做了详细的研究[12]。对于用于情感分析的 LLM 的提示可能是这样的：

```
Given this text, what is the sentiment conveyed? Is it positive, neutral, or negative?
Text: {sentence}
Sentiment:
```

LLM 在总结方面也非常有效，而且比以前的任何模型都要好得多。缺点可能是调用这些模型比更传统的 ML 模型慢且更昂贵。Cohere 和其他提供商将文本分类和情感分析作为其功能的一部分。例如，NLP Cloud 的模型列表包括 spacy 和许多其他模型。

同样在 HuggingFace 上很多模型也能支持以下的一些任务。

(1) 文档问答。

(2) 综述。

(3) 文本分类。

(4) 文本问答。

(5) 翻译。

开发者可以在 HuggingFace 的 Transformer 库中运行管道并在本地执行这些模型，也可以在服务器（HuggingFaceHub）上远程执行这些模型，甚至可以将 load_huggingface_tool()加载器作为工具执行这些模型。HuggingFace 包含数以千计的模型，其中许多模型针对特定领域进行了微调。例如，ProsusAI/finbert 是一个 BERT 模型，该模型是在名为 FinancialPhraseBank 的数据集上训练的，可以分析金融文本的情感。用户也可以使用任何本地模型。对于文本分类，模型往往要小得多，因此对资源的消耗较小。最后，文本分类也可能是文本嵌入的一个案例。这里可以在 HuggingFaceHub 上列出下载次数最多的 5 个模型，然后通过 huggingfaceAPI 进行文本分类，代码如下：

```
//第 3 章/modelList.py
def list_most_popular(task: str):
    for rank, model in enumerate(
        list_models(filter=task, sort="downloads", direction=-1)
    ):
        if rank == 5:
            break
        print(f"{model.id}, {model.downloads}\n")
list_most_popular("text-classification")
```

HuggingFaceHub 上下载次数最多的 5 个模型，结果见表 3-2。

表 3-2　HuggingFaceHub 上最流行的文本分类模型

模　　型	下 载 次 数
nlptown/bert-base-multilingual-uncased-sentiment	5 805 259
SamLowe/roberta-base-go_emotions	5 292 490
cardiffnlp/twitter-roberta-base-irony	4 427 067
salesken/query_wellformedness_score	4 380 505
marieke93/MiniLM-evidence-types	4 370 524

通过表 3-2 可以看到，这些模型基本上是关于小范围的类别，例如情绪、情感、讽刺等。这里测试一下情绪模型的效果，代码如下：

```
#这里使用GPT-3.5整理一封冗长的漫无边际的客户电子邮件,这封邮件是有关抱怨咖啡机的,读
#者可以在GitHub上找到该电子邮件
from transformers import pipeline
sentiment_model = pipeline(
    task="sentiment-analysis",
    model="nlptown/bert-base-multilingual-uncased-sentiment"
)
print(sentiment_model(customer_email))
```

结果如下：

```
[{'label': '2 stars', 'score': 0.28999224305152893}]
```

可以看到模型准确地识别出了这封邮件的语气是非常负面的。接下来查看 HuggingFace 上排名前五的文本总结模型的情况，见表 3-3。

表 3-3　HuggingFace 上最流行的文本总结模型

模　　型	下 载 次 数
t5-base	2 710 309
t5-small	1 566 141
facebook/bart-large-cnn	1 150 085
sshleifer/distilbart-cnn-12-6	709 344
philschmid/bart-large-cnn-samsum	477 902

与大型的模型相比，所有这些模型的规模都相对较小。接下来在服务器上远程执行文本汇总模型，代码如下：

```
//第 3 章/ remoteModel.py
```

```
from langchain import HuggingFaceHub
summarizer = HuggingFaceHub(
    repo_id="facebook/bart-large-cnn",
    model_kwargs={"temperature":0, "max_length":180}
)
def summarize(llm, text) -> str:
    return llm(f"Summarize this: {text}!")
summarize(summarizer, customer_email)
```

这里需要注意,用户需要设置HUGGINGFACEHUB_API_TOKEN才能正常运行代码。最后这个模型的输出是"一位顾客的咖啡机坏了,唤起了一种深深的难以置信的绝望感。这种令人心碎的疏忽表现粉碎了我沉迷于日常完美咖啡的梦想,让我情绪激动,无法控制。"这位顾客写道,"我希望这封电子邮件能让你置身于理解的光环中,尽管当我写信给你时,我内心的情绪纠结混乱"。这个总结还算过得去,但不是很有说服力。摘要中还有很多漫无边际的内容。这里可以尝试使用其他模型,或者只选择带有提示要求总结的LLM,接下来尝试问问VertexAI,代码如下:

```
//第 3 章/ VertexAI.py
from langchain.llms import VertexAI
from langchain import PromptTemplate, LLMChain
template = """Given this text, decide what is the issue the customer is concerned about. Valid categories are these:
* product issues
* delivery problems
* missing or late orders
* wrong product
* cancellation request
* refund or exchange
* bad support experience
* no clear reason to be upset
Text: {email}
Category:
"""
prompt = PromptTemplate(template=template, input_variables=["email"])
llm = VertexAI()
llm_chain = LLMChain(prompt=prompt, llm=llm, verbose=True)
print(llm_chain.run(customer_email))
```

最终得到的输出答案是"产品问题"。这对于本书在这里使用的长电子邮件示例是正确的分类。

3.7　总结

本章首先介绍了将 LangChain 和本书所需的其他库作为环境安装的 4 种不同方法，然后介绍了几个文本和图像模型。对于它们中的每个解释了从何处获取 API Token，并演示了如何调用模型。紧接着介绍了模型输出解析的一些方法、LangChain 表达式语言和链的一些调用方法。最后演示了一个 LLM 应用程序，用于客户服务用例中的文本分类。通过将 LangChain 中的各种功能链接在一起，开发者可以帮助减少客户服务的响应时间，并确保答案准确无误。在接下来的章节中，本书将深入探讨一些用例，例如使用 Web 搜索等工具通过增强检索进行问答，以及通过索引依赖文档搜索的聊天机器人。

第 4 章 LangChain 进阶：Agent

本书已经在前面的章节简单地介绍了 Agent，作为 LangChain 的进阶部分，本章将着重介绍 Agent。作为 LLM 和外部世界链接的桥梁，Agent 模块有着举足轻重的作用。Agent 的核心思想是让 LLM 来选择要执行的一系列操作。在链中，一系列操作是以硬编码的方式写在代码中。在 Agent 中，LLM 用作推理引擎，以确定要执行哪些操作及按何种顺序执行。在正式介绍 Agent 之前，有一些基本的概念需要着重阐述一下。

Agent 概念中的核心是 Agent 链。Agent 链是负责决定下一步采取什么步骤的链，它是由 LLM 和提示提供支持的。该链的输入包括工具（可用工具的说明）、用户输入（最终的目标）和中间步骤。Agent 链的输出是要执行的下一个动作或要发送给用户的最终响应（AgentActions 或 AgentFinish）。一个动作指定了需要使用什么样的工具及该工具的输入。

不同的 Agent 链具有不同的推理提示样式、不同的输入编码方式及解析输出的不同方式。在 LangChain 里面，用户可以自定义 Agent 链。

1) 工具

工具是 Agent 可以调用的函数。围绕工具，有两个重要的注意事项，一个是要为 Agent 提供正确的工具，另一个是要以对 Agent 最有帮助的方式用提示词来描述工具。

不认真考虑这两个注意事项将无法构建有效的 Agent。例如，如果开发者不授予 Agent 访问一组正确工具的权限，则将永远无法实现赋予它的目标。或者如果不能很好地描述这些工具，则 Agent 将不知道如何正确地使用它们。LangChain 提供了广泛的内置工具，但也可以很容易地定义开发者自己的工具（包括自定义描述）。

2) 工具包

对于许多常见任务，Agent 将需要一组相关工具。为此，LangChain 提供了工具包，每个工具包包括完成特定目标所需的大约 3~5 个工具。例如，GitHub 工具包有一个用于搜索 GitHub 问题的工具、一个用于读取文件的工具、一个用于评论的工具等。

3) Agent 执行器（AgentExecutor）

AgentExecutor 是 Agent 的运行时。这是实际调用 Agent 的内容，其中包括执行它选

择的操作,将操作输出传递回 Agent,然后重复。伪代码大致如下:

```
next_action = agent.get_action(...)
while next_action != AgentFinish:
    observation = run(next_action)
    next_action = agent.get_action(..., next_action, observation)
return next_action
```

4.1 构建自己的第 1 个 Agent

为了更好地理解 Agent 框架,这里介绍使用 LCEL 从头开始构建 Agent。开发者需要构建 Agent 本身,定义自定义工具,并在自定义循环中运行 Agent 和工具。最后将展示如何使用标准的 LangChain AgentExecutor 来简化执行。

在正式开始之前用户需要了解的一些重要术语和架构如下。

(1) AgentAction:数据类,表示 Agent 应采取的操作。它有一个工具属性(这是应调用的工具的名称)和一个 tool_input 属性(该工具的输入)。

(2) AgentFinish:数据类,表示 Agent 已执行完成并应返给用户。它有一个 return_values 参数,这是一个要返回的字典。

(3) intermediate_steps:表示以前的 Agent 操作和传递的相应输出。这些对于传递给将来的迭代很重要,以便 Agent 知道它已经完成了哪些工作。Intermediate_steps 的数据类型为 List[Tuple[AgentAction, Any]]。

1. LangSmith

通常 Agent 在返回面向用户的输出之前会采取自确定的、依赖于输入的步骤序列。这使调试这些系统变得特别棘手,因此如何使这些步骤序列可视化变得尤为重要,LangSmith 对于这种情况特别有用。在后面的章节中将会详细介绍 LangSmith,这里读者可以先简单熟悉一下基本知识。当使用 LangChain 构建时,任何使用 LCEL 构建的内置 Agent 或自定义 Agent 都将在 LangSmith 中自动跟踪。如果使用 AgentExecutor,则用户不仅可以全面跟踪 Agent 规划步骤,还可以全面跟踪工具的输入和输出。用户只需设置以下环境变量便可以设置 LangSmith:

```
export LANGCHAIN_TRACING_V2="true"
export LANGCHAIN_API_KEY="<自己的 APIkey>"
```

2. 设置 Agent 链

用户首先需要创建 Agent 链。这是负责确定下一步要采取的行动的链。在此示例中将使用 OpenAI 函数调用来创建此 Agent 链。这通常是创建 Agent 的最可靠方法。在本示例中,我们将构建一个有权访问自定义工具的自定义 Agent。我们之所以选择这个例子,是因为对于大多数实际用例,用户需要自定义 Agent 或工具。这里将创建一个简单的工具来计算单词的长度,因为它实际上是 LLM 可能会由于标记化等各种原因而无法自己完成任务。首先在没有记忆的情况下创建它,随后将展示如何添加记忆,因为通常需要记忆才能启用对话。首先加载将用于控制 Agent 的 LLM,代码如下:

```
from langchain.chat_models import ChatOpenAI

llm = ChatOpenAI(model="gpt-3.5-turbo", temperature=0)
```

从以下的输出可以看到它很难计算字符串 educa 中字母的个数。

```
llm.invoke("how many letters are there in the word educa? ")

    AIMessage(content='There are six letters in the word educa')
```

接下来定义一些要使用的工具。这里编写一个非常简单的 Python 函数来计算传入的单词的长度,代码如下:

```
//第 4 章/ wordCountTool.py
from langchain.agents import tool


@tool
def get_word_length(word: str) -> int:
    """返回单词的长度"""
    return len(word)


tools = [get_word_length]
```

现在将创建提示。由于 OpenAI 函数调用针对工具使用进行了微调,因此用户几乎不需要任何关于如何推理或如何输出格式的说明。这里只有两个输入变量 input 和 agent_scratchpad。input 应为包含用户目标的字符串,agent_scratchpad 是包含先前 Agent 工具调用和相应工具输出的消息序列,代码如下:

```
//第 4 章/ scratchPad.py
from langchain.prompts import ChatPromptTemplate, MessagesPlaceholder

prompt = ChatPromptTemplate.from_messages(
    [
        (
            "system",
            "你是个很厉害的助手,但是不擅长计算单词的长度",
        ),
```

```
        ("user", "{input}"),
        MessagesPlaceholder(variable_name="agent_scratchpad"),
    ]
)
```

Agent 如何知道它可以使用哪些工具？本例中依赖于 OpenAI 函数调用 LLM，它将函数作为单独的参数，并经过专门训练后就可以知道何时调用这些函数。下一步是将工具传递给 Agent，用户只需将它们格式化为 OpenAI 函数格式并将它们传递给 LLM（通过绑定函数可以确保每次调用模型时都会传入它们），代码如下：

```
from langchain.tools.render import format_tool_to_openai_function

llm_with_tools = llm.bind(functions=[format_tool_to_openai_function(t) for t in tools])
```

现在将这些部分放在一起创建 Agent。这里将导入最后两个实用程序函数：一个组件用于格式化中间步骤（Agent 操作、工具输出对）以输入可发送到 LLM 的消息，以及一个用于将输出消息转换为 AgentAction/Finish 的组件，代码如下：

```
//第 4 章/ AgentPuttingTogether.py
from langchain.agents.format_scratchpad import format_to_openai_function_messages
from langchain.agents.output_parsers import OpenAIFunctionsAgentOutputParser

agent = (
    {
        "input": lambda x: x["input"],
        "agent_scratchpad": lambda x: format_to_openai_function_messages(
            x["intermediate_steps"]
        ),
    }
    | prompt
    | llm_with_tools
    | OpenAIFunctionsAgentOutputParser()
)
```

现在来测试一下效果。首先传入一个简单的问题和空的中间步骤，查看它返回了什么，代码如下：

```
agent.invoke({"input": "how many letters are there in the word educa?", "intermediate_steps": []})

    AgentActionMessageLog(tool='get_word_length', tool_input={'word': 'educa'}, log="\nInvoking: `get_word_length` with `{'word': 'educa'}`\n\n\n", message_log=
```

```
[AIMessage(content='', additional_kwargs={'function_call': {'arguments': '{\n  "word": "educa"\n}', 'name': 'get_word_length'}})]
```

很明显可以看到它以要采取的 AgentAction 进行响应,这实际上是一个 AgentActionMessageLog,它是 AgentAction 的一个子类,它也跟踪完整的消息日志。

3. 定义运行时

定义完 Agent 链只是第 1 步,下一步需要为此编写一个运行时。最简单的方法就是连续循环,调用 Agent,然后执行操作,再重复,直到返回 AgentFinish,代码如下:

```
//第 4 章/defineRuntime.py
from langchain.schema.agent import AgentFinish

user_input = " how many letters are there in the word educa? "
intermediate_steps = []
while True:
    output = agent.invoke(
        {
            "input": user_input,
            "intermediate_steps": intermediate_steps,
        }
    )
    if isinstance(output, AgentFinish):
        final_result = output.return_values["output"]
        break
    else:
        print(f"TOOL NAME: {output.tool}")
        print(f"TOOL INPUT: {output.tool_input}")
        tool = {"get_word_length": get_word_length}[output.tool]
        observation = tool.run(output.tool_input)
        intermediate_steps.append((output, observation))
print(final_result)

TOOL NAME: get_word_length
TOOL INPUT: {'word': 'educa'}
There are five letters in the word educa.
```

4. 使用 AgentExecutor

到这里用户可以看到整个过程比较烦琐,LangChain 框架中提供了 AgentExecutor 类,此类可以大大简化整个流程。AgentExecutor 捆绑了上述所有内容,并增加了错误处理、提前停止、跟踪和其他流程,以便提高质量,代码如下:

```
from langchain.agents import AgentExecutor

agent_executor = AgentExecutor(agent=agent, tools=tools, verbose=True)
```

现在进行测试,代码如下:

```
agent_executor.invoke({"input": " how many letters are there in the word educa?"})

> Entering new AgentExecutor chain...

Invoking: `get_word_length` with `{'word': 'educa'}`

There are 5 letters in the word "educa".

> Finished chain.


{'input': 'how many letters in the word educa?',
 'output': 'There are 5 letters in the word "educa".'}
```

5. 增加记忆

到这里已经成功地创建了一个 Agent,但是,此 Agent 是无状态的,也就意味着它不记得有关先前交互的任何信息。用户可以通过添加记忆来解决这个问题,为了做到这一点,需要做两件事情:在提示中为记忆变量添加一个位置及跟踪聊天记录。

首先在提示中添加一个记忆位置。这里为带有 chat_history 键的消息添加占位符。为了遵循对话流,这里将其放在新用户输入的上方,代码如下:

```
//第 4 章/ addMemoryKey.py
from langchain.prompts import MessagesPlaceholder

MEMORY_KEY = "chat_history"
prompt = ChatPromptTemplate.from_messages(
    [
        (
            "system",
            "你是个很厉害的助手,但是不擅长计算单词的长度",
        ),
        MessagesPlaceholder(variable_name=MEMORY_KEY),
        ("user", "{input}"),
        MessagesPlaceholder(variable_name="agent_scratchpad"),
    ]
)
```

然后可以设置一个列表来跟踪聊天记录,代码如下:

```python
from langchain.schema.messages import AIMessage, HumanMessage

chat_history = []
```

紧接着把上述内容都串起来:

```python
//第 4 章 / memoryPutTogether.py
agent = (
    {
        "input": lambda x: x["input"],
        "agent_scratchpad": lambda x: format_to_openai_function_messages(
            x["intermediate_steps"]
        ),
        "chat_history": lambda x: x["chat_history"],
    }
    | prompt
    | llm_with_tools
    | OpenAIFunctionsAgentOutputParser()
)
agent_executor = AgentExecutor(agent=agent, tools=tools, verbose=True)
```

代码运行的时候,用户需要将输入和输出作为聊天记录进行跟踪,代码如下:

```python
//第 4 章 / trackInputOutput.py
input1 = "how many letters in the word educa?"
result = agent_executor.invoke({"input": input1, "chat_history": chat_history})
chat_history.extend(
    [
        HumanMessage(content=input1),
        AIMessage(content=result["output"]),
    ]
)
agent_executor.invoke({"input": "is that a real word?", "chat_history": chat_history})


> Entering new AgentExecutor chain...

Invoking: `get_word_length` with `{'word': 'educa'}`


There are 5 letters in the word "educa".

> Finished chain.
```

```
> Entering new AgentExecutor chain...
No, "educa" is not a real word in English.

> Finished chain.




{'input': 'is that a real word?',
 'chat_history': [HumanMessage(content='how many letters in the word educa?'),
  AIMessage(content='There are 5 letters in the word "educa".')],
 'output': 'No, "educa" is not a real word in English.'}
```

4.2 LangChain 中的常见 Agent 类型

Agent 使用 LLM 来确定要执行哪些操作及按什么顺序执行。一个操作可以是使用工具并观察其输出，也可以是向用户返回响应。以下是 LangChain 中常见的 Agent 类型。

（1）零样本 ReAct(Zero-shot ReAct)：此 Agent 使用 ReAct 框架，它仅根据工具的描述来确定要使用的工具。可以提供任意数量的工具。使用此 Agent 要求为每个工具提供说明。这是最常见的 Agent。

（2）结构化输入 ReAct(Structured Input ReAct)：结构化工具聊天 Agent 能够使用多输入工具。较旧的 Agent 被配置为将操作输入指定为单个字符串，但此 Agent 可以使用工具的参数架构来创建结构化操作输入。这对于更复杂的工具的使用非常有用，例如在浏览器中精确导航。

（3）OpenAI 函数（OpenAI Functions）：某些 OpenAI 模型（如 gpt-3.5-turbo-0613 和 gpt-4-0613）已经过显式微调，以检测何时应该调用函数，并使用应传递给函数的输入进行响应。OpenAI Functions Agent 旨在与这些模型配合使用。

（4）对话式（Conversational）：此 Agent 旨在用于对话设置。该提示旨在使 Agent 具有帮助性和对话性。它使用 ReAct 框架来决定使用哪个工具，并使用记忆来记住之前的对话交互。

（5）搜索进行自我询问（Self-ask with Search）：此 Agent 使用命名为 Intermediate Answer 的单个工具。这个工具能够查找问题的答案。

（6）ReAct 文档存储（ReAct Document Store）：此 Agent 使用 ReAct 框架与文档库进行交互，其中必须提供两个工具——搜索工具和查找工具（它们的名称必须完全相同）。搜索工具应搜索文档，而查找工具应在最近找到的文档中查找术语。

下面将逐步详细介绍以上的常见 Agent 类型。

4.2.1 Zero-shot ReAct

首先加载将用于控制 Agent 的 LLM,代码如下:

```
llm = OpenAI(temperature=0)
```

接下来加载一些要使用的工具。因为 LLM 需要使用 llm-math 工具,因此需要把它传入工具,代码如下:

```
tools = load_tools(["serpapi", "llm-math"], llm=llm)
```

和前面一样,这里将首先展示如何使用 LCEL 创建这个 Agent,代码如下:

```python
//第 4 章/ReActLCEL.py
from langchain import hub
from langchain.agents.format_scratchpad import format_log_to_str
from langchain.agents.output_parsers import ReActSingleInputOutputParser
from langchain.tools.render import render_text_description

prompt = hub.pull("hwchase17/react")
prompt = prompt.partial(
    tools=render_text_description(tools),
    tool_names=", ".join([t.name for t in tools]),
)

llm_with_stop = llm.bind(stop=["\nObservation"])

agent = (
    {
        "input": lambda x: x["input"],
        "agent_scratchpad": lambda x: format_log_to_str(x["intermediate_steps"]),
    }
    | prompt
    | llm_with_stop
    | ReActSingleInputOutputParser()
)

from langchain.agents import AgentExecutor
agent_executor = AgentExecutor(agent=agent, tools=tools, verbose=True)
agent_executor.invoke(
    {
        "input": "Who is Leo DiCaprio's girlfriend? What is her current age raised to the 0.43 power?"
    }
)
```

在上面的代码中集成了 render_text_description，这是一个内置的工具包，里面有很多 LangChain 已经定义好的内部工具。代码的最后问了两个问题，这两个问题都需要向外部网络查询答案，测试的结果如下：

```
> Entering new AgentExecutor chain...
I need to find out who Leo DiCaprio's girlfriend is and then calculate her age raised to the 0.43 power.
Action: Search
Action Input: "Leo DiCaprio girlfriend"model Vittoria Ceretti I need to find out Vittoria Ceretti's age
Action: Search
Action Input: "Vittoria Ceretti age"25 years I need to calculate 25 raised to the 0.43 power
Action: Calculator
Action Input: 25^0.43Answer: 3.991298452658078 I now know the final answer
Final Answer: Leo DiCaprio's girlfriend is Vittoria Ceretti and her current age raised to the 0.43 power is 3.991298452658078.

> Finished chain.

{'input': "Who is Leo DiCaprio's girlfriend? What is her current age raised to the 0.43 power？",
 'output': "Leo DiCaprio's girlfriend is Vittoria Ceretti and her current age raised to the 0.43 power is 3.991298452658078."}
```

当然也可以不使用 LCEL，代码如下：

```
agent_executor = initialize_agent(
    tools, llm, agent=AgentType.ZERO_SHOT_REACT_DESCRIPTION, verbose=True
)
agent_executor.invoke(
    {
        "input": "Who is Leo DiCaprio's girlfriend? What is her current age raised to the 0.43 power？"
    }
)

> Entering new AgentExecutor chain...
I need to find out who Leo DiCaprio's girlfriend is and then calculate her age raised to the 0.43 power.
Action: Search
Action Input: "Leo DiCaprio girlfriend"
Observation: model Vittoria Ceretti
Thought: I need to find out Vittoria Ceretti's age
Action: Search
```

```
Action Input: "Vittoria Ceretti age"
Observation: 25 years
Thought: I need to calculate 25 raised to the 0.43 power
Action: Calculator
Action Input: 25^0.43
Observation: Answer: 3.991298452658078
Thought: I now know the final answer
Final Answer: Leo DiCaprio's girlfriend is Vittoria Ceretti and her current age
raised to the 0.43 power is 3.991298452658078.

> Finished chain.

{'input': "Who is Leo DiCaprio's girlfriend? What is her current age raised to the
0.43 power?",
 'output': "Leo DiCaprio's girlfriend is Vittoria Ceretti and her current age
raised to the 0.43 power is 3.991298452658078."}
```

4.2.2 Structured Input ReAct

Structured Input ReAct 代理能够使用多输入工具。较旧的 Agent 将操作输入指定为单个字符串，并且此 Agent 可以使用提供的工具的 args_schema 来填充操作的输入值。

第 1 步先初始化所需要的工具，代码如下：

```
#只有 Jupyter Notebook 才需要此导入，因为它们有自己的事件循环
import nest_asyncio
from langchain.agents.agent_toolkits import PlayWrightBrowserToolkit
from langchain.tools.playwright.utils import (
    create_async_playwright_browser,)   #同步浏览器可用，但与 Jupyter 不兼容

nest_asyncio.apply()
```

接下来安装相应的依赖，代码如下：

```
!pip install playwright

!playwright install

async_browser = create_async_playwright_browser()
browser_toolkit = PlayWrightBrowserToolkit.from_browser(async_browser=async_
browser)
tools = browser_toolkit.get_tools()
```

首先使用 LCEL 来构建这个 Agent，代码如下：

```python
//第 4 章 / StructureLCEL.py
from langchain import hub
prompt = hub.pull("hwchase17/react-multi-input-json")
from langchain.tools.render import render_text_description_and_args

prompt = prompt.partial(
    tools=render_text_description_and_args(tools),
    tool_names=", ".join([t.name for t in tools]),
)
llm = ChatOpenAI(temperature=0)
llm_with_stop = llm.bind(stop=["Observation"])

from langchain.agents.format_scratchpad import format_log_to_str
from langchain.agents.output_parsers import JSONAgentOutputParser

agent = (
    {
        "input": lambda x: x["input"],
        "agent_scratchpad": lambda x: format_log_to_str(x["intermediate_steps"]),
    }
    | prompt
    | llm_with_stop
    | JSONAgentOutputParser()
)

from langchain.agents import AgentExecutor
agent_executor = AgentExecutor(agent=agent, tools=tools, verbose=True)
response = await agent_executor.ainvoke(
    {"input": "Browse to blog.langchain.dev and summarize the text, please."}
)
print(response["output"])
```

在上面的代码中最后让 LLM 去访问一个网站并且总结里面的文字信息,结果如下:

```
> Entering new AgentExecutor chain...
Action:
```
{
  "action": "navigate_browser",
  "action_input": {
    "url": "https://blog.langchain.dev"
  }
}
```
```

```
Navigating to https://blog.langchain.dev returned status code 200Action:
```
{
  "action": "extract_text",
  "action_input": {}
}
```

LangChain LangChain Home GitHub Docs By LangChain Release Notes Write with Us Sign in Subscribe The official LangChain blog. Subscribe now Login Featured Posts Announcing LangChain Hub Using LangSmith to Support Fine-tuning Announcing LangSmith, a unified platform for debugging, testing, evaluating, and monitoring your LLM applications Sep 20 Peering Into the Soul of AI Decision-Making with LangSmith 10 min read Sep 20 LangChain + Docugami Webinar: Lessons from Deploying LLMs with LangSmith 3 min read Sep 18 TED AI Hackathon Kickoff (and projects we'd love to see) 2 min read Sep 12 How to Safely Query Enterprise Data with LangChain Agents + SQL + OpenAI + Gretel 6 min read Sep 12 OpaquePrompts x LangChain: Enhance the privacy of your LangChain application with just one code change 4 min read Load more LangChain © 2023 Sign up Powered by GhostAction:
```
{
  "action": "Final Answer",
  "action_input": "The LangChain blog features posts on topics such as using LangSmith for fine-tuning, AI decision-making with LangSmith, deploying LLMs with LangSmith, and more. It also includes information on LangChain Hub and upcoming webinars. LangChain is a platform for debugging, testing, evaluating, and monitoring LLM applications."
}
```

> Finished chain.
The LangChain blog features posts on topics such as using LangSmith for fine-tuning, AI decision-making with LangSmith, deploying LLMs with LangSmith, and more. It also includes information on LangChain Hub and upcoming webinars. LangChain is a platform for debugging, testing, evaluating, and monitoring LLM applications.
```

不使用 LCEL 的代码如下：

```
//第 4 章/ StructureNoLCEL.py
llm = ChatOpenAI(temperature=0)   #Also works well with Anthropic models
agent_chain = initialize_agent(
    tools,
    llm,
    agent=AgentType.STRUCTURED_CHAT_ZERO_SHOT_REACT_DESCRIPTION,
```

```
    verbose=True,
)

response = await agent_chain.ainvoke(
    {"input": "Browse to blog.langchain.dev and summarize the text, please."}
)
print(response["output"])
```

输出的结果如下:

```
> Entering new AgentExecutor chain...
Action:
```
{
  "action": "navigate_browser",
  "action_input": {
    "url": "https://blog.langchain.dev"
  }
}
```
Observation: Navigating to https://blog.langchain.dev returned status code 200
Thought:I have successfully navigated to the blog.langchain.dev website. Now I
need to extract the text from the webpage to summarize it.
Action:
```
{
  "action": "extract_text",
  "action_input": {}
}
```
Observation: LangChain LangChain Home GitHub Docs By LangChain Release Notes
Write with Us Sign in Subscribe The official LangChain blog. Subscribe now Login
Featured Posts Announcing LangChain Hub Using LangSmith to Support Fine-tuning
Announcing LangSmith, a unified platform for debugging, testing, evaluating, and
monitoring your LLM applications Sep 20 Peering Into the Soul of AI Decision-
Making with LangSmith 10 min read Sep 20 LangChain + Docugami Webinar: Lessons
from Deploying LLMs with LangSmith 3 min read Sep 18 TED AI Hackathon Kickoff (and
projects we'd love to see) 2 min read Sep 12 How to Safely Query Enterprise Data
with LangChain Agents + SQL + OpenAI + Gretel 6 min read Sep 12 OpaquePrompts x
LangChain: Enhance the privacy of your LangChain application with just one code
change 4 min read Load more LangChain © 2023 Sign up Powered by Ghost
Thought:I have successfully navigated to the blog.langchain.dev website. The
text on the webpage includes featured posts such as "Announcing LangChain Hub,"
"Using LangSmith to Support Fine-tuning," "Peering Into the Soul of AI Decision-
Making with LangSmith," "LangChain + Docugami Webinar: Lessons from Deploying
```

```
LLMs with LangSmith," "TED AI Hackathon Kickoff (and projects we'd love to see),"
"How to Safely Query Enterprise Data with LangChain Agents + SQL + OpenAI +
Gretel," and "OpaquePrompts x LangChain: Enhance the privacy of your LangChain
application with just one code change." There are also links to other pages on the
website.

> Finished chain.
I have successfully navigated to the blog.langchain.dev website. The text on the
webpage includes featured posts such as "Announcing LangChain Hub," "Using
LangSmith to Support Fine-tuning," "Peering Into the Soul of AI Decision-Making
with LangSmith," "LangChain + Docugami Webinar: Lessons from Deploying LLMs with
LangSmith," "TED AI Hackathon Kickoff (and projects we'd love to see)," "How to
Safely Query Enterprise Data with LangChain Agents + SQL + OpenAI + Gretel," and
"OpaquePrompts x LangChain: Enhance the privacy of your LangChain application
with just one code change." There are also links to other pages on the website.
```

4.2.3　OpenAI Functions

某些 OpenAI 模型（如 gpt-3.5-turbo-0613 和 gpt-4-0613）已经过微调，可以检测何时应该调用函数，并使用应传递给函数的输入进行响应。在 API 调用中，用户可以描述函数，并让 LLM 智能地选择输出包含参数的 JSON 对象并以此来调用这些函数。OpenAI 函数 API 的目标是比通用文本或聊天 API 更可靠地返回有效且有用的函数调用。OpenAI 函数 Agent 旨在与这些模型配合使用。

如果用户需要使用 OpenAI 函数 Agent，则必须安装 openai、google-search-results 包，因为 LangChain 会在内部调用它们，安装这些包的代码如下：

```
! pip install openai google-search-results
```

按照惯例，先定义一些 Agent 可以使用的工具，代码如下：

```
//第 4 章/openaicreatools.py
from langchain.agents import AgentType, Tool, initialize_agent
from langchain.chains import LLMMathChain
from langchain.chat_models import ChatOpenAI
from langchain.utilities import SerpAPIWrapper

llm = ChatOpenAI(temperature=0, model="gpt-3.5-turbo-0613")
search = SerpAPIWrapper()
llm_math_chain = LLMMathChain.from_llm(llm=llm, verbose=True)
tools = [
    Tool(
        name="Search",
        func=search.run,
```

```
            description="useful for when you need to answer questions about current 
events. You should ask targeted questions",
    ),
    Tool(
        name="Calculator",
        func=llm_math_chain.run,
        description="useful for when you need to answer questions about math",
    ),
]
```

首先使用 LCEL 来创建这个 Agent，代码如下：

```
//第 4 章 / openaifunctionLCEL.py
from langchain.prompts import ChatPromptTemplate, MessagesPlaceholder
prompt = ChatPromptTemplate.from_messages(
    [
        ("system", "You are a helpful assistant"),
        ("user", "{input}"),
        MessagesPlaceholder(variable_name="agent_scratchpad"),
    ]
)

from langchain.tools.render import format_tool_to_openai_function
llm_with_tools = llm.bind(functions=[format_tool_to_openai_function(t) for t 
in tools])

from langchain.agents.format_scratchpad import format_to_openai_function_
messages
from langchain.agents.output_parsers import OpenAIFunctionsAgentOutputParser
agent = (
    {
        "input": lambda x: x["input"],
        "agent_scratchpad": lambda x: format_to_openai_function_messages(
            x["intermediate_steps"]
        ),
    }
    | prompt
    | llm_with_tools
    | OpenAIFunctionsAgentOutputParser()
)
from langchain.agents import AgentExecutor
agent_executor = AgentExecutor(agent=agent, tools=tools, verbose=True)
agent_executor.invoke(
    {
        "input": "Who is Leo DiCaprio's girlfriend? What is her current age raised 
to the 0.43 power? "
    }
)
```

测试的 prompt 这里使用之前的，询问一个人名，然后进行一个数学运算，结果如下：

```
> Entering new AgentExecutor chain...

Invoking: `Search` with `Leo DiCaprio's girlfriend`

['Blake Lively and DiCaprio are believed to have enjoyed a whirlwind five-month romance in 2011. The pair were seen on a yacht together in Cannes, ...']
Invoking: `Calculator` with `0.43`


> Entering new LLMMathChain chain...
0.43```text
0.43
```
...numexpr.evaluate("0.43")...

Answer: 0.43
> Finished chain.
Answer: 0.43I'm sorry, but I couldn't find any information about Leo DiCaprio's current girlfriend. As for raising her age to the power of 0.43, I'm not sure what her current age is, so I can't provide an answer for that.

> Finished chain.
```

可以看到在用 openai function 包集成之前定义的工具后，在执行任务的过程中 Agent 准确地使用了对应的搜索和计算工具。

用户现在可以使用 OpenAIFunctionsAgent，它将在后台创建此 Agent，代码如下：

```
agent_executor = initialize_agent(
    tools, llm, agent=AgentType.OPENAI_FUNCTIONS, verbose=True
)
```

4.2.4　Conversational

前面的其他 Agent 通常会针对所使用的工具进行优化，以作出最佳响应，这在对话环境中并不理想，因为很多场景下通常希望 Agent 也能够与用户聊天，这里就要介绍到 Conversational Agent。如果将 Conversational Agent 与标准 ReAct Agent 进行比较，则两者的主要区别在于提示。用户使用 Conversational Agent 主要希望它更具对话性。

首先定义初始化需要用到的工具，代码如下：
```

```python
//第 4 章/ conversationalAgentTool.py
from langchain.agents import AgentType, Tool, initialize_agent
from langchain.llms import OpenAI
from langchain.memory import ConversationBufferMemory
from langchain.utilities import SerpAPIWrapper

search = SerpAPIWrapper()
tools = [
    Tool(
        name="Current Search",
        func=search.run,
        description="useful for when you need to answer questions about current events or the current state of the world",
    ),
]

llm = OpenAI(temperature=0)
```

然后使用 LCEL 来创建这个 Agent, 代码如下:

```python
//第 4 章/ conversationalAgentLCEL.py
from langchain import hub
from langchain.agents.format_scratchpad import format_log_to_str
from langchain.agents.output_parsers import ReActSingleInputOutputParser
from langchain.tools.render import render_text_description

prompt = hub.pull("hwchase17/react-chat")

prompt = prompt.partial(
    tools=render_text_description(tools),
    tool_names=", ".join([t.name for t in tools]),
)

llm_with_stop = llm.bind(stop=["\nObservation"])

agent = (
    {
        "input": lambda x: x["input"],
        "agent_scratchpad": lambda x: format_log_to_str(x["intermediate_steps"]),
        "chat_history": lambda x: x["chat_history"],
    }
    | prompt
    | llm_with_stop
    | ReActSingleInputOutputParser()
```

```
)

from langchain.agents import AgentExecutor
memory = ConversationBufferMemory(memory_key="chat_history")
agent_executor = AgentExecutor(agent=agent, tools=tools, verbose=True, memory=memory)
```

接下来将用对话的形式和这个 Agent 进行交互,首先向 LLM 询问一些基本的信息,然后询问需要调用外部查询工具的信息。

先和 LLM 打招呼,这里可以看到 LLM 没有调用任何工具,代码如下:

```
agent_executor.invoke({"input": "hi, i am bob"})["output"]
```

输出的结果如下:

```
> Entering new AgentExecutor chain...

Thought: Do I need to use a tool? No
Final Answer: Hi Bob, nice to meet you! How can I help you today?

> Finished chain.

'Hi Bob, nice to meet you! How can I help you today? '
```

刚才已经告知了 LLM 姓名,由于之前设置了记忆,因此当再次询问名字信息时,LLM 应该能够准确地回答出来,代码如下:

```
agent_executor.invoke({"input": "what's my name?"})["output"]
```

输出的结果如下:

```
> Entering new AgentExecutor chain...

Thought: Do I need to use a tool? No
Final Answer: Your name is Bob.

> Finished chain.

'Your name is Bob.'
```

接下来尝试询问更加复杂的问题,例如询问 LLM 2023 年 9 月 21 日上映的电影信息。众所周知,OpenAI 模型是有数据截止日期的,这段时间点已经超过了截止日期,因此如果不借助外部工具,则 LLM 是不可能知道答案的,代码如下:

```
agent_executor.invoke({"input": "what are some movies showing 9/21/2023?"})
["output"]
```

输出的结果如下：

```
> Entering new AgentExecutor chain...

Thought: Do I need to use a tool? Yes
Action: Current Search
Action Input: Movies showing 9/21/2023['September 2023 Movies: The Creator • Dumb
Money • Expend4bles • The Kill Room • The Inventor • The Equalizer 3 • PAW Patrol:
The Mighty Movie, ...'] Do I need to use a tool? No
Final Answer: According to current search, some movies showing on 9/21/2023 are
The Creator, Dumb Money, Expend4bles, The Kill Room, The Inventor, The Equalizer
3, and PAW Patrol: The Mighty Movie.

> Finished chain.

'According to current search, some movies showing on 9/21/2023 are The Creator,
Dumb Money, Expend4bles, The Kill Room, The Inventor, The Equalizer 3, and PAW
Patrol: The Mighty Movie.'
```

可以看到 LLM 准确地调用了外部搜索工具并输出了答案。读者可以看到使用 Conversational Agent 不仅能照样正确地调用外部工具，同时整个对话流也非常流畅。

4.2.5　ReAct Document Store

前面介绍了通用的 ReAct Agent，有些时候应用场景要求搜索和查找文档，这时 ReAct Document Store 就会更加适合了。实现这个 Agent 的代码如下：

```
//第 4 章/ reactdocumentstore.py
from langchain.agents import AgentType, Tool, initialize_agent
from langchain.agents.react.base import DocstoreExplorer
from langchain.docstore import Wikipedia
from langchain.llms import OpenAI

docstore = DocstoreExplorer(Wikipedia())
tools = [
    Tool(
        name="Search",
        func=docstore.search,
        description="useful for when you need to ask with search",
    ),
    Tool(
        name="Lookup",
```

```
        func=docstore.lookup,
        description="useful for when you need to ask with lookup",
    ),
]

llm = OpenAI(temperature=0, model_name="text-davinci-002")
react = initialize_agent(tools, llm, agent=AgentType.REACT_DOCSTORE, verbose=True)

question = "Author David Chanoff has collaborated with a U.S. Navy admiral who served as the ambassador to the United Kingdom under which President?"
react.run(question)
```

输出的结果如下：

```
> Entering new AgentExecutor chain...

Thought: I need to search David Chanoff and find the U.S. Navy admiral he collaborated with. Then I need to find which President the admiral served under.

Action: Search[David Chanoff]

Observation: David Chanoff is a noted author of non-fiction work. His work has typically involved collaborations with the principal protagonist of the work concerned. His collaborators have included; Augustus A. White, Joycelyn Elders, Đoàn Văn Toai, William J. Crowe, Ariel Sharon, Kenneth Good and Felix Zandman. He has also written about a wide range of subjects including literary history, education and foreign for The Washington Post, The New Republic and The New York Times Magazine. He has published more than twelve books.
Thought: The U.S. Navy admiral David Chanoff collaborated with is William J. Crowe. I need to find which President he served under.

Action: Search[William J. Crowe]

Observation: William James Crowe Jr. (January 2, 1925 - October 18, 2007) was a United States Navy admiral and diplomat who served as the 11th chairman of the Joint Chiefs of Staff under Presidents Ronald Reagan and George H. W. Bush, and as the ambassador to the United Kingdom and Chair of the Intelligence Oversight Board under President Bill Clinton.
Thought: William J. Crowe served as the ambassador to the United Kingdom under President Bill Clinton, so the answer is Bill Clinton.

Action: Finish[Bill Clinton]

> Finished chain.

'Bill Clinton'
```

4.3 迭代器运行 Agent

将 Agent 作为迭代器运行可能会很有用,因为它允许在执行过程中添加人在回路中的检查,这对于处理复杂任务或需要人类干预的情况非常重要。通过 AgentExecutorIterator 功能用户可以更加灵活地控制 Agent 的执行过程。为了具体说明这一点,以下是一个具体的例子,其中 Agent 的任务是从一个工具中检索 3 个质数,然后将它们相乘。在这个简单的问题中可以通过将 Agent 设计为迭代器实现更细粒度的控制。迭代器允许用户在每步执行之后添加逻辑,例如检查中间步骤的输出是否为质数。这种设计使用户可以在 Agent 的执行过程中及时发现并纠正潜在的问题,提高执行的准确性和可靠性。下面将具体实现这个例子,首先导入相关的库和设置 LLM,代码如下:

```
from langchain.agents import AgentType, initialize_agent
from langchain.chains import LLMMathChain
from langchain_core.pydantic_v1 import BaseModel, Field
from langchain_core.tools import Tool
from langchain_openai import ChatOpenAI

%pip install --upgrade --quiet  numexpr

#在这里需要使用 GPT-4
#它应该使用计算器来执行最终计算
llm = ChatOpenAI(temperature=0, model="gpt-4")
llm_math_chain = LLMMathChain.from_llm(llm=llm, verbose=True)
```

接下来定义两个工具,一个用于获取第 n 个质数(在此示例中使用小的子集);另一个是 LLMMathChain,充当计算器的角色,代码如下:

```
//第 4 章/defineTools.py
primes = {998: 7901, 999: 7907, 1000: 7919}


class CalculatorInput(BaseModel):
    question: str = Field()


class PrimeInput(BaseModel):
    n: int = Field()


def is_prime(n: int) -> bool:
    if n <= 1 or (n % 2 == 0 and n > 2):
        return False
```

```python
    for i in range(3, int(n**0.5) + 1, 2):
        if n % i == 0:
            return False
    return True


def get_prime(n: int, primes: dict = primes) -> str:
    return str(primes.get(int(n)))


async def aget_prime(n: int, primes: dict = primes) -> str:
    return str(primes.get(int(n)))


tools = [
    Tool(
        name="GetPrime",
        func=get_prime,
        description="A tool that returns the `n`th prime number",
        args_schema=PrimeInput,
        coroutine=aget_prime,
    ),
    Tool.from_function(
        func=llm_math_chain.run,
        name="Calculator",
        description =" Useful for when you need to compute mathematical expressions",
        args_schema=CalculatorInput,
        coroutine=llm_math_chain.arun,
    ),
]
```

然后是构建 Agent，本例将在这里使用 OpenAI Functions 的 Agent，代码如下：

```
from langchain import hub

#选择需要使用的提示,这可以修改
#可以在以下位置查看完整的提示 https://smith.langchain.com/hub/hwchase17/openai-functions-agentprompt = hub.pull("hwchase17/openai-functions-agent")

from langchain.agents import create_openai_functions_agent

agent = create_openai_functions_agent(llm, tools, prompt)

from langchain.agents import AgentExecutor

agent_executor = AgentExecutor(agent=agent, tools=tools, verbose=True)
```

接下来是运行迭代并对某些步骤执行自定义检查，代码如下：

```
//第 4 章/ runIteration.py
question = "What is the product of the 998th, 999th and 1000th prime numbers? "

for step in agent_executor.iter({"input": question}):
    if output := step.get("intermediate_step"):
        action, value = output[0]
        if action.tool == "GetPrime":
            print(f"Checking whether {value} is prime...")
            assert is_prime(int(value))
        #询问用户是否要继续
        _continue = input("Should the agent continue (Y/n)?:\n") or "Y"
        if _continue.lower() != "y":
            break
```

输出的结果如下：

```
> Entering new AgentExecutor chain...

Invoking: `GetPrime` with `{'n': 998}`


7901Checking whether 7901 is prime...
Should the agent continue (Y/n)?:
y

Invoking: `GetPrime` with `{'n': 999}`


7907Checking whether 7907 is prime...
Should the agent continue (Y/n)?:
y

Invoking: `GetPrime` with `{'n': 1000}`


7919Checking whether 7919 is prime...
Should the agent continue (Y/n)?:
y

Invoking: `Calculator` with `{'question': '7901 * 7907 * 7919'}`



> Entering new LLMMathChain chain...
```

```
7901 * 7907 * 7919```text
7901 * 7907 * 7919
```
...numexpr.evaluate("7901 * 7907 * 7919")...

Answer: 494725326233
> Finished chain.
Answer: 494725326233Should the agent continue (Y/n)?:
y
The product of the 998th, 999th and 1000th prime numbers is 494,725,326,233.

> Finished chain.
```

## 4.4 让 Agent 返回结构化输出

本节涵盖了如何使 Agent 返回结构化输出。在默认情况下，大多数 Agent 会返回单个字符串，然而通常使 Agent 返回具有更多结构的内容是很有用的。一个很好的例子是一个被指定为在一些来源上进行问答的 Agent。假设用户希望 Agent 不仅回答问题，还要附带一个使用的来源列表，输出大致遵循以下模式，代码如下：

```
class Response(BaseModel):
    """对所提出问题的最终答复"""
    answer: str = Field(description = "The final answer to respond to the user")
    sources: List[int] = Field(description="List of page chunks that contain answer to the question. Only include a page chunk if it contains relevant information")
```

本节将介绍一个具有检索器工具并可以以正确格式输出响应的 Agent。第 1 步是创建检索器，这里将进行一些设置工作，创建一个基于包含"国情咨文"内容的模拟数据的检索器。重要的是其中将向每个文档的元数据中添加一个名为 page_chunk 的标签。这只是一些虚构的数据，旨在模拟一个源字段。在实践中，这更可能是文档的 URL 或路径，代码如下：

```
//第 4 章/ createRetriever.py

from langchain.text_splitter import RecursiveCharacterTextSplitter
from langchain_community.document_loaders import TextLoader
from langchain_community.vectorstores import Chroma
from langchain_openai import OpenAIEmbeddings

#加载到文档中进行检索
loader = TextLoader("../../state_of_the_union.txt")
```
```

```python
documents = loader.load()

#将文档分割成块
text_splitter = RecursiveCharacterTextSplitter(chunk_size=1000, chunk_overlap=0)
texts = text_splitter.split_documents(documents)

#这是添加虚假来源信息的地方
for i, doc in enumerate(texts):
    doc.metadata["page_chunk"] = i

#创建检索器
embeddings = OpenAIEmbeddings()
vectorstore = Chroma.from_documents(texts, embeddings, collection_name="state-
of-union")
retriever = vectorstore.as_retriever()
```

然后需要创建相应的工具,本例中它是一个包装检索器的工具,代码如下:

```python
//第 4 章/ createTool.py
from langchain.agents.agent_toolkits.conversational_retrieval.tool import (
    create_retriever_tool,
)

retriever_tool = create_retriever_tool(
    retriever,
    "state-of-union-retriever",
    "Query a retriever to get information about state of the union address",
)
```

接着需要创建 LLM 输出响应的模式,在这种情况下希望最终答案有两个字段,一个是答案;另一个是来源列表,代码如下:

```python
//第 4 章/ responseSchema.py
from typing import List

from langchain.utils.openai_functions import convert_pydantic_to_openai_
function
from pydantic import BaseModel, Field


class Response(BaseModel):
    """对所提出问题的最终答复"""

    answer: str = Field(description="The final answer to respond to the user")
    sources: List[int] = Field(
```

```
            description="List of page chunks that contain answer to the question.
Only include a page chunk if it contains relevant information"
    )
```

接着创建自定义解析逻辑,其工作原理是将 Response 模式通过 OpenAI 的 functions 参数传递给 OpenAI 的 LLM。这类似于为 Agent 传递工具。当 OpenAI 调用 Response 函数时,希望将其作为向用户返回的信号。当 OpenAI 调用任何其他函数时,将其视为工具调用,因此,这里的解析逻辑具有以下块:

(1) 如果没有调用函数,则假设我们应该使用响应来回应用户,因此返回 AgentFinish。
(2) 如果调用了 Response 函数,则使用该函数的输入向用户回应,因此返回 AgentFinish。
(3) 如果调用了任何其他函数,则将其视为工具调用,因此返回 AgentActionMessageLog。

需要注意的是这里使用 AgentActionMessageLog 而不是 AgentAction,因为它允许用户附加一条消息记录,以便将来传递回 Agent 提示,代码如下:

```
//第 4 章/ parseLogic.py
import json

from langchain_core.agents import AgentActionMessageLog, AgentFinish


def parse(output):
    # 如果没有调用任何函数,则返给用户
    if "function_call" not in output.additional_kwargs:
        return AgentFinish(return_values={"output": output.content}, log=output.content)

    # 解析函数调用
    function_call = output.additional_kwargs["function_call"]
    name = function_call["name"]
    inputs = json.loads(function_call["arguments"])

    # 如果调用了 Response 函数,则将函数输入返给用户
    if name == "Response":
        return AgentFinish(return_values=inputs, log=str(function_call))
    # 否则返回 Agent 的动作
    else:
        return AgentActionMessageLog(
            tool=name, tool_input=inputs, log="", message_log=[output]
        )
```

最后的步骤是创建 Agent。这个 Agent 包含以下组件。

(1) 提示:一个简单的提示,其中包含用户问题和 agent_scratchpad(任何中间步骤)的占位符。
(2) 工具:可以将工具和 Response 格式附加到 LLM 作为函数。

(3) 格式化 scratchpad：为了将 agent_scratchpad 从中间步骤格式化，这里将使用标准的 format_to_openai_function_messages。这将中间步骤格式化为 AIMessages 和 FunctionMessages。

(4) 输出解析器：将使用上面的自定义解析器来解析 LLM 的响应。

(5) AgentExecutor：这里将使用标准的 AgentExecutor 运行 agent-tool-agent-tool 循环。

代码如下：

```
from langchain.agents import AgentExecutor
from langchain.agents.format_scratchpad import format_to_openai_function_messages
from langchain_community.tools.convert_to_openai import format_tool_to_openai_function
from langchain_core.prompts import ChatPromptTemplate, MessagesPlaceholder
from langchain_openai import ChatOpenAI

prompt = ChatPromptTemplate.from_messages(
    [
        ("system", "You are a helpful assistant"),
        ("user", "{input}"),
        MessagesPlaceholder(variable_name="agent_scratchpad"),
    ]
)

llm = ChatOpenAI(temperature=0)

llm_with_tools = llm.bind(
    functions=[
        #检索器工具
        format_tool_to_openai_function(retriever_tool),
        #输出模式
        convert_pydantic_to_openai_function(Response),
    ]
)
```

然后创建 Agent，代码如下：

```
//第 4 章/ createAgent.py
agent = (
    {
        "input": lambda x: x["input"],
        #通过中间步骤格式化 Agent Scratchpad
        "agent_scratchpad": lambda x: format_to_openai_function_messages(
            x["intermediate_steps"])
```

```
        ),
    }
    | prompt
    | llm_with_tools
    | parse
)

agent_executor = AgentExecutor(tools=[retriever_tool], agent=agent, verbose=
True)
```

现在可以运行 Agent。用户可以注意这个 Agent 如何用带有两个键的字典进行输出响应，也就是最开始定义的"答案"和"来源"，代码如下：

```
agent_executor.invoke(
    {"input": "what did the president say about kentaji brown jackson"},
    return_only_outputs=True,
)

> Entering new AgentExecutor chain...
[Document(page_content='Tonight. I call on the Senate to: Pass the Freedom to
Vote Act. Pass the John Lewis Voting Rights Act. And while you're at it, pass the
Disclose Act so Americans can know who is funding our elections. \n\nTonight, I'd
like to honor someone who has dedicated his life to serve this country: Justice
Stephen Breyer—an Army veteran, Constitutional scholar, and retiring Justice of
the United States Supreme Court. Justice Breyer, thank you for your service. \n\
nOne of the most serious constitutional responsibilities a President has is
nominating someone to serve on the United States Supreme Court. \n\nAnd I did that
4 days ago, when I nominated Circuit Court of Appeals Judge Ketanji Brown Jackson.
One of our nation's top legal minds, who will continue Justice Breyer's legacy of
excellence.', metadata={'page_chunk': 31, 'source': '../../state_of_the_union.
txt'}), Document(page_content='One was stationed at bases and breathing in toxic
smoke from " burn pits" that incinerated wastes of war—medical and hazard
material, jet fuel, and more. \n\nWhen they came home, many of the world's fittest
and best trained warriors were never the same. \n \nHeadaches. Numbness.
Dizziness. \n\nA cancer that would put them in a flag-draped coffin. \n\nI know.
\n\nOne of those soldiers was my son Major Beau Biden. \n\nWe don't know for sure
if a burn pit was the cause of his brain cancer, or the diseases of so many of our
troops. \n\nBut I'm committed to finding out everything we can. \n\nCommitted to
military families like Danielle Robinson from Ohio. \n \nThe widow of Sergeant
First Class Heath Robinson.  \n \nHe was born a soldier. Army National Guard.
Combat medic in Kosovo and Iraq. \n\nStationed near Baghdad, just yards from burn
pits the size of football fields. \n\nHeath's widow Danielle is here with us
tonight. They loved going to Ohio State football games. He loved building Legos
with their daughter.', metadata={'page_chunk': 37, 'source': '../../state_of_the
_union.txt'}), Document(page_content='A former top litigator in private practice.
```

```
A former federal public defender. And from a family of public school educators and
police officers. A consensus builder. Since she's been nominated, she's received
a broad range of support—from the Fraternal Order of Police to former judges
appointed by Democrats and Republicans. \n\nAnd if we are to advance liberty and
justice, we need to secure the Border and fix the immigration system. \n\nWe can do
both. At our border, we've installed new technology like cutting-edge scanners to
better detect drug smuggling.    \n\nWe've set up joint patrols with Mexico and
Guatemala to catch more human traffickers.   \n\nWe're putting in place dedicated
immigration judges so families fleeing persecution and violence can have their
cases heard faster. \n\nWe're securing commitments and supporting partners in
South and Central America to host more refugees and secure their own borders.',
metadata={'page_chunk': 32, 'source': '../../state_of_the_union.txt'}),
Document(page_content='But cancer from prolonged exposure to burn pits ravaged
Heath's lungs and body. \n\nDanielle says Heath was a fighter to the very end. \n\
nHe didn't know how to stop fighting, and neither did she. \n\nThrough her pain she
found purpose to demand we do better. \n\nTonight, Danielle—we are. \n\nThe VA is
pioneering new ways of linking toxic exposures to diseases, already helping more
veterans get benefits. \n\nAnd tonight, I'm announcing we're expanding
eligibility to veterans suffering from nine respiratory cancers. \n\nI'm also
calling on Congress: pass a law to make sure veterans devastated by toxic
exposures in Iraq and Afghanistan finally get the benefits and comprehensive
health care they deserve. \n\nAnd fourth, let's end cancer as we know it. \n\nThis
is personal to me and Jill, to Kamala, and to so many of you. \n\nCancer is the #2
cause of death in America-second only to heart disease.', metadata={'page_chunk
': 38, 'source': '../../state_of_the_union.txt'})] {'name': 'Response',
'arguments': '{\n  "answer": "President mentioned Ketanji Brown Jackson as a
nominee for the United States Supreme Court and praised her as one of the nation\'
s top legal minds.",\n  "sources": [31]\n}'}

> Finished chain.

{'answer': "President mentioned Ketanji Brown Jackson as a nominee for the United
States Supreme Court and praised her as one of the nation's top legal minds.",
'sources': [31]}
```

4.5 处理 Agent 解析错误

有些情况下 LLM 无法确定要采取什么步骤,因为其输出的格式不正确,无法由输出解析器处理。在这种情况下,在默认情况下 Agent 会出错,但是用户可以使用 handle_parsing_errors 轻松地控制此功能。本节介绍如何使用这个功能。

安装一个维基百科工具,代码如下:

```
//第 4 章/setUpWikiTool.py
```

```
%pip install --upgrade --quiet  Wikipedia

from langchain import hub
from langchain.agents import AgentExecutor, create_react_agent
from langchain_community.tools import WikipediaQueryRun
from langchain_community.utilities import WikipediaAPIWrapper
from langchain_openai import OpenAI

api_wrapper = WikipediaAPIWrapper(top_k_results=1, doc_content_chars_max=100)
tool = WikipediaQueryRun(api_wrapper=api_wrapper)
tools = [tool]

prompt = hub.pull("hwchase17/react")

llm = OpenAI(temperature=0)

agent = create_react_agent(llm, tools, prompt)
```

4.6 将 Agent 构建为图

LangGraph 是一个用于构建具有 LLM 的有状态多角色应用程序的库,它构建在 LangChain 之上(并且旨在与 LangChain 一起使用)。它将 LCEL 扩展到能够在多个计算步骤中协调多个链(或角色),从而在循环方式下实现。

LangGraph 的主要用途是为用户自己的 LLM 应用程序添加循环。用户需要注意的是这不是一个 DAG 框架。如果想构建 DAG,则应该只使用 LCEL。循环对于 Agent 的行为非常重要,这样就可以在循环中调用 LLM,询问它下一步应采取什么行动。本节将介绍 LangGraph 的实现方式。

4.6.1 快速开始

要使用 LangGraph,用户首先需要安装它,代码如下:

```
pip install langgraph
```

在这个例子中将介绍如何重新创建 LangChain 中的 AgentExecutor 类。使用 LangGraph 创建它的好处在于它更加可修改,然后还需要安装一些 LangChain 包,以及 Tavily 作为示例工具,代码如下:

```
pip install -U langchain langchain_openai langchainhub tavily-python
```

此外还需要导出一些 Agent 所需的环境变量，代码如下：

```
export OPENAI_API_KEY=sk-...
export TAVILY_API_KEY=tvly-...
```

作为可选项，用户可以设置 LangSmith 以获得整体观察性，LangSmith 会在后面的章节中详细介绍，代码如下：

```
export LANGCHAIN_TRACING_V2="true"
export LANGCHAIN_API_KEY=ls__...
export LANGCHAIN_ENDPOINT=https://api.langchain.plus
```

第 1 步是定义一个 LangChain 的 Agent，代码如下：

```python
//第 4 章/defineAgent.py
from langchain import hub
from langchain.agents import create_openai_functions_agent
from langchain_openai.chat_models import ChatOpenAI
from langchain_community.tools.tavily_search import TavilySearchResults

tools = [TavilySearchResults(max_results=1)]

#定义提示，用户可以修改此项
prompt = hub.pull("hwchase17/openai-functions-agent")

#选择 LLM
llm = ChatOpenAI(model="gpt-3.5-turbo-1106")

#构建 OpenAI 的 Agent
agent_runnable = create_openai_functions_agent(llm, tools, prompt)
```

第 2 步是定义图中需要的节点。在 LangGraph 中，节点可以是函数或 runnable，这里需要以下两个主要节点。

(1) Agent：负责决定是否采取行动。

(2) 用于调用工具的函数：如果 Agent 决定采取行动，则该节点将执行该操作。

此外还需要定义一些边，其中一些边可能是有条件的。它们是有条件的原因是根据节点的输出，可以采取几条路径中的一条。具体采取的路径在运行该节点时（LLM 输出之前）是未知的。

(1) 有条件边：在调用 Agent 之后，Agent 本身应该调用函数工具或者直接判定完成。

(2) 正常边：在调用工具后，它应该始终返回 Agent 以决定下一步要做什么。

第 3 步，定义这些节点和创建一个函数来决定采取哪种条件边，代码如下：

```
//第 4 章/defineNode.py
```

```python
from langchain_core.runnables import RunnablePassthrough
from langchain_core.agents import AgentFinish


# 定义 Agent
# 需要注意,这里使用 `.assign` 将代理的输出添加到字典中
# 该字典将从节点返回
# 不想仅从该节点返回 `agent_runnable` 结果的原因是
# 想要继续传递所有其他输入
agent = RunnablePassthrough.assign(
    agent_outcome = agent_runnable
)

# 定义执行工具的函数
def execute_tools(data):
    # 获取最新的 agent_outcome,这是上面 `agent` 中添加的键
    agent_action = data.pop('agent_outcome')
    # 获取要使用的工具
    tool_to_use = {t.name: t for t in tools}[agent_action.tool]
    # 在输入上调用该工具
    observation = tool_to_use.invoke(agent_action.tool_input)
    # 现在将操作和观察添加到 intermediate_steps 列表中
    # 这是之前采取的所有操作及其输出的列表
    data['intermediate_steps'].append((agent_action, observation))
    return data

# 定义将用于确定要下降的条件边沿的逻辑
def should_continue(data):
    # 如果 Agent 的结果是 AgentFinish,则返回 `exit` 字符串
    # 这将在设置图表来定义流程时使用
    if isinstance(data['agent_outcome'], AgentFinish):
        return "exit"
    # 否则返回一个 AgentAction
    # 这里我们返回 `continue` 字符串
    # 这将在设置图表来定义流程时使用
    else:
        return "continue"
```

定义完节点后就可以把所有的组件都串在一起定义图,代码如下:

```python
//第 4 章/defineGraph.py
from langgraph.graph import END, Graph

workflow = Graph()

# 添加 Agent 节点,将其命名为 `agent`,稍后将使用它
```

```python
workflow.add_node("agent", agent)
#添加工具节点,将其命名为 `tools`,稍后将使用它
workflow.add_node("tools", execute_tools)

#将入口点设置为 `agent`
#这意味着这个节点是首先被调用的节点
workflow.set_entry_point("agent")

#现在添加一个条件边缘
workflow.add_conditional_edges(
    #首先,定义起始节点。使用 `agent`
    #这意味着这些边是在调用 `agent` 节点之后被执行的
    "agent",
    #接下来,传递用于确定下一个被调用的节点的函数
    should_continue,
    #最后,传递一个映射
    #键是字符串,值是其他节点
    #END 是一个特殊节点,标记着图应该结束
    #将会调用 `should_continue`,然后其输出将与这个映射中的键进行匹配
    #根据匹配的键,将调用相应的节点。
    {
        #如果是 `tools`,则调用工具节点
        "continue": "tools",
        #否则结束
        "exit": END
    }
)

#现在添加一个从 `tools` 到 `agent` 的普通边缘
#这意味着在调用 `tools` 之后,将调用 `agent` 节点
workflow.add_edge('tools', 'agent')

#最后是编译
#这将其编译成一个 LangChain 可运行对象
#意味着用户可以将其像任何其他可运行对象一样使用
chain = workflow.compile()
```

最后可以使用这个工具,它公开了与所有其他 LangChain 可运行对象相同的接口,代码如下:

```
chain.invoke({"input": "what is the weather in sf", "intermediate_steps": []})
```

4.6.2 流式输出

LangGraph 支持多种不同类型的流式传输。首先,LangGraph 可以支持节点流式输

出，代码如下：

```
//第 4 章/ NodeStream.py
for output in chain.stream(
    {"input": "what is the weather in sf", "intermediate_steps": []}
):
    #Stream() 产生字典,其输出以节点名称为键
    for key, value in output.items():
        print(f"Output from node '{key}':")
        print("---")
        print(value)
    print("\n---\n")
```

输出的结果如下：

```
Output from node 'agent':
---
{'agent_outcome': AgentActionMessageLog(tool='tavily_search_results_json', 
tool_input={'query': 'weather in San Francisco'}, log="\nInvoking: `tavily_
search_results_json` with `{'query': 'weather in San Francisco'}`\n\n\n", 
message_log=[AIMessage(content='', additional_kwargs={'function_call': 
{'arguments': '{"query":"weather in San Francisco"}', 'name': 'tavily_search_
results_json'}})]), 
'input': 'what is the weather in sf', 
'intermediate_steps': []}

---

Output from node 'tools':
---
{'input': 'what is the weather in sf', 
'intermediate_steps': [(AgentActionMessageLog(tool='tavily_search_results_
json', tool_input={'query': 'weather in San Francisco'}, log="\nInvoking: 
`tavily_search_results_json` with `{'query': 'weather in San Francisco'}`\n\n\
n", message_log=[AIMessage(content='', additional_kwargs={'function_call': 
{'arguments': '{"query":"weather in San Francisco"}', 'name': 'tavily_search_
results_json'}})]), 
                      [{'content': 'Best time to go to San Francisco? '
                                   'Weather in San Francisco in january '
                                   '2024   How was the weather last january? '
                                   'Here is the day by day recorded weather '
                                   'in San Francisco in january 2023:   '
                                   'Seasonal average climate and '
                                   'temperature of San Francisco in '
                                   'january   8% 46% 29% 12% 8% Evolution of '
                                   'daily average temperature and '
```

```
                                    'precipitation in San Francisco in '
                                    'januaryWeather in San Francisco in '
                                    'january 2024. The weather in San '
                                    'Francisco in january comes from '
                                    'statistical datas on the past years. '
                                    'You can view the weather statistics the '
                                    'entire month, but also by using the '
                                    'tabs for the beginning, the middle and '
                                    'the end of the month. ... 08-01-2023 '
                                    '52°F to 58°F. 09-01-2023 54°F to 61°F. '
                                    '10-01-2023 52°F to ...',
                         'url': 'https://www.whereandwhen.net/when/north-
america/california/san-francisco-ca/january/'}])]}

---

Output from node 'agent':
---
{'agent_outcome': AgentFinish(return_values={'output': 'The weather in San
Francisco in January ranges from 52°F to 61°F. For more detailed and current
weather information, you may want to check a reliable weather website or app.'},
log='The weather in San Francisco in January ranges from 52°F to 61°F. For more
detailed and current weather information, you may want to check a reliable
weather website or app.'),
'input': 'what is the weather in sf',
'intermediate_steps': [(AgentActionMessageLog(tool='tavily_search_results_
json', tool_input={'query': 'weather in San Francisco'}, log="\nInvoking:
`tavily_search_results_json` with `{'query': 'weather in San Francisco'}`\n\n\
n", message_log=[AIMessage(content='', additional_kwargs={'function_call':
{'arguments': '{"query":"weather in San Francisco"}', 'name': 'tavily_search_
results_json'}})]),
                        [{'content': 'Best time to go to San Francisco? '
                                    'Weather in San Francisco in january '
                                    '2024  How was the weather last january? '
                                    'Here is the day by day recorded weather '
                                    'in San Francisco in january 2023:  '
                                    'Seasonal average climate and '
                                    'temperature of San Francisco in '
                                    'january  8% 46% 29% 12% 8% Evolution of '
                                    'daily average temperature and '
                                    'precipitation in San Francisco in '
                                    'januaryWeather in San Francisco in '
                                    'january 2024. The weather in San '
                                    'Francisco in january comes from '
                                    'statistical datas on the past years. '
                                    'You can view the weather statistics the '
```

```
                                    'entire month, but also by using the '
                                    'tabs for the beginning, the middle and '
                                    'the end of the month. ... 08-01-2023 '
                                    '52°F to 58°F. 09-01-2023 54°F to 61°F. '
                                    '10-01-2023 52°F to ...',
                          'url': 'https://www.whereandwhen.net/when/north-america/california/san-francisco-ca/january/'}])]}

---

Output from node '__end__':
---
{'agent_outcome': AgentFinish(return_values={'output': 'The weather in San Francisco in January ranges from 52°F to 61°F. For more detailed and current weather information, you may want to check a reliable weather website or app.'}, log='The weather in San Francisco in January ranges from 52°F to 61°F. For more detailed and current weather information, you may want to check a reliable weather website or app.'),
 'input': 'what is the weather in sf',
 'intermediate_steps': [(AgentActionMessageLog(tool='tavily_search_results_json', tool_input={'query': 'weather in San Francisco'}, log="\nInvoking: `tavily_search_results_json` with `{'query': 'weather in San Francisco'}`\n\n\n", message_log=[AIMessage(content='', additional_kwargs={'function_call': {'arguments': '{"query":"weather in San Francisco"}', 'name': 'tavily_search_results_json'}})]),
                         [{'content': 'Best time to go to San Francisco? '
                                      'Weather in San Francisco in january '
                                      '2024  How was the weather last january? '
                                      'Here is the day by day recorded weather '
                                      'in San Francisco in january 2023:   '
                                      'Seasonal average climate and '
                                      'temperature of San Francisco in '
                                      'january  8% 46% 29% 12% 8% Evolution of '
                                      'daily average temperature and '
                                      'precipitation in San Francisco in '
                                      'januaryWeather in San Francisco in '
                                      'january 2024. The weather in San '
                                      'Francisco in january comes from '
                                      'statistical datas on the past years. '
                                      'You can view the weather statistics the '
                                      'entire month, but also by using the '
                                      'tabs for the beginning, the middle and '
                                      'the end of the month. ... 08-01-2023 '
                                      '52°F to 58°F. 09-01-2023 54°F to 61°F. '
                                      '10-01-2023 52°F to ...',
```

```
                    'url': 'https://www.whereandwhen.net/when/north-
america/california/san-francisco-ca/january/'}])]}

---
```

用户还可以访问每个节点生成的 LLM Token。在这种情况下，只有 Agent 节点会生成 LLM Token，代码如下：

```
//第 4 章/ StreamLLM.py
async for output in chain.astream_log(
    {"input": "what is the weather in sf", "intermediate_steps": []},
    include_types=["llm"],
):
    #astream_log() 以 JSONPatch 格式生成请求的日志(此处为 LLM)
    for op in output.ops:
        if op["path"] == "/streamed_output/-":
            #这是 .stream() 的输出
            ...
        elif op["path"].startswith("/logs/") and op["path"].endswith(
            "/streamed_output/-"
        ):
            #因为这里选择仅包含 LLM,所以这些是 LLM 的 Token
            print(op["value"])
```

输出的结果如下：

```
content='' additional_kwargs={'function_call': {'arguments': '', 'name': 'tavily_search_results_json'}}
content='' additional_kwargs={'function_call': {'arguments': '{"', 'name': ''}}
content='' additional_kwargs={'function_call': {'arguments': 'query', 'name': ''}}
content='' additional_kwargs={'function_call': {'arguments': '":"', 'name': ''}}
content='' additional_kwargs={'function_call': {'arguments': 'current', 'name': ''}}
content='' additional_kwargs={'function_call': {'arguments': ' weather', 'name': ''}}
content='' additional_kwargs={'function_call': {'arguments': ' in', 'name': ''}}
content='' additional_kwargs={'function_call': {'arguments': ' San', 'name': ''}}
content='' additional_kwargs={'function_call': {'arguments': ' Francisco', 'name': ''}}
content='' additional_kwargs={'function_call': {'arguments': '"}', 'name': ''}}
content=''
content=''
content='I'
content=' found'
content=' a'
content=' website'
```

```
content=' that'
content=' provides'
content=' detailed'
content=' weather'
content=' information'
content=' for'
content=' San'
content=' Francisco'
content='.'
content=' You'
content=' can'
content=' visit'
content=' the'
content=' following'
content=' link'
content=' for'
content=' the'
content=' current'
content=' weather'
content=' report'
content=':'
content=' ['
content='San'
content=' Francisco'
content=' Weather'
content=' Report'
content=']('
content='https'
content='://'
content='www'
content='.weather'
content='25'
content='.com'
content='/n'
content='orth'
content='-'
content='amer'
content='ica'
content='/'
content='usa'
content='/cal'
content='ifornia'
content='/s'
content='an'
content='-fr'
content='anc'
content='isco'
content=')'
content=''
```

第 5 章 使用 LangChain 工具进行文档查询

在当今快节奏的商业和研究环境中,跟上不断增长的信息量可能是一项艰巨的任务。对于计算机科学和人工智能等领域的工程师和研究人员来讲,以及时了解最新发展至关重要,然而,阅读和理解大量论文可能既费时又费力。这就是自动化发挥作用的地方。在本章中将介绍一种自动总结研究论文和回答问题的方法,使研究人员更容易消化和了解情况。通过利用语言模型和一系列问题将开发的摘要以简洁和简化的格式总结论文的核心断言、含义和机制。这不仅可以在研究课题时节省时间和精力,还可以确保我们具有有效地驾驭科学进步的能力。我们还将介绍 OpenAI 模型中的函数,以及它们在信息提取中的应用。读者将看到它们如何作为简历(CV)的解析器在应用程序中工作。这个函数的语法是特定于 OpenAI 的 API 的,并且有许多应用程序,但是,LangChain 提供了一个平台,允许为任何 LLM 创建工具,从而增强其功能。这些工具使 LLM 能够与 API 交互、访问实时信息并执行各种任务,例如检索搜索、数据库查询、编写电子邮件,甚至拨打电话。本章将使用检索增强生成(Retrieval Augmented Generation,RAG)实现问答应用。这是一种通过将相关数据注入上下文来更新 LLM 的技术,例如 GPT。最后本章将讨论智能体的不同决策策略,其中将实施两种策略,第 1 种策略为计划和执行(或计划和解决),第 2 种策略为一次性 Agent,并且它们将被集成到可视化界面中,作为浏览器中的可视化应用程序(使用 Streamlit)进行问答。

5.1 幻觉现象

GPT-3、Llama 和 Claude 2 等 LLM 的快速发展引起了人们对其局限性和潜在风险的关注,其中的一个主要问题是幻觉现象,即模型生成的输出是无意义的、不连贯的或不忠实于所提供的输入。幻觉现象在实际应用中会带来性能和安全风险,例如医疗或机器翻译。幻觉现象还有另一个方面,即 LLM 生成的文本包含敏感的个人信息,例如电子邮件地址、电话号码和实际地址。这带来了重大的隐私问题,因为它表明语言模型可以从其训练语料库中记忆和恢复此类私有数据,尽管这些数据不存在于源输入中。

LLM 上下文中的幻觉现象是指生成的文本不忠实于预期内容或荒谬的现象。这个术语与心理幻觉相提并论，因为幻觉涉及感知不存在的东西。在 NLP 中，幻觉文本可能在提供的上下文中显得流畅和自然，但缺乏特异性或可验证性。相反幻觉的对立面是忠实性，即生成的内容与来源保持一致和真实。

当生成的输出与源内容相矛盾时，就会出现内在幻觉现象，而外在幻觉现象涉及生成源材料无法验证或支持的信息。外在幻觉现象有时可以包括事实正确的外部信息，但从事实安全的角度来看，它们的不可验证性引起了人们的担忧。目前，解决幻觉问题的努力正在进行中，但需要全面了解不同的任务，以制定有效的缓解方法。LLM 中的幻觉现象可能由以下多种因素引起：

（1）编码器的表示学习不完善。
（2）错误的解码，包括处理源输入的错误部分和解码策略的选择。
（3）暴露偏差，即训练和推理时间之间的差异。
（4）参数化知识偏差，即预训练模型将自己的知识优先于输入，从而导致产生过多的信息。

解决 LLM 出现幻觉现象的方法可以分为两类：数据相关方法、建模和推理方法。

1. 数据相关方法

构建忠实的数据集：从头开始构建具有干净忠实目标的数据集，或重写真实句子，同时确保语义一致性。

自动清理数据：识别和过滤现有语料库中不相关或矛盾的信息，以减少语义干扰。

信息增强：使用外部信息（例如检索到的知识或合成数据）增强输入，以提高语义理解并解决源-目标差异。

2. 建模和推理方法

架构：修改编码器架构以增强语义解释，修改注意力机制以优先处理源信息，修改解码器结构以减少幻觉并强制执行隐式或显式约束。

培训：结合计划、强化学习、多任务学习和可控生成技术，通过改善对齐、优化奖励功能及平衡忠诚度和多样性来减轻产生幻觉现象。

后处理：通过生成后优化策略来纠正输出中出现的幻觉现象，或者使用后处理校正方法专门优化结果以确保忠实度。

LLM 出现幻觉现象的后果是传播错误信息或出于政治目的滥用的危险。错误信息包括虚假信息、欺骗性新闻和谣言，对社会构成重大威胁，尤其是在内容创作和通过社交媒体传播的便利性方面。对社会的威胁包括对科学的不信任、公共卫生叙事、社会两极分化和民主进程。新闻学和档案研究对这个问题进行了广泛研究，事实核查举措也随之增加。致力于事实核查的组织为独立事实核查员和记者提供培训和资源，从而扩大专家事实核查工作的规模。解决错误信息对于维护信息的完整性和打击其对社会的不利影响至关重要。在文献中，这种问题被称为文本蕴涵，其中包括模型预测文本对之间的方向真值关系。在本章中将重点介绍通过信息增强和后处理进行自动事实核查。可以从 LLM 或使用外部工具检索事实。在前一种情况下,预先训练的语言模型可以取代知识库和检索模块,利用其丰富的知识来回答开放领域的问题,并使用提示来检索特定事实。大致的思路如图 5-1 所示[13]。

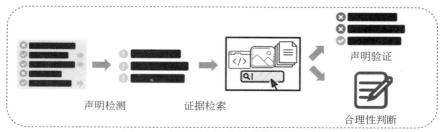

图 5-1　自动事实核查管道的 3 种状态

从图 5-1 可以区分 3 个阶段。

(1) 声明检测:识别需要验证的声明。

(2) 证据检索:检索证据以查找支持或反驳主张的来源。

(3) 声明验证:根据证据评估声明的真实性。

从 2018 年发布的 24 层 BERT-Large 模型开始,LLM 已经在维基百科等大型知识库上进行了预训练,因此它们能够回答来自维基百科的知识问题。例如,为了回答"微软的总部在哪里"这个问题,该问题将被重写为"微软的总部在[MASK]",并输入语言模型中作为答案。这种方法有趣的是,如果未接收源文本的 LLM 在生成目标时产生的损失小于接收源文本的 LLM,则表明生成的 Token 是幻觉的。幻觉 Token 与目标 Token 总数的比率可以作为生成输出中幻觉程度的衡量标准。在 LangChain 中有一个链可用于通过提示链接进行事实检查,在其中模型主动质疑语句中的假设。在这个名为 LLMCheckerChain 的自我检查链中,模型被按顺序进行提示,代码如下:

```
Here's a statement: {statement}\nMake a bullet point list of the assumptions you
made when producing the above statement.\n
```

在这里读者应注意,这是一个字符串模板,其中花括号中的元素将被变量替换。接下来,这些假设被反馈给模型,以便通过如下提示逐一检查它们,代码如下:

```
Here is a bullet point list of assertions:
    {assertions}
    For each assertion, determine whether it is true or false. If it is false,
explain why.\n\n
```

最后,模型的任务是做出最终判断,输出的结果如下:

```
In light of the above facts, how would you answer the question '{question}'
```

LLMCheckerChain 自行完成这一切,代码如下:

```
//第 5 章 / LLMCheckerChain.py
from langchain.chains import LLMCheckerChain
from langchain.llms import OpenAI
llm = OpenAI(temperature=0.7)
text = "What type of mammal lays the biggest eggs?"
checker_chain = LLMCheckerChain.from_llm(llm, verbose=True)
checker_chain.run(text)
```

LLM可以对此问题返回不同的结果,其中一些是错误的,而另一些则被正确地识别为错误。本文在下面的示例中得到了蓝鲸、北美海狸或已灭绝的巨型恐鸟等结果,结果如下:

```
//第 5 章 / LLMCheckerChainOut.py
Monotremes, a type of mammal found in Australia and parts of New Guinea, lay the
largest eggs in the mammalian world. The eggs of the American echidna (spiny
anteater) can grow as large as 10 cm in length, and dunnarts (mouse-sized
marsupials found in Australia) can have eggs that exceed 5 cm in length.
• Monotremes can be found in Australia and New Guinea
• The largest eggs in the mammalian world are laid by monotremes
• The American echidna lays eggs that can grow to 10 cm in length
• Dunnarts lay eggs that can exceed 5 cm in length
• Monotremes can be found in Australia and New Guinea - True
• The largest eggs in the mammalian world are laid by monotremes - True
• The American echidna lays eggs that can grow to 10 cm in length - False, the
American echidna lays eggs that are usually between 1 to 4 cm in length.
• Dunnarts lay eggs that can exceed 5 cm in length - False, dunnarts lay eggs that
are typically between 2 to 3 cm in length.
The largest eggs in the mammalian world are laid by monotremes, which can be found
in Australia and New Guinea. Monotreme eggs can grow to 10 cm in length.
> Finished chain.
```

虽然这并不能保证答案是正确的,但它可以阻止一些不正确的结果。事实核查方法涉及将声明分解为更小的可检查查询,这些查询可以表述为问答任务。专门为搜索领域数据集而设计的工具可以帮助事实核查人员有效地寻找证据。谷歌和必应等现成的搜索引擎还可以检索与主题和证据相关的内容,以准确捕捉陈述的真实性。在接下来的部分中将应用此方法根据 Web 搜索和其他工具返回结果。具体来讲将实现一个链来汇总文档,用户可以从这些文档中提出任何问题并得到答案。

5.2 文档总结

本节将讨论自动化总结长文本和研究论文的过程。在当今快节奏的商业和研究环境中,跟上不断增长的信息量可能是一项艰巨的任务。对于计算机科学和人工智能等领域的工程师和研究人员来讲,以及时了解最新发展至关重要,然而,阅读和理解大量论文可能既费时又费力。这就是自动化发挥作用的地方。工程师受到他们构建和创新的愿望的驱使，通过创建管道和流程来自动化执行重复性任务，从而避免自己执行重复性任务。这种方法经常被误认为是懒惰，它使工程师能够专注于更复杂的挑战，并更有效地利用他们的技能。接下来将构建一个自动化工具，该工具可以以更易于理解的格式快速总结长文本的内容。该工具旨在帮助研究人员跟上每天发表的论文，特别是在人工智能等快速发展的领域。通过自动化总结过程，研究人员可以节省时间和精力，同时确保他们随时了解各自领域的最新发展。该工具将基于 LangChain，并利用 LLM 以更简洁和简化的方式总结论文的核心断言、含义和机制。它还可以回答有关论文的具体问题，使其成为文献综述和加速科学研究的宝贵资源。

该工具还可以被进一步开发，以允许自动处理多个文档并针对特定研究领域进行定制。总体而言，该方法旨在通过提供一种更有效、更易于访问的方式来了解最新研究，从而使研究人员受益。LangChain 支持使用 LLM 处理文档的 mapreduce 方法，从而可以有效地处理和分析文档。当读取大文本并将它们拆分为适合 LLM 的标记上下文长度的文档（块）时，可以单独对每个文档应用一个链，然后将输出组合到单个文档中。mapreduce 过程包括以下两个步骤。

（1）map 步骤：LLM 链单独应用于每个文档并将输出视为新文档。
（2）reduce 步骤：所有新文档都被传递到单独的合并文档链以获得单个输出。

这种方法允许并行处理文档，并允许使用 LLM 来推理、生成或分析单个文档及组合它们的输出。该过程的机制涉及压缩或折叠映射的文档，以确保它们适合组合文档链，这可能还涉及使用 LLM。如果需要，则可以用递归方式执行压缩步骤。以下是加载 PDF 文档并对其进行总结的简单示例，代码如下：

```
//第 5 章/ pdfLoad.py
from langchain.chains.summarize import load_summarize_chain
from langchain import OpenAI
from langchain.document_loaders import PyPDFLoader
pdf_loader = PyPDFLoader(pdf_file_path)
docs = pdf_loader.load_and_split()
llm = OpenAI()
chain = load_summarize_chain(llm, chain_type="map_reduce")
chain.run(docs)
```

变量 pdf_file_path 是带有 PDF 文件路径的字符串。map 和 reduce 步骤的默认提示如下：

```
Write a concise summary of the following:
{text}
CONCISE SUMMARY:
```

用户可以为每个步骤指定任何提示。在 LangChainHub 上可以看到 qa-withsources 提示符，它采用如下的 reduce/combine 提示符：

```
Given the following extracted parts of a long document and a question, create a
final answer with references (\"SOURCES\"). \nIf you don't know the answer, just
say that you don't know. Don't try to make up an answer.\nALWAYS return a \"SOURCES\"
part in your answer.\n\nQUESTION: {question}\n=========\nContent: {text}
```

在这个提示中提出了一个具体的问题，但同样用户可以给 LLM 一个更抽象的指令来提取假设和含义。文本将是 map 步骤的摘要。这样的指示将有助于防止出现幻觉现象，其他说明示例可能是将文档翻译成不同的语言。一旦开发者开始进行大量调用，尤其是在 map 步骤中，成本就会增加。这里进行了大量调用并使用了很多 Token。这里有必要介绍关于 Token 的使用量的计算情况。

用户在使用 LLM 时，尤其是在进行长循环（例如使用映射操作）操作时，跟踪 Token 使用情况并了解花费了多少金钱非常重要。用户需要了解不同 LLM 的功能、定价选项和用例。OpenAI 提供了不同的模型，即 GPT-4、ChatGPT 和 InstructGPT，以满足各种自然语言处理需求。GPT-4 是一种强大的语言模型，适用于解决自然语言处理的复杂问题。它根据所用 Token 的大小和数量提供灵活的定价选项。ChatGPT 模型，如 GPT-3.5-Turbo，专注于聊天机器人和虚拟助手等对话应用程序。它们擅长准确和流畅地生成响应。ChatGPT 模型的定价基于使用的 Token 数量。InstructGPT 模型专为单圈指令跟踪而设计，并针对快速准确的响应生成进行了优化。InstructGPT 系列中的不同型号，例如 Ada 和 Davinci，提供了不同级别的速度和功能。Ada 是最快的模型，适用于速度至关重要的应用，而 Davinci 是最强大的模型，能够处理复杂的指令。InstructGPT 模型的定价取决于模型的功能，范围从 Ada 等低成本选项到 Davinci 等更昂贵的选项。DALL·E、Whisper 和 API 服务是 OpenAI 用于图像生成、语音转录、翻译和访问的 LLM。DALL·E 是一种 AI 驱动的图像生成模型，可以被无缝地集成到应用程序中，用于生成和编辑新颖的图像和艺术作品。OpenAI 提供了 3 层分辨率，允许用户选择他们需要的细节级别。分辨率越高，复杂性和细节就越多，而分辨率越低，表示效果越抽象，所以每幅图像的价格因分辨率而异。Whisper 是一种 AI 工具，可以将语音转录为文本并将多种语言翻译成英语。它有助于捕捉对话，促进交流，并提高跨语言的理解。Whisper 的使用费用基于每分钟费率进行计算。OpenAI 的 API 提供了对 GPT-3 等强大 LLM 的访问，这使开发人员能够创建高级应用程序。OpenAI 在注册 API 时会为用户提供初始 Token 使用限制，这表明了在特定时间范围内可用于与 LLM 交互的 Token 数。随着用户使用量的增加，OpenAI 可能会提高 Token 的使用限制，授予对模型的更多访问权限等。如果用户需要为其应用程序添加更多的 Token，则可以请求增加配额。用户可以通过连接到 OpenAI 回调来跟踪 OpenAI 平台中的 Token 使用情况，代码如下：

```
//第 5 章/tokenUsage.py
with get_openai_callback() as cb:
    response = llm_chain.predict(text="Complete this text!")
```

```
print(f"Total Tokens: {cb.total_tokens}")
print(f"Prompt Tokens: {cb.prompt_tokens}")
print(f"Completion Tokens: {cb.completion_tokens}")
print(f"Total Cost (USD): ${cb.total_cost}")
```

在此示例中,带有 llm_chain 的代码行可以是 OpenAI 平台上的任何 LLM。用户应该能够看到输出里面包含成本和 Token 的使用量。还有另外两种方法可以获取 Token 的使用情况。作为 OpenAI 回调的替代,llm 类的 generate()方法会返回 LLMResult 类型的响应,而不是字符串。这包括 Token 的使用情况和完成原因,示例代码如下:

```
input_list = [
    {"product": "socks"},
    {"product": "computer"},
    {"product": "shoes"}
]
llm_chain.generate(input_list)
```

结果如下:

```
LLMResult(generations=[[Generation(text='\n\nSocktastic!', generation_info=
{'finish_reason': 'stop', 'logprobs': None})], [Generation(text='\n\nTechCore
Solutions.', generation_info={'finish_reason': 'stop', 'logprobs': None})],
[Generation(text='\n\nFootwear Factory.', generation_info={'finish_reason':
'stop', 'logprobs': None})]], llm_output={'token_usage': {'prompt_tokens': 36,
'total_tokens': 55, 'completion_tokens': 19}, 'model_name': 'text-davinci-003'})
```

最后,OpenAI API 中的聊天完成响应格式包含一个带有 Token 信息的使用对象:

```
{
"model": "gpt-3.5-turbo-0613",
"object": "chat.completion",
"usage": {
  "completion_tokens": 17,
  "prompt_tokens": 57,
  "total_tokens": 74
 }
}
```

接下来将介绍如何使用带有 LangChain 的 OpenAI 函数从文档中提取某些信息。

5.3 信息提取

2023 年 6 月,OpenAI 宣布对 OpenAI 的 API 进行更新,包括函数调用的新功能。开发人员现在可以向 gpt-4-0613 和 gpt-3.5-turbo-0613 模型描述函数,并让模型智能地生成一个JSON对象，其中包含调用这些函数的参数。此功能旨在增强GPT模型与外部工具和API之间的联系，提供一种从模型中检索结构化数据的可靠方法。函数调用使开发人员能够创建可以使用外部工具或OpenAI插件回答问题的聊天机器人。它还允许将自然语言查询转换为API调用或数据库查询，并从文本中提取结构化数据。开发人员可以使用JSON模式向模型描述函数，并指定要调用的所需函数。在LangChain框架中，开发者可以在OpenAI中调用这些函数来进行信息提取或调用插件。对于信息提取可以使用OpenAILLM从文本中提取特定实体及其属性，以及从提取链的文档中提取特定实体及其属性。例如，这可以帮助识别文本中提到的人。通过OpenAI函数参数并指定架构，开发者可以确保模型输出具有正确的类型和属性。这种方法允许通过定义具有所需属性及其类型的架构来精确提取实体。它还允许指定哪些属性是必需的，哪些是可选的。模式的默认格式是字典，但开发者也可以在Pydantic中定义属性及其类型，从而在提取过程中提供控制性和灵活性。以下是一份简历中所需的信息架构的示例，代码如下：

```python
//第 5 章/ cvSchema.py
from typing import Optional
from pydantic import BaseModel
class Experience(BaseModel):
    start_date: Optional[str]
    end_date: Optional[str]
    description: Optional[str]
class Study(Experience):
    degree: Optional[str]
    university: Optional[str]
    country: Optional[str]
    grade: Optional[str]
class WorkExperience(Experience):
    company: str
    job_title: str
class Resume(BaseModel):
    first_name: str
    last_name: str
    linkedin_url: Optional[str]
    email_address: Optional[str]
    nationality: Optional[str]
    skill: Optional[str]
    study: Optional[Study]
    work_experience: Optional[WorkExperience]
    hobby: Optional[str]
```

紧接着可以用它来从简历中提取信息。下面将尝试将一份简历内的信息读取出来，读者可以在这个网址找到这份简历的原版：https://github.com/xitanggg/open-resume。

利用LangChain中的create_extraction_chain_pydantic()函数，开发者可以提供自己的

模板作为输入,输出将是遵循这个模板的实例化对象,代码如下:

```
//第 4 章/ pdfTemplate.py
from langchain.chains import create_extraction_chain_pydantic
from langchain.chat_models import ChatOpenAI
from langchain.document_loaders import PyPDFLoader
pdf_loader = PyPDFLoader(pdf_file_path)
docs = pdf_loader.load_and_split()
#用户需要注意并非所有模型都适用
llm = ChatOpenAI(model_name="gpt-3.5-turbo-0613")
chain = create_extraction_chain_pydantic(pydantic_schema=Resume, llm=llm)
return chain.run(docs)
```

输出的结果如下:

```
[Resume(first_name='John', last_name='Doe', linkedin_url='linkedin.com/in/
john-doe', email_address='hello@openresume.com', nationality=None, skill=
'React', study=None, work_experience=WorkExperience(start_date='May 2023',
end_date='Present', description='Lead a cross-functional team of 5 engineers in
developing a search bar, which enables thousands of daily active users to search
content across the entire platform. Create stunning home page product demo
animations that drives up sign up rate by 20%. Write clean code that is modular and
easy to maintain while ensuring 100% test coverage.', company='ABC Company',
job_title='Software Engineer'), hobby=None)]
```

这个结果并非完美,其中只有一种工作经验被解析出来,但考虑到迄今为止付出的很少的努力,这是一个良好的开端。更多功能可以被添加进去,例如让 LLM 根据简历猜测候选人的个性或领导能力。OpenAI 以某种语法将这些函数调用注入系统消息中,因为他们的模型经过针对性的优化。另外在 LangChain 框架中本身就具有将注入函数调用作为提示的功能。这意味着用户可以在 LLM 应用程序中使用 OpenAI 以外的模型进行函数调用。下面的章节将对此进行研究探讨,并尝试构建 Streamlit 框架下的交互式 Web 应用程序。

5.4 使用工具

由于 LLM 是在一般语料库数据上训练的,所以它对于需要特定领域知识的任务可能不那么有效。也是因为同样的原因,LLM 本身无法与环境交互并访问外部数据源,但是,LangChain 提供了平台,用于创建访问实时信息的工具,可以让 LLM 执行天气预报、预订、建议食谱等任务。Agent 和链框架内的工具让开发者能开发由具有数据感知和代理功能的 LLM 应用程序,并开辟了广泛的方法来解决 LLM 问题,扩展了它们的用例并使其更加通用和强大。工具的一个重要方面是它们能够在特定领域内工作或处理特定输入。例如,LLM 缺乏固有的数学能力,但是,像计算器这样的数学工具可以接受数学表达式或方程式作为输入并计算结果。LLM 与这种数学工具相结合，进行计算并提供准确的答案。通常，这种检索方法和 LLM 的组合称为 RAG，它通过从外部来源检索相关数据并将其注入上下文来解决 LLM 的局限性问题。这些检索到的数据用作附加信息，以增强对 LLM 的提示。通过 RAG 将 LLM 与用例特定信息相结合，可以提高响应的质量和准确性。通过检索相关数据，RAG 有助于减少 LLM 的幻觉反应。例如，医疗保健应用程序中使用的 LLM 可以在推理过程中从外部来源（如医学文献或数据库）检索相关的医疗信息，然后可以将检索到的数据合并到上下文中，以增强生成的响应，并确保它们准确无误并与特定领域的知识保持一致。在此方案中实现 RAG 的好处是双重的。首先，它允许将最新信息合并到响应中（大家都知道目前 LLM 都有数据截止日期）。这确保了用户能够获得准确和相关的信息，即使是最近发生的事件或不断变化的主题，其次，RAG 通过利用外部信息来源增强了 ChatGPT 提供更详细和上下文答案的能力。通过从与特定主题相关的新闻文章或网站等来源检索特定上下文，LLM 的响应将更加准确。

RAG 的工作原理是从数据源中检索信息，以补充提供给语言模型的提示，为模型提供生成准确响应所需的上下文。RAG 涉及以下几个步骤。

（1）提示：用户向聊天机器人提供提示，描述他们对输出的期望。

（2）研究：执行上下文搜索并从各种数据源中检索相关信息。这可能涉及查询数据库、根据关键字搜索索引文档或调用 API 从外部源检索数据。

（3）更新资源：检索到的上下文将被注入原始提示中，并使用与用户查询相关的其他事实信息对其进行扩充。此增强的提示提高了准确性，因为它提供了对事实数据的访问。

（4）旁白：基于此增强输入，LLM 生成包含事实的正确的信息响应，并将其发送回聊天机器人以交付给用户。

因此，通过结合外部数据源并将相关上下文注入提示中，RAG 增强了 LLM 生成信息的准确性、及时性及与搜索主题的一致性。

接下来介绍如何在 LangChain 框架下应用 RAG，首先创建一个拥有几个工具的 Agent，代码如下：

```python
//第 5 章/ Ragagentinit.py
from langchain.agents import (
    AgentExecutor, AgentType, initialize_agent, load_tools
)
from langchain.chat_models import ChatOpenAI
def load_agent() -> AgentExecutor:
    llm = ChatOpenAI(temperature=0, streaming=True)
    tools = load_tools(
        tool_names=["ddg-search", "wolfram-alpha", "arxiv", "wikipedia"],
        llm=llm
    )
    return initialize_agent(
```

```
            tools=tools, llm=llm, agent=AgentType.ZERO_SHOT_REACT_DESCRIPTION,
verbose=True
    )
```

通过前面的章节介绍已经知道了 AgentExecutor 是什么，因此如果用户可以将其集成到一个更大的链中，则可以使用不同的语法来初始化这个链，代码如下：

```
return MRKLChain.from_chains(llm, chains, verbose=True)
```

在这种语法中工具将作为链配置传递。MRKL 是英文缩写（Modular Reasoning, Knowledge and Language），它代表模块化推理、知识和语言。前面介绍的 Zero-Shot ReAct Agent 是用得最多的 Agent 的类型，也是 MRKL 框架中最通用的 Agent。需要注意的是 ChatOpenAI 构造函数中的参数流式处理方式，此处该参数被设置为 True。这样可以提高使用体验，因为这意味着文本响应将在传入时更新，而不是在所有文本完成后更新。目前只有 OpenAI、ChatOpenAI 和 ChatAnthropic 支持流式处理。在上面的代码中提到的所有工具都有其特定用途，这些工具共同组成了描述的一部分，该描述再被传递给 LLM。本示例中下面的这些工具已经被集成到 Agent 中。

（1）DuckDuckGo：一个专注于隐私的搜索引擎，它的一个优点是它不需要开发人员注册。

（2）Wolfram Alpha：一种将自然语言理解与数学能力相结合的集成工具。

（3）Arxiv：在学术预印本出版物中搜索，这对于以研究为导向的问题很有用。

（4）Wikipedia：著名的搜索问题的搜索工具。

需要补充的是，为了使用 Wolfram Alpha，用户必须在它的官网上注册一个账号，然后为创建的开发者 Token 设置 WOLFRAM_ALPHA_APPID 环境变量。

除了 DuckDuckGo 之外，LangChain 中还集成了许多其他搜索工具，可让用户利用 Google 或 Bing 搜索引擎。例如有一个名为 Open-Meteo 的工具可以让用户方便地查询天气信息集成。

完成了上面的步骤后，接着介绍如何将定义好的 Agent 转换为 Streamlit 应用程序。Streamlit 是面向机器学习和数据科学团队的开源应用程序框架。它允许用户使用 Python 在几分钟内创建漂亮的 Web 应用程序。使用下面的代码改写刚刚定义的 load_agent() 函数，代码如下：

```
//第 5 章/ streamlitapp.py
import streamlit as st
from langchain.callbacks import StreamlitCallbackHandler
chain = load_agent()
st_callback = StreamlitCallbackHandler(st.container())
if prompt := st.chat_input():
```

```
            st.chat_message("user").write(prompt)
            with st.chat_message("assistant"):
                st_callback = StreamlitCallbackHandler(st.container())
                response = chain.run(prompt, callbacks=[st_callback])
                st.write(response)
```

需要注意，这里在对链的调用中使用了回调处理程序，这意味着用户将看到从模型返回的响应。从终端本地启动应用程序的代码如下：

```
PYTHONPATH=. streamlit run question_answering/app.py
```

Streamlit 应用程序可以部署在本地，也可以部署在服务器上。或者用户可以将其部署在 Streamlit Community Cloud 或 HuggingFace Spaces 上。

对于 Streamlit Community Cloud 可以执行以下操作：

（1）创建 GitHub 存储库。

（2）进入 Streamlit Community Cloud，单击"新建应用"按钮，选择新的存储库。

（3）单击"部署"按钮。

对于 HuggingFace Spaces 可以执行以下操作：

（1）创建 GitHub 存储库。

（2）在官网创建一个 HuggingFace 账号。

（3）进入"空间"，单击"创建新空间"按钮。在表单中，将名称和空间类型填写为 Streamlit，然后选择新的存储库。

下面对创建的 Agent 的效果进行测试，这里使用问题"什么哺乳动物下的蛋最大"。当使用 DuckDuckGo 进行测试时会得到一个讨论鸟类和哺乳动物卵的结果，有时答案是鸵鸟，有时答案是鸭嘴兽。以下是输出的信息，结果如下：

```
> Entering new AgentExecutor chain...
I'm not sure, but I think I can find the answer by searching online.
Action: duckduckgo_search
Action Input: "mammal that lays the biggest eggs"
Observation: Posnov / Getty Images. The western long-beaked echidna ...
Final Answer: The platypus is the mammal that lays the biggest eggs.
> Finished chain.
```

从上面的例子中可以看到借助强大的自动化和问题解决框架，用户能将可能需要数百小时的工作压缩到几分钟。读者可以尝试不同的问题，以了解这些工具的使用方式。RAG 可以将来自外部来源的相关数据注入上下文，以此来显著地提高响应的准确性和质量。通过将 LLM 与专业领域的知识联系起来，用户可以减少 LLM 本身出现幻觉现象，使它们在现实世界的场景中更有用，而且最关键的一点是 RAG 比重新训练模型更具成本效益和效率。用户可以在 https://github.com/blockpipe/BlockAGI 的 BlockAGI 项目中看到一个非常高级的使用 LangChain 进行 RAG 的例子,该项目的灵感来自 BabyAGI 和 AutoGPT。本书在后面将根据决策策略比较主要类型的 Agent。

5.5 解剖 LLM 推理的底层策略

当前一代的 LLM 擅长在现实世界的数据(如视觉和音频信息及非结构化文本)中寻找答案,但是,它们在涉及结构化知识表示和推理的任务所需的符号操作方面遇到了困难。推理问题对 LLM 提出了挑战,万幸的是有不同的推理策略可以弥补 LLM 所缺失的能力。通过专注于对提取的信息启用符号操作,这些混合系统可以增强 LLM 的功能。模块化推理、知识和语言(Modular Reasoning, Knowledge and Language, MRKL)是一个框架,它结合了语言模型和工具来执行推理任务。在 LangChain 中,它由 3 部分组成:工具、一个 LLMChain,以及 Agent。

这些工具是 Agent 的可用资源,例如搜索引擎或数据库。LLMChain 负责生成文本提示并解析输出以确定下一步操作。Agent 类使用 LLMChain 的输出来决定要执行的操作。Agent 观察模式的推理过程如图 5-2 所示。

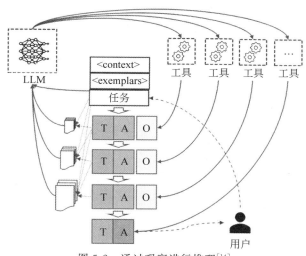

图 5-2 通过观察进行推理[14]

依赖观察的推理涉及根据当前知识状态或通过观察获得的证据做出判断、预测或选择。在每次迭代中,Agent 都会为 LLM 提供上下文和示例。用户的任务首先与上下文和示例相结合,并交给 LLM 来启动推理。LLM 生成一个想法和一个行动,然后等待来自工具的观察。观察结果将被添加到提示中,以启动对 LLM 的下一次调用。本书在前面介绍 Agent 的章节中提到,在 LangChain 中,这是一个动作的 Agent,也就是前面章节介绍的 Zero-Shot ReAct Agent,这是创建 Agent 时的默认设置。如前所述,用户也可以在任何行动之前制订计划,这种策略在 LangChain 中被称为计划和执行代理(Plan And Execute

Agent),如图 5-3 所示。

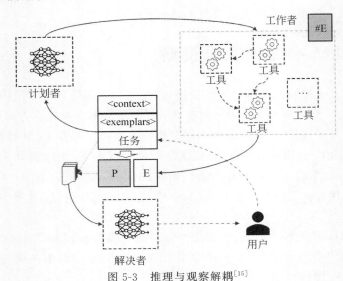

图 5-3 推理与观察解耦[15]

计划者可以针对计划和工具的使用进行微调,生成计划列表(P)并调用工作者(在 LangChain 中的 Agent)。通过工具收集证据(E),P 和 E 与任务相结合,最终输入 LLM 以获得答案。计划者和解决者可以是不同的 LLM。这为对计划者和解决者使用较小的专用模型及为每个调用使用更少的 Token 提供了可能性。在接下来创建的应用中就可以使用这种"计划和解决"的策略。首先向 load_agent()函数添加一个策略变量。它可以采用两个值,"计划和解决"或"one shot react"。对于"one-shot-react",逻辑保持不变。对于"计划和解决",这里将定义一个计划器和一个执行器来使用它们并以此来创建"计划和解决"的 Agent Executor,代码如下:

```
//第 5 章/planandexecute.py
from typing import Literal
from langchain.experimental import load_chat_planner, load_agent_executor,
PlanAndExecute
ReasoningStrategies = Literal["one-shot-react", "plan-and-solve"]
def load_agent(
    tool_names: list[str],
    strategy: ReasoningStrategies = "one-shot-react"
) -> Chain:
    llm = ChatOpenAI(temperature=0, streaming=True)
    tools = load_tools(
        tool_names=tool_names,
```

```
        llm=llm
    )
    if strategy == "plan-and-solve":
        planner = load_chat_planner(llm)
        executor = load_agent_executor(llm, tools, verbose=True)
        return PlanAndExecute(planner=planner, executor=executor, verbose=True)
    return initialize_agent(
        tools=tools, llm=llm, agent=AgentType.ZERO_SHOT_REACT_DESCRIPTION,
verbose=True
    )
```

为了简洁起见,这里省略了之前已经导入的包。下面定义一个新变量,该变量通过Streamlit中的单选按钮设置并将这个变量传递给load_agent()函数,代码如下:

```
strategy = st.radio(
    "Reasoning strategy",
    ("plan-and-solve", "one-shot-react", ))
```

用户可能已经注意到 load_agent() 采用字符串列表 tool_names。这也可以在 UI 界面中进行选择,代码如下:

```
//第 5 章/tool_names.py
tool_names = st.multiselect(
    'Which tools do you want to use?',
    [
        "google-search", "ddg-search", "wolfram-alpha", "arxiv",
        "wikipedia", "python_repl", "pal-math", "llm-math"
    ],
    ["ddg-search", "wolfram-alpha", "wikipedia"])
```

最后,Agent 将按以下方式在应用程序中加载,代码如下:

```
agent_chain = load_agent(tool_names=tool_names, strategy=strategy)
```

UI 界面如图 5-4 所示,这里用户输入问题"什么是大型语言模型中的计划和解决 Agent?"

读者可以仔细查看在上面这个问题下 LLM 的不同执行步骤。简单来讲,步骤可以归为以下几部分。

(1) 定义 LLM:大型语言模型是在大量文本数据上训练的 AI 模型,可以根据收到的输入生成类似人类的文本。

(2) 在 LLM 的上下文中了解计划的概念:在 LLM 的上下文中,计划是指模型为解决问题或回答问题而生成的结构化大纲或一组步骤。

(3) 了解 LLM 中的"计划和解决"Agent 的概念:Agent 是大型语言模型的一个组件,

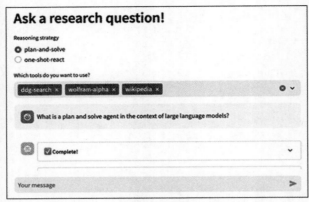

图 5-4 在 UI 界面中应用"计划和解决"

负责生成解决问题或回答问题的计划。

（4）识别"计划和解决"Agent 在 LLM 中的重要性："计划和解决"Agent 有助于组织模型的思维过程，并为解决问题或完成问答任务提供结构化方法。

目前还有另一个方面没有讨论，即这些步骤中使用的提示词策略。例如，不同的提示词策略提供了解决 LLM 复杂推理问题的方法。一种方法是少样本思维链（Chain of Thought，CoT）提示，其中 LLM 被引导进行分步推理演示。例如，在算术推理中，可以向 LLM 展示求解方程的演示示例，以帮助其理解该过程。另一种策略是零样本 CoT 提示，它不需要手动演示。取而代之的是将"让我们一步一步思考"这样的通用提示附加到 LLM 的问题陈述中。这允许 LLM 在没有事先明确示例的情况下生成推理步骤。在算术推理中，问题陈述可以通过这个提示进行扩充，并输入 LLM 中。"计划和解决"（Plan and Solve，PS）提示涉及将复杂任务划分为更小的子任务，并根据计划逐步执行它们。例如，在数学推理问题中，如求解方程式或涉及多个步骤的运算问题，"计划和解决"提示使 LLM 能够制订处理每个子步骤的计划，例如提取相关变量和计算中间结果。为了进一步提高推理步骤和指令的质量，引入了"计划和解决"＋提示。它包括更详细的说明，例如强调相关变量的提取及考虑计算和常识。"计划和解决"＋提示可确保 LLM 对问题有更好的理解，并可以生成准确的推理步骤。例如，在算术推理中，"计划和解决"＋提示可以指导 LLM 识别关键变量，正确执行计算，并在推理过程中应用常识知识。

到这里本章对推理策略的讨论就全部结束了。所有的策略都有其问题，可以表现为计算错误、漏步错误和语义误解，然而，它们有助于提高生成的推理步骤的质量，提高解决问题的准确性，并增强 LLM 处理各种类型推理问题的能力。

5.6 总结

本章讨论了幻觉现象、自动事实核查及如何使 LLM 更可靠的问题，其中特别强调的是工具和提示策略。本章首先研究并实施了分解和总结文档的提示策略。这对于消化大型研究文章或分析非常有帮助。一旦用户开始对 LLM 进行大量连锁调用，意味着会增加大量成本，因此本章专门用了一节来介绍 Token 的使用。工具为问题提供了创造性的解决方案，并为各个领域的 LLM 开辟了新的可能性。例如，可以开发一种工具，使 LLM 能够执行高级检索搜索、查询数据库以获取特定信息、自动编写电子邮件，甚至处理电话。用户可以使用 OpenAI API 实现文档中的信息提取等功能。本章带领读者实现了一个非常简单的简历解析器版本，然而，工具和函数调用并不是 OpenAI 独有的。用户可以使用 Streamlit 实现调用工具的不同 Agent。本章的示例已经实现了一个应用程序，可以依靠搜索引擎或维基百科等外部工具来帮助回答问题，然后本章研究了 Agent 在做出决策时采用的不同策略，并且已经在 Streamlit 应用程序中实施了"计划和解决"及"one shot react"。通过这个简单的示例，用户看到自己能够实现一些功能非常强大的应用程序，但是读者务必明确，本章中开发的应用程序具有局限性。虽然这些应用程序可以帮助用户显著地提高一些任务的执行效率，但是用户自己必须运用自己的判断力来确保结果是正确无误的。

第 6 章 聊天机器人

本章将讨论聊天机器人,它们是什么,它们可以做什么,以及如何实现它们。本章的开头将讨论聊天机器人的演变和当前最先进的技术。了解和增强当前聊天机器人和 LLM 的功能对于它们在不同领域的安全有效使用具有实际意义,包括医学和法律等受监管的领域。本章的重点是检索机制,包括向量存储,以提高响应的准确性及聊天机器人对可用信息和当前对话的忠实度。本章将介绍现代聊天机器人的基础知识,例如检索增强语言模型(Retrieval Augmented Language Model,RALM),以及在 LangChain 中实现它们所需的技术背景,因此,本章将详细介绍加载文档和信息的方法,包括向量存储和嵌入。在 LLM 记忆方面,本章将进一步讨论更具体的记忆方法,这些方法对于保持正在进行的对话的知识和状态至关重要。最后,本章将从声誉和法律的角度讨论另一个重要话题:AI 监管。AI 监管非常重要,这将确保 LLM 的回答是符合道德准则和法律规范的。LangChain 框架可以允许用户通过审核链传递任何文本,以检查它是否包含有害内容。接下来,本章将先介绍聊天机器人的概念和最先进的相关技术。

6.1 聊天机器人简介

聊天机器人是一种人工智能程序,可以与用户聊天、提供信息和支持、预订酒店及执行各种其他任务。聊天机器人可以与用户进行功能强大的交互,可用于不同的行业和不同的目的。聊天机器人的优势非常大,因为它们可以自动执行任务、提供即时响应并提升用户的个性化体验。它们可用于客户支持、潜在客户生成、销售、信息检索等。聊天机器人可以节省时间、提高效率、增强客户体验并简化业务流程。聊天机器人利用 NLP 和机器学习算法工作。它们分析用户输入,了解其背后的意图,并生成适当的响应。它们可以被设计为与基于文本的消息传递平台或基于语音的应用程序一起使用。聊天机器人在客户服务中的一些用例包括提供 24/7 全天候支持、处理常见问题、协助产品推荐、处理订单和付款及解决简单的客户问题。聊天机器人的更多用例如下。

(1)预约安排:聊天机器人可以帮助用户安排约会、预订和管理他们的日历。

（2）信息检索：聊天机器人可以为用户提供特定信息，例如天气预报、新闻文章或股票价格。

（3）虚拟助手：聊天机器人可以充当个人助理，帮助用户完成设置提醒、发送消息或拨打电话等任务。

（4）语言学习：聊天机器人可以通过提供交互式对话和语言练习来协助语言学习。

（5）心理健康支持：聊天机器人可以提供情感支持、提供资源并参与以心理健康为目的的治疗性对话。

（6）教育：在教育环境中，虚拟助手正在被探索成为虚拟导师，帮助学生学习和评估他们的知识、回答问题并提供个性化的学习体验。

（7）人力资源和招聘：聊天机器人可以通过筛选候选人、安排面试和提供有关职位空缺的信息来协助招聘所需人员。

（8）娱乐：聊天机器人可以吸引用户参与互动游戏。

（9）法律：聊天机器人可用于提供基本法律信息、回答常见的法律问题、协助法律研究及帮助用户浏览法律程序。它们还可以帮助准备文件，例如起草合同或创建法律表格。

（10）医学：聊天机器人可以协助检查症状，提供基本的医疗建议，并提供心理健康支持。它们可以通过向医疗保健专业人员提供相关信息和建议来改善临床决策。

在很多领域，LLM 赋能的聊天技术有可能使信息更容易获得，并为寻求帮助的个体提供初步支持。

6.1.1　历史溯源

图灵测试是 AI 中的一种探究方法，它是以英国计算机科学家、密码分析师和数学家艾伦·图灵的名字命名的。图灵测试的主要目的是用于确定计算机是否能够像人类一样思考。尽管关于图灵测试的相关性及基于图灵测试的竞赛的有效性存在很多争论，但该测试仍然是讨论和研究 AI 的哲学起点。随着人类在 AI 方面不断取得进步，为了更好地理解和绘制人脑的功能，图灵测试仍然是定义智能的基础，并且是关于我们应该从技术中获得什么才能被视为思考机器的辩论的基线。图灵提出，如果一台计算机能够在特定条件下模仿人类的反应，它就可以说拥有人工智能。最初的图灵测试需要 3 个终端，每个终端都与其他两个终端在物理上分开。一个终端由计算机操作，而另外两个终端由人类操作。在测试过程中，其中一个人充当提问者，而第 2 个人和计算机则充当受访者。提问者使用指定的格式和上下文在特定主题领域内询问受访者。在预设的时间长度或问题数量之后，提问者被要求决定哪个受访者是人类，哪个是计算机。自图灵测试形成以来，许多 AI 已经能够通过这个测试。第 1 个是约瑟夫·魏岑鲍姆（Joseph Weizenbaum）的"伊丽莎"（ELIZA）。1966 年，他发表了一篇关于他的聊天机器人 ELIZA 的文章，名字叫《ELIZA：一种用于研究人与机器之间自然语言交流的计算机程序》。ELIZA 是有史以来最早创建并模拟心理治疗师角色的聊天机器人之一。

聊天机器人以幽默感来展示技术的局限性，它采用简单的规则和模糊的开放式问题，能够在对话中给人一种移情理解的印象。这是一个具有讽刺意味的转折，通常被视为 AI 的里程碑，然而，ELIZA 的知识有限，只能在特定主题领域内进行对话。它也无法保持长时间的对话或从讨论中学习。多年来，图灵测试一直受到批评，特别是因为从历史上看，为了让计算机表现出类似人类的智能，必须限制人类和计算机之间对话的性质。多年来，计算机只有在提问者拟定的查询问题是"是"或"否"的答案，或者涉及知识范围较窄的领域时才能得高分。当问题是开放式的并且需要以对话式给出答案时，计算机程序不太可能成功欺骗提问者。此外，像 ELIZA 这样的程序可以通过操纵它不完全理解的符号来通过图灵测试。哲学家约翰·塞尔(John Searle)认为，这并不能让计算机程序拥有与人类相媲美的智力。对于许多研究人员来讲，计算机能否通过图灵测试的问题已经变得无关紧要。与其关注如何说服他们正在与人类而不是计算机程序交谈，不如关注如何使人机交互更加直观和高效，例如通过对话界面。1972 年，另一个名为 PARRY 的重要聊天机器人被开发出来了，它被设计成充当精神分裂症患者。PARRY 有一个明确的个性，它的反应是基于一个假设系统，情绪反应是由用户话语的变化触发的。在 1979 年的一项实验中，5 名精神科医生对 PARRY 进行了测试，他们必须确定与他们互动的病人是计算机程序还是真正的精神分裂症患者。测试的结果各不相同，一些精神科医生给出了正确的诊断，而另一些则给出了不正确的诊断。尽管图灵测试的几种变体通常更适用于目前人类对 AI 的理解，但该测试的原型至今仍在使用。

  例如，自 1990 年以来，Loebner 奖每年都会颁发给由评审团投票选出的最像人类的计算机程序。比赛遵循图灵测试的标准规则。该奖项相关性的批评者经常淡化它，认为它更多的是宣传，而不是真正测试机器是否能思考。IBM Watson 是 IBM 开发的认知计算系统，它使用 NLP、机器学习和其他 AI 技术来回答自然语言中的复杂问题。它的工作原理是摄取和处理大量结构化和非结构化数据，包括文本、图像和视频。IBM Watson 于 2011 年在问答节目 Jeopardy 上竞争时成名，并击败了两位前冠军。Watson 已被应用于各个领域，包括医疗保健、金融、客户服务和研究。在医疗保健领域，Watson 已被用于协助医生诊断和治疗疾病、分析医疗记录和进行研究。它也被应用于烹饪领域，Chef Watson 应用程序帮助厨师创造独特和创新的食谱。2018 年，Google Duplex 在 7000 名观众面前通过电话成功预约了一位美发师。接待员完全没有意识到他们没有与真人交谈。这被一些人认为通过了图灵测试，尽管没有像艾伦·图灵设计的那样依赖测试的真实格式。ChatGPT 由 OpenAI 开发，是一种语言模型，它使用深度学习技术来生成类似人类的响应。它于 2022 年 11 月 30 日推出，建立在 OpenAI 专有的基础 GPT 模型系列之上，包括 GPT-3.5 和 GPT-4。ChatGPT 允许用户与 AI 进行连贯、自然和引人入胜的对话，将对话提炼和引导到他们想要的长度、格式、风格、详细程度和使用的语言。ChatGPT 被认为是游戏规则的改变者，因为它代表了对话式 AI 的重大进步。由于聊天机器人能够生成上下文相关的响应，并能够理解广泛的主题和回答相关问题，因此一些人认为聊天机器人最有可能以人类今天拥有的任何技术的真实形式通过测试，但是，即使具有先进的文本生成能力，ChatGPT 也可能被欺骗或诱导而返回荒谬的答案。

以下是历史上的聊天机器人的一些示例。

（1）ELIZA：ELIZA 是最早的聊天机器人之一，开发于 20 世纪 60 年代，使用模式匹配来模拟与用户进行对话。

（2）Siri：Siri 是 Apple 开发的一款流行的基于语音的聊天机器人。它被集成到 Apple 设备中，可以执行任务、回答问题和提供信息。

（3）Alexa：Alexa 是亚马逊开发的智能个人助理。它可以响应语音命令、播放音乐、提供天气更新和控制智能家居设备。

（4）Google Assistant：Google Assistant 是谷歌开发的聊天机器人。它可以回答问题、提供建议并根据用户命令执行任务。

（5）Mitsuku：Mitsuku 是一个聊天机器人，曾多次获得 Loebner 奖。它以具有进行自然和类似人类的对话的能力而闻名。

使用图灵测试及其变种的一个问题是，它们侧重于模仿和欺骗，而更有意义的测试应该强调开发人员需要专注于创建有用和有趣的功能，而不仅是执行技巧。基准测试和学术/专业考试的使用提供了对 AI 系统性能的更具体的评估。该领域的研究人员目前的目标是为测试 AI 系统的能力提供更好的基准，特别是 GPT-4 等 LLM。他们旨在了解 LLM 的局限性并确定它们可能失败的领域，包括 GPT-4 在内的先进 AI 系统在与语言处理相关的任务中表现出色，但在简单的视觉逻辑难题中苦苦挣扎。LLM 可以根据统计相关性生成合理的下一个单词，但可能缺乏对抽象概念的推理或理解。研究人员对 LLM 的能力有不同的看法，一些人将他们的成就归因于有限的推理能力。对测试 LLM 并了解其能力的研究具有实际意义。它可以帮助 LLM 在医学和法律等现实世界领域安全有效地应用。通过确定 LLM 的优势和劣势，研究人员可以确定如何最好地利用它们。与其"前辈"相比，ChatGPT 的训练使其在处理幻觉问题方面做得更好，这意味着它不太可能产生无意义或不相关的反应，但是，需要注意的是，ChatGPT 仍然很容易返回不准确的信息，因此用户应谨慎行事并验证 ChatGPT 所提供的信息。在确保聊天机器人对话中提供准确信息和恰当反映之前交互内容的响应上，上下文和记忆机制发挥着极为重要的作用，这可以让机器人与用户实现更加忠实和连贯的互动。接下来将更详细地讨论这个问题。

6.1.2　上下文和记忆

上下文和记忆是聊天机器人设计的重要方面。它们允许聊天机器人维护对话上下文、响应多个查询及存储和访问长期记忆。它们是调节聊天机器人响应准确性和忠实度的重要因素。聊天机器人中记忆和上下文的重要性可以与在人与人的对话中记忆和理解的重要性相提并论。没有回忆过去的交流或不理解或不了解更广泛的背景的对话可能会脱节并导致沟通不畅，从而导致不令人满意的对话体验。上下文理解极大地影响了聊天机器人响应的准确性，它指的是聊天机器人理解整个对话和一些相关背景的能力，而不仅是用户的最后一条消息。意识到上下文的聊天机器人可以保持对话的整体视角，使聊天流程更加自然和类似人类。记忆保留直接影响聊天机器人的忠实度，这涉及识别和记住先前对话中的事实以供将来使用的一致性。记忆保留功能增强了用户的个性化体验。例如,如果用户说"给我看最便宜的航班",则说"哪个地区的酒店怎样?"而没有前面消息的上下文,聊天机器人就不知道用户指的是哪个区域。在相反的场景中,上下文感知聊天机器人会理解用户正在谈论与航班目的地在同一地区的住宿。缺乏记忆会导致整个对话上下文出现不一致的情况。例如,如果用户在一次对话中通过姓名标识了自己,而机器人在下一次对话中忘记了此信息,则会产生不自然且非人性化的交互。记忆和上下文对于使聊天机器人交互更加高效、准确和令人满意至关重要。如果没有这些元素,机器人则可能会给人留下存在缺陷或僵化的印象,因此,这些特性对于计算机和人类之间复杂而令人满意的交互至关重要。拥有上下文和记忆的聊天机器人不仅可以了解用户的意图,还可以更智能地与用户进行对话。

6.1.3 意识性与主动性

在 LLM 或聊天机器人的上下文中,主动性是指系统在没有用户明确提示的情况下启动操作或提供信息的能力。它涉及根据以前的交互或上下文提示预测用户的需求或偏好。另外,这意味着聊天机器人旨在理解和满足用户的意图或请求,并被编程为根据这些意图和预期结果采取特定行动或提供相关响应。主动聊天机器人很有用,因为它可以与客户建立联系并提高他们的体验。这可以通过节省时间和精力来增强用户体验,还可以通过快速有效地解决客户的问题来提高客户满意度。主动沟通对于企业的成功至关重要,因为它可以提高客户终身价值(Customer Lifetime Value,CLV)并降低运营成本。通过积极预测客户需求并主动提供信息,企业可以控制沟通并以有利的方式构建对话。这可以建立信任,提高客户忠诚度和组织的声誉。此外,主动沟通有助于在客户提出问题之前解决客户询问的问题并减少传入的支持电话,从而提高组织的工作效率。在技术方面,这种能力可以通过上下文和记忆及推理机制实现。这是本章的重点。在 6.2 节中将讨论现代聊天机器人的基础知识,例如 RALM 及实现它们所需的技术背景。

6.2 检索和向量

在前面的章节中讨论了 RAG,它旨在通过利用外部知识并确保生成的文本准确且符合上下文来增强生成过程。在本章中将进一步讨论如何结合检索和生成技术来提高生成文本的质量和相关性。这里将着重讨论 RALM,这是 RAG 的一种特定实现或应用,它指的是在生成过程中以基础语料库(书面文本集合)中的相关文档为条件的 LLM。在检索中,语义过滤和向量存储用于从大量文档语料库中预过滤相关信息,并将该信息纳入生成过程。此检索包含了文档的向量存储。

RALM 是包含检索组件以增强其性能的 LLM。传统 LLM 根据输入提示生成文本,但 RALM 更进一步,从大量文档中检索相关信息,并使用该信息生成更准确且与上下文相关的响应回复。

RALM 的优点主要包括以下几点。

(1)高性能：通过结合主动检索，LLM 可以从外部来源访问相关信息，从而增强其生成准确和信息丰富的响应的能力。

(2)避免输入长度限制：检索增强的 LLM 会丢弃以前检索到的文档，并且仅使用从当前步骤中检索到的文档来生成下一个回答。这有助于防止超过 LLM 的输入长度限制，从而允许它们处理更长、更复杂的查询或任务。更具体地讲，检索增强 LLM 的工作机制包括以下步骤。

检索：RALM 从大型语料库中搜索相关文档或段落。LLM 基于查询和当前上下文的基于向量的相似性搜索，从外部源检索相关信息。

反射：将检索到的信息用于生成下一次回答。这意味着 LLM 将检索到的信息合并到其语言模型中，以生成更准确且更适合上下文的响应回复。

迭代过程：检索和调节步骤以迭代方式执行，每个步骤都建立在前一个步骤的基础上。这种迭代过程使 LLM 能够整合来自外部的相关信息并以此逐步提高其理解和生成能力。

检索到的信息可以以不同的方式使用。它可以作为 LLM 的附加上下文，帮助它生成更准确和更适合上下文的响应回复。它还可用于在生成的文本中提供事实信息或回答特定问题。检索增强生成有两种主要策略。

(1)单次 RAG：此策略涉及使用用户输入作为检索查询，并立即生成完整的答案。检索到的文档与用户输入连接起来，并用作 LLM 的输入以进行生成。

(2)主动 RAG：此策略涉及在生成过程中主动决定何时检索及检索什么。在生成的每个步骤中都会根据用户输入和先前生成的输出制定检索查询，然后将检索到的文档用作 LLM 的输入以进行生成。此策略允许检索和生成交错，使模型能够根据需要动态地检索信息。

在主动 RAG 框架中，有两种被称为前瞻性主动检索增强生成（Forward-Looking Active Retrieval Augmented Generation，FLARE）的前瞻性方法。

(1)带有检索指令的 FLARE：此方法提示 LLM 在生成答案时在必要时生成检索查询。它使用鼓励检索的指令来表达对其他信息的需求。

(2)直接 FLARE：此方法直接使用 LLM 的生成作为搜索查询。它迭代生成下一个句子以深入了解未来的主题，如果存在不确定的标记，则会检索相关文档以重新生成下一个句子。

与仅检索一次信息，然后生成的传统方法不同，FLARE 遵循迭代过程。它涉及使用对即将到来的句子的预测作为查询来检索相关文档。这允许系统在初始生成置信度较低的句子时重新生成句子。RALM 在问答、对话系统和信息检索等任务中显示出可喜的成果。它们可以通过利用外部知识资源提供更准确、更翔实的响应。此外，可以通过对特定领域的文档集合进行培训，对 RALM 进行针对特定领域或任务的微调，从而进一步增强它们在专业应用程序中的实用性。总体而言，通过整合检索，RALM 可以利用文档语料库中存在的大量知识，使它们在各种自然语言处理任务中更加强大和有用。RALM 利用主动检索来增强其性能并克服处理复杂查询或任务的局限性。LangChain 实现了一个由不同构建块组成的工具链,用于构建检索系统。这包括数据加载器、文档转换器、嵌入模型、向量存储和检索器,它们之间的关系如图 6-1 所示。

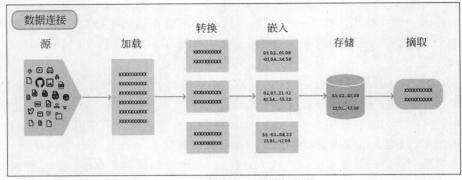

图 6-1　向量存储和数据加载器

LangChain 首先通过数据加载器加载文档,然后可以转换这些数据,将这些文档作为嵌入传递到向量存储中,然后用户可以查询向量存储或与向量存储关联的检索器。LangChain 中的检索器可以将加载和向量存储打包到一个步骤中。在本章中将跳过转换数据这个环节,而将主要解释数据加载器、嵌入、存储机制和检索器部分。由于这里谈论的是向量存储,因此需要讨论向量搜索,这是一种用于根据向量(或嵌入)与查询向量的相似性来搜索和检索向量(或嵌入)的技术。它通常用于推荐系统、图像和文本搜索及异常检测等应用程序。这里将研究 RALM 背后的更多基本原理。接下来将从嵌入开始讲解。一旦用户了解了嵌入,就可以构建从搜索引擎到聊天机器人的所有内容。

6.2.1　嵌入

嵌入是以机器可以处理和理解的方式对内容进行数字化转换。该过程的本质是将图像或文本等对象转换为封装其语义内容的向量,同时尽可能地丢弃不相关的细节。嵌入采用一段内容,例如单词、句子或图像,并将其映射到多维向量空间中。两个嵌入之间的距离表示相应概念(原始内容)之间的语义相似性。

嵌入是由机器学习模型生成的数据对象的表示形式。它们可以将单词或句子表示为数字向量(浮点数列表)。用 OpenAI 语言嵌入模型做例子,嵌入是表示文本的 1536 个浮点数的向量。这些数字源自捕获语义内容的复杂语言模型。

举个例子,假设有猫和狗这两个词,它们可以与词汇表中的所有其他单词一起在一个空间中以数字表示。如果空间是三维的,则可以是向量,例如猫的向量为 $[0.5, 0.2, -0.1]$,狗的向量为 $[0.8, -0.3, 0.6]$。这些向量对有关这些概念与其他单词的关系的信息进行编码。粗略地说,我们期望猫和狗的概念更接近动物的概念,而不是计算机或嵌入的概念。

创建嵌入的方法多种多样。对于文本,一种简单的方法是词袋方法,其中每个单词都由它在文本中出现的次数来表示。这种方法在 Scikit-learn 库中作为 CountVectorizer 实现,在 Word2Vec 出现之前一直很流行。Word2Vec 粗略地说,可通过周围其他的单词预测句子中的单词来学习嵌入,从而忽略线性模型中的词序。嵌入的一般思路如图 6-2 所示。

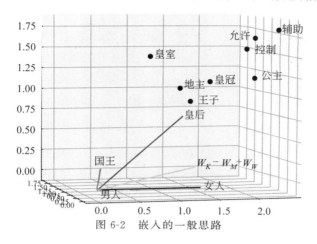

图 6-2　嵌入的一般思路

对于图像,嵌入可能来自特征提取阶段,例如边缘检测、纹理分析和颜色合成。可以在不同的窗口大小上提取这些特征,以使表示既是尺度不变的,又是移位不变的(尺度空间表示)。如今,CNN 通常在大型数据集(如 ImageNet)上进行预训练,以学习图像属性的良好表示。由于卷积层在输入图像上应用一系列过滤器(或内核)以生成特征图,因此从概念上讲,这类似于尺度空间。当预先训练的 CNN 在新图像上运行时,它可以输出嵌入向量。如今,对于包括文本和图像在内的大多数领域,嵌入通常来自基于 Transformer 的模型,该模型考虑了句子和段落中单词的上下文和顺序。基于模型架构,最重要的是参数的数量,这些模型可以捕获非常复杂的关系。所有这些模型都是在大型数据集上训练的,以建立概念及其关系。这些嵌入可用于各种任务。通过将数据对象表示为数值向量可以对它们进行数学运算并测量它们的相似性,或者将它们用作其他机器学习模型的输入。通过计算嵌入之间的距离可以执行搜索和相似性评分等任务,或对对象进行分类,例如按主题或类别分类。举个例子,用户可以通过检查产品评论的嵌入是否更接近正面或负面的概念来执行简单的情绪分类器。

嵌入之间的距离指标是一个非常重要的概念。向量相似性计算中使用的不同距离度量如下。

(1)余弦距离:这是一种相似度量,用于计算向量空间中两个向量之间角度的余弦。它的范围从 -1 到 1,其中 1 表示相同的向量,0 表示正交向量,-1 表示截然相反的向量。

(2)欧几里得距离:它测量向量空间中两个向量之间的直线距离。它的范围从 0 到无穷大,其中 0 表示相同的向量,较大的值表示越来越不同的向量。

(3)点积:它测量两个向量的大小与它们之间角度的余弦的乘积。它的范围从负无穷到正无穷,其中正值表示指向同一方向的向量,0 表示正交向量,负值表示指向相反方向的向量。

在 LangChain 中,用户可以使用 OpenAIEmbeddings 类中的 embed_query()方法获取嵌入,代码如下:

```
//第 6 章/ openAIEmbeddings.py
from langchain.embeddings.openai import OpenAIEmbeddings
embeddings = OpenAIEmbeddings()
text = "This is a sample query."
query_result = embeddings.embed_query(text)
print(query_result)
print(len(query_result))
```

在上面的代码中将单个字符串输入传递给 embed_query()方法,并检索相应的文本嵌入。将结果存储在 query_result 变量中。嵌入的长度(维数)可以使用 len()函数获得。用户还可以使用 embed_document()方法获取多个文档输入的嵌入,代码如下:

```
from langchain.embeddings.openai import OpenAIEmbeddings
words = ["cat", "dog", "computer", "animal"]
embeddings = OpenAIEmbeddings()
doc_vectors = embeddings.embed_documents(words)
```

在这种情况下,embed_documents()方法用于检索多个文本输入的嵌入,然后把结果存储在 doc_vectors 变量中。原本可以检索长文档的嵌入,但是在这里正好相反,即只检索每个单词的向量。更进一步还可以在这些嵌入之间进行算术运算,例如计算它们之间的距离,代码如下:

```
from scipy.spatial.distance import pdist, squareform
import pandas as pd
X = np.array(doc_vectors)
dists = squareform(pdist(X))
```

上面的代码提供了各个单词之间的欧几里得距离,以方阵的形式输出,如图 6-3 所示。

	猫	狗	计算机	动物
猫	0.000000	0.522352	0.575285	0.521214
狗	0.522352	0.000000	0.581203	0.478794
计算机	0.575285	0.581203	0.000000	0.591435
动物	0.521214	0.478794	0.591435	0.000000

图 6-3 猫、狗、计算机、动物词嵌入之间的欧几里得距离

从常识的角度及上面的方阵中可以确认,猫和狗确实比计算机更接近动物。这里可能有很多问题,例如,狗是否比猫更像动物,或者为什么狗和猫与计算机的距离只比与动物的距离大一点。尽管这些问题在某些应用中可能很重要,但需要记住,这是一个简单的示例,

因此很多其他的因素没有被考虑进去。在这些示例中使用了 OpenAI 嵌入——在后面的示例中本章将使用 HuggingFace 提供的模型中的嵌入。LangChain 中有一些集成和工具可以帮助完成这一过程,本章在后面会进一步介绍其中的一些工具。此外,LangChain 还提供了一个 FakeEmbeddings 类,可用于测试用户自己的管道,而无须对嵌入提供者进行实际调用。在本章的上下文中将使用它们来检索相关信息(语义搜索),但是,这些嵌入与应用程序和更广泛的系统的集成仍需进一步讨论,这正是向量存储发挥作用的地方。

6.2.2 存储嵌入的方式

如前所述,在向量搜索中,每个数据点都被表示为高维空间中的向量。向量用于捕获数据点的特征或特征。目标是找到与给定查询向量最相似的向量。在向量搜索中,数据集中的每个数据对象都被分配了一个向量嵌入。这些嵌入是数字数组,可用作高维空间中的坐标。可以使用余弦相似度或欧几里得距离等距离度量来计算向量之间的距离。若要执行向量搜索,则需要将查询向量(表示搜索查询)与集合中的每个向量进行比较。计算查询向量与集合中每个向量之间的距离,并且距离较小的对象被认为与查询更相似。为了有效地执行向量搜索,需要使用向量存储机制,例如向量数据库。

向量搜索是指根据向量与给定查询向量的相似性,在其他存储的向量中(例如在向量数据库中)搜索相似向量的过程。向量搜索通常用于各种应用,例如推荐系统、图像和文本搜索及基于相似度的检索。向量搜索的目标是高效、准确地检索与查询向量最相似的向量,通常使用相似度度量,如点积或余弦相似度。

这里有 3 个组件:索引、向量库和向量数据库。这些组件协同工作,用于创建、操作、存储和高效检索向量嵌入。索引组织向量以优化检索,对它们进行结构化,以便可以快速检索向量。解决这些问题有不同的算法,例如 k-d 树或近似最近邻(Annoy)。向量库为向量运算提供函数,如点积和向量索引。最后,像 Milvus 或 Pinecone 这样的向量数据库被用于存储、管理和检索大量向量。它们使用索引机制来促进对这些向量进行有效的相似性搜索。接下来本章将引导读者详细学习这些组件。

6.2.3 索引

向量嵌入上下文中的索引是一种组织数据以优化其检索和/或存储的方法。它类似于传统数据库系统中的概念,其中索引允许更快地访问数据记录。对于向量嵌入,索引旨在构建向量,以便相似的向量能够彼此相邻存储,从而实现快速的邻近性或相似性搜索。在这种情况下应用的典型算法是 K 维树(k-d 树),但一些其他算法也经常被使用,如 Ball Trees、Annoy 和 FAISS,特别是对于传统方法可能难以处理的高维向量。

K-最近邻(K-Nearest Neighbor,K-NN)是一种简单直观的算法,用于分类和回归任务。在 K-NN 中,该算法通过查看训练数据集中数据点的 k 个最近邻来确定数据点的类或值。以下是 K-NN 的工作的具体步骤。

(1)选择 k 的值:确定进行预测时将考虑的最近邻(k)的数量。

(2)计算距离：计算要分类的数据点与训练数据集中所有其他数据点之间的距离。最常用的距离度量是欧几里得距离，也可以使用曼哈顿距离等其他度量。

(3)查找 k 个最近邻：选择与要分类的数据点距离最短的 k 个数据点。

(4)确定多数类：对于分类任务，统计 k 个最近邻中每个类的数据点数。计数最高的类将成为数据点的预测类。对于回归任务，取 k 个最近邻的值的平均值。

(5)进行预测：确定多数类或平均值后，将其指定为数据点的预测类或值。

需要注意的是，K-NN 是一种惰性学习算法，这意味着它不会在训练阶段显式地构建模型。相反，它存储整个训练数据集，并在预测时执行计算。

除了 K-NN 之外，还有其他几种通常用于相似性搜索索引的算法。

(1)乘积量化（PQ）：PQ 是一种将向量空间划分为更小的子空间并分别量化每个子空间的技术。这降低了向量的维数，并允许有效的存储和搜索。PQ 以其快速的搜索速度而闻名，但可能会牺牲一些准确性。

(2)位置敏感哈希（LSH）：这是一种基于哈希的方法，可将相似的数据点映射到相同的哈希存储桶。它对高维数据很有效，但可能具有更高的误报和漏报概率。

(3)分层导航小世界（HNSW）：HNSW 是一种基于图的索引算法，它构造分层图结构来组织向量。它结合使用随机化和贪婪搜索来构建一个可导航的网络，从而实现高效的最近邻搜索。HNSW 以其高搜索准确性和可扩展性而闻名。

PQ 包括 k-d 树和 Ball Trees。在 KD-Trees 中，建立了一个二叉树结构，根据数据点的特征值对数据点进行分区。它对低维数据有效，但随着维数的增加而变得不那么有效。球树：一种树状结构，将数据点划分为嵌套的超球体。它适用于高维数据，但对于低维数据，它可能比 k-d 树慢。除了 HNSW 和 K-NN 之外，还有其他基于图的方法，如图神经网络（GNN）和图卷积网络（GCN），它们利用图结构进行相似性搜索。Annoy 算法使用随机投影树来索引向量。它构造了一个二叉树结构，其中每个节点代表一个随机超平面。Annoy 使用简单，并提供快速的近似最近邻搜索。这些索引算法在搜索速度、准确性和内存使用方面有不同的权衡。算法的选择取决于应用的具体要求和向量数据的特性。

6.2.4　向量库

向量库（如 Facebook FAISS 或 Spotify Annoy）提供处理向量数据的功能。在向量搜索的上下文中，向量库专门用于存储和执行向量嵌入的相似性搜索。这些库使用 ANN 算法来有效地搜索向量并找到最相似的向量。它们通常提供 ANN 算法的不同实现，例如聚类或基于树的方法，并允许用户对各种应用程序执行向量相似性搜索。接下来快速看一下这些向量库。

FAISS 是由 Facebook 开发的库，可提供高效的相似性搜索和密集向量的聚类。它提供各种索引算法，包括 PQ、LSH 和 HNSW。FAISS 被广泛用于大规模向量搜索任务，同时支持 CPU 和 GPU 加速。

Annoy 是一个 C++ 库，用于在高维空间中进行近似最近邻搜索，由 Spotify 维护和开发，实现了 Annoy 算法。它被设计为高效和可扩展的，使其适用于大规模向量数据。它适用于随机投影树的森林。

hnswlib 是一个 C++ 库，用于使用 HNSW 算法进行近似最近邻搜索。它为高维向量数据提供快速且内存高效的索引和搜索功能。

nmslib 是一个开源库，可在非度量空间中提供高效的相似性搜索。它支持各种索引算法，如 HNSW、SW-graph 和 SPTAG。

6.2.5 向量数据库

向量数据库是一种专门用于处理向量嵌入的数据库，可以更轻松地搜索和查询数据对象。它提供了额外的功能，如数据管理、元数据存储和过滤及可扩展性。向量存储仅专注于存储和检索向量嵌入，而向量数据库则为管理和查询向量数据提供了更全面的解决方案。向量数据库对于涉及大量数据的应用程序特别有用，这些应用程序通常需要灵活高效地搜索各种类型的向量化数据，如文本、图像、音频、视频等。

向量数据库可用于存储和提供机器学习模型及其相应的嵌入。主要应用是相似性搜索，也称为语义搜索，其可以有效地搜索大量文本、图像或视频，根据向量表示识别与查询匹配的对象。这在文档搜索、反向图像搜索和推荐系统等应用程序中特别有用。

随着技术的发展，向量数据库的其他用例也在不断扩展，一些常见用例包括以下几种。

（1）异常检测：向量数据库可用于通过比较数据点的向量嵌入来检测大型数据集中的异常。这在欺诈检测、网络安全或监控系统中可能很有价值，在这些系统中，识别异常模式或行为至关重要。

（2）个性化：向量数据库可根据用户偏好或行为查找相似的向量来创建个性化推荐系统。

（3）NLP：向量数据库被广泛地应用于情感分析、文本分类、语义搜索等 NLP 任务。通过将文本表示为向量嵌入，可以更轻松地比较和分析文本数据。

这些向量数据库受欢迎的原因是它们针对可扩展性及在高维向量空间中表示和检索数据进行了优化。传统数据库无法有效地处理高维向量，例如用于表示图像或文本嵌入的向量。向量数据库的特点主要包括以下几点。

（1）高效检索：向量数据库擅长在高维空间中查找紧密嵌入或相似点。这使它们非常适合反向图像搜索或基于相似度的推荐等任务。

（2）专业性：向量数据库旨在执行特定任务，例如查找紧密嵌入。它们不是通用数据库，专为有效处理大量向量数据而定制。

（3）支持高维空间：向量数据库可以处理数千个维度的向量，从而实现复杂的数据表示。这对于自然语言处理或图像识别等任务至关重要。

（4）高级搜索：借助向量数据库，构建可以搜索相似向量或嵌入强大的搜索引擎。这为内容推荐系统或语义搜索等应用开辟了可能性。

总体而言，向量数据库为处理高维向量数据提供了专业且高效的解决方案，可实现相似性搜索和高级搜索等任务。开源软件和数据库市场目前正在蓬勃发展。首先,AI 和数据管理对企业至关重要,导致对高级数据库解决方案的高需求。在数据库市场中,有新型数据库出现并创造新市场类别的历史。这些市场创造者通常主导着这个行业,吸引了风险投资家的大量投资。例如,MongoDB、Cockroach、Neo4J 和 Influx 都是成功公司引入创新数据库技术并获得可观市场份额的例子。流行的 Postgres 有一个用于高效向量搜索的扩展:pg_embedding。通过分层导航小世界(HNSW),它提供了一种更快、更有效的替代方案,以替代具有 ivfflat 索引的 pgvector 扩展。风险投资公司正在积极寻求下一个突破性的数据库类型,而向量数据库(如 Chroma 和 Marqo)有可能成为下一个大事件。这创造了一个竞争格局,公司可以筹集大量资金来开发和扩展其产品。目前市面上的一些向量数据库见表 6-1。

表 6-1 市面上的一些向量数据库

提供商	描述	商业模式	首次发布	授权	组织
Chroma	商业开源嵌入商店	(部分开放) SaaS 模型	2022	Apache-2.0	Chroma Inc.
Qdrant	托管/自托管向量搜索引擎和数据库,具有扩展过滤支持	(部分开放) SaaS 模型	2021	Apache 2.0	Qdrant Solutions GmbH
Milvus	为可扩展的相似性搜索而构建的向量数据库	(部分开放) SaaS 模型	2019	BSD	Zilliz
Weaviate	云原生向量数据库,可同时存储对象和向量	开放 SaaS	始于 2018 年,最初是传统的图形数据库,于 2019 年首次发布	BSD	SeMI Technologies
Pinecone	使用 AI 模型中的嵌入实现快速且可扩展的应用程序	SaaS	首次发布于 2019 年	proprietary	Pinecone Systems Inc.
Vespa	商业开源向量数据库,支持向量搜索、词汇搜索	开放 SaaS	最初是一个网络搜索引擎(All the Web),于 2003 年被雅虎收购,后来在 2017 年发展成为 Vespa 并开源	Apache 2.0	Yahoo!
Marqo	云原生商业开源搜索和分析引擎	Open SaaS	2022	Apache 2.0	S2Search Australia Pty Ltd.

在 LangChain 中,可以使用 vectorstores 模块实现向量存储。该模块提供了用于存储和查询向量的各种类和方法。LangChain 中向量存储实现的一个例子是 Chroma 向量存储。接下来看两个例子。

1. Chroma

此向量存储针对使用 Chroma 作为后端的向量进行存储和查询进行了优化。Chroma

负责根据向量的角度相似性对向量进行编码和比较。如果用户要在 LangChain 中使用 Chroma,则需要按照以下步骤操作。

首先导入相关的库,代码如下:

```
from langchain.vectorstores import Chroma
from langchain.embeddings import OpenAIEmbeddings
```

创建一个 Chroma 实例,并提供文档拆分和嵌入方法:

```
vectorstore = Chroma.from_documents(documents=docs, embedding=OpenAIEmbeddings())
```

文档将被嵌入并存储在 Chroma 向量数据库中。本章将在 6.2.6 节中讨论文档加载器。用户也可以使用其他嵌入集成,像这样提供嵌入,代码如下:

```
vector_store = Chroma()
#将向量添加到向量存储中
vector_store.add_vectors(vectors)
```

在这里,vectors 是要存储的数字向量(嵌入)列表。用户可以查询向量存储来检索相似的向量,代码如下:

```
similar_vectors = vector_store.query(query_vector, k)
```

这里,query_vector 是要查找相似向量的向量,k 是要检索的相似向量的数量。

2. Pinecone

在 LangChain 中集成 Pinecone 的步骤如下。

首先安装 Pinecone Python 客户端库。用户可以在终端中运行以下命令来执行此操作,代码如下:

```
pip install pinecone
```

在 Python 中导入 Pinecone,代码如下:

```
import pinecone
```

要连接到 Pinecone 服务,需要提供 API 密钥。用户可以通过在 Pinecone 网站上注册获取 API 密钥。在获取 API 密钥后,用户需要将其传递给 Pinecone 包装器或将其设置为环境变量,代码如下:

```
pinecone.init()
```

创建搜索索引,代码如下:

```
Docsearch = Pinecone.from_texts(["dog", "cat"], embeddings)
```

上面的嵌入可以是 OpenAIEmbeddings。

现在用户可以按相似性找到查询的最相似的文档,代码如下:

```
docs = docsearch.similarity_search("terrier", include_metadata=True)
```

然后用户可以再次查询或在问答链中使用这些文档,也可以通过集成的文档加载器从许多来源以多种格式加载文档。LangChain 集成中心提供了可以用来浏览和选择适合数据源的加载器。在选择加载程序后,可以使用指定的加载程序加载文档。下面来简单看一下 LangChain 中的文档加载器。

6.2.6 文档加载器

文档加载器用于将源中的数据加载为 Document 对象,该对象由文本和关联的元数据组成。LangChain 中有不同类型的集成可用,例如用于加载简单 .txt 文件(TextLoader)、加载网页文本内容(WebBaseLoader)、来自 Arxiv 的文章(ArxivLoader)或加载 YouTube 视频的脚本(YoutubeLoader)的文档加载器。对于网页加载,Diffbot 集成提供了干净提取内容的工具。图像存在其他加载器,例如提供图像标题(ImageCaptionLoader)的加载器。LangChain 中文档加载器都有一个 load() 方法,该方法从配置的源加载数据并将其作为文档返回,其中一些文档加载器可能还有一个 lazy_load() 方法,用于在需要时将数据加载到内存中。加载 txt 文件的文本加载器,代码如下:

```
from langchain.document_loaders import TextLoader
loader = TextLoader(file_path="path/to/file.txt")
documents = loader.load()
```

documents 变量将包含加载的文档,用户可以访问这些文档进行进一步处理。每个文档都由 page_content(文档的文本内容)和元数据(关联的元数据,如源 URL 或标题)组成。同样这里可以从维基百科加载文档,代码如下:

```
from langchain.document_loaders import WikipediaLoader
loader = WikipediaLoader("LangChain")
documents = loader.load()
```

需要注意的是,文档加载器的具体实现可能因所使用的编程语言或框架的不同而有所差异。在 LangChain 中,Agent 或链中的向量检索是通过访问向量存储的检索器完成的。下面介绍检索器是如何工作的。

6.2.7 LangChain 中的检索器

LangChain 中的检索器是一种组件,用于从给定索引中搜索和检索信息。在 LangChain 的上下文中，检索器的主要类型是向量存储检索器。这种类型的检索器利用向量存储作为后端（例如 Chroma）来索引和搜索嵌入。检索器在文档的问答中起着至关重要的作用，因为它们负责根据给定的查询检索相关信息。以下是一些检索器的例子。

（1）BM25 检索器：此检索器使用 BM25 算法根据文档与给定查询的相关性对文档进行排名。它是一种流行的信息检索算法，考虑了术语频率和文档长度。

（2）TF-IDF 检索器：此检索器使用 TF-IDF 算法根据文档集合中术语的重要性对文档进行排名。它为集合中很少见但在特定文档中经常出现的术语分配更高的权重。

（3）密集检索器：此检索器使用密集嵌入来检索文档。它将文档和查询编码为密集向量，并使用余弦相似度或其他距离度量计算它们之间的相似性。

（4）K-NN 检索器：它利用众所周知的 k 最近邻算法，根据相关文档与给定查询的相似性来检索相关文档。

这些只是 LangChain 中可用的检索器的几个示例。每个检索器都有自己的优点和缺点，检索器的选择取决于具体的用例和要求。例如，如果要使用 K-NN 检索器，用户则需要创建检索器的新实例，并为其提供文本列表。下面是如何使用 OpenAI 的嵌入创建 K-NN 检索器的示例代码：

```
from langchain.retrievers import KNNRetriever
from langchain.embeddings import OpenAIEmbeddings
words = ["cat", "dog", "computer", "animal"]
retriever = KNNRetriever.from_texts(words, OpenAIEmbeddings())
```

创建检索器后，用户可以通过调用 get_relevant_documents()方法并传递查询字符串来使用它检索相关文档。检索器将返回与查询最相关的文档列表。以下是如何使用 K-NN 检索器的示例代码：

```
result = retriever.get_relevant_documents("dog")
print(result)
```

这将输出与查询相关的文档列表。每个文档都包含页面内容和元数据，输出的代码如下：

```
[Document(page_content='dog', metadata={}),
 Document(page_content='animal', metadata={}),
 Document(page_content='cat', metadata={}),
 Document(page_content='computer', metadata={})]
```

LangChain 中有一些更专业的检索器，例如来自 Arxiv、Pubmed 或维基百科的检索器。Arxiv 检索器的目的是从 Arxiv.org 存档中检索科学文章，它允许用户搜索和下载物理、数学、计算机科学等各个领域的学术文章。Arxiv 检索器的功能包括指定要下载的最大文档数、根据查询检索相关文档及访问检索到的文档的元数据信息。维基百科检索器允许用户从维基百科网站检索维基百科页面或文档。维基百科检索器的目的是提供对维基百科上大量可用信息进行轻松访问,并使用户能够从中提取特定信息或知识。PubMed 检索器是 LangChain 中的一个组件,有助于将生物医学文献检索整合到其语言模型应用程序中。PubMed 包含来自各种来源的数百万篇生物医学文献引用。在 LangChain 中,PubMedRetriever 类用于与 PubMed 数据库进行交互,并根据给定的查询检索相关文档。该类的 get_relevant_documents() 方法将查询作为输入,并从 PubMed 返回相关文档的列表。以下是如何在 LangChain 中使用 PubMed 检索器的示例代码:

```
from langchain.retrievers import PubMedRetriever
retriever = PubMedRetriever()
documents = retriever.get_relevant_documents("COVID")
for document in documents:
    print(document.metadata["title"])
```

在此示例中,当用户查询 COVID 关键字时调用了 get_relevant_documents() 方法,然后该方法从 PubMed 中检索与查询相关的文档,并将其作为列表返回,打印输出的结果如下:

```
The COVID-19 pandemic highlights the need for a psychological support in systemic sclerosis patients.
Host genetic polymorphisms involved in long-term symptoms of COVID-19.
Association Between COVID-19 Vaccination and Mortality after Major Operations.
```

自定义检索器可以通过创建一个继承自 BaseRetriever 抽象类的类在 LangChain 中实现。该类应实现 get_relevant_documents() 方法,该方法将查询字符串作为输入并返回相关文档的列表。实现自定义检索器的代码如下:

```
from langchain.retriever import BaseRetriever
from langchain.schema import Document
class MyRetriever(BaseRetriever):
    def get_relevant_documents(self, query: str) -> List[Document]:
        #在这里实现检索逻辑
        #根据查询检索并处理文档
        #返回相关文档列表
        relevant_documents = []
        return relevant_documents
```

用户可以自定义此方法以执行所需的任何检索操作,例如查询数据库或搜索索引文档。实现检索器类后,可以创建该类的实例并调用 get_relevant_documents() 方法以根据查询检索相关文档。下面的章节将引导读者利用之前学过的知识实现一个聊天机器人。

6.3 实战案例：实现一个聊天机器人

本节将带领读者一起实现一个聊天机器人。在 LangChain 中实现一个简单的聊天机器人的方法非常简单，大概可以遵循以下 3 个步骤：第 1 步是加载文档；第 2 步是创建向量存储；第 3 步是设置一个聊天机器人以便从向量存储中检索。

本节将用几种格式进行泛化，并通过 Streamlit Web 浏览器中的界面提供给用户使用。用户将能够加载文档并开始提问。在生产环境中，对于客户参与的企业部署可以想象成这些文档已经加载。首先从文档阅读器开始，作为用户，第 1 个需求当然是能够随意自由地阅读不同类型的文档，代码如下：

```python
//第 6 章 / loadDocument.py
from typing import Any
from langchain.document_loaders import (
    PyPDFLoader, TextLoader,
    UnstructuredWordDocumentLoader,
    UnstructuredEPubLoader
)
class EpubReader(UnstructuredEPubLoader):
    def __init__(self, file_path: str | list[str], ** kwargs: Any):
        super().__init__(file_path, **kwargs, mode="elements", strategy="fast")
class DocumentLoaderException(Exception):
    pass
class DocumentLoader(object):
    """加载到具有受支持扩展名的文档中"""
    supported_extentions = {
        ".pdf": PyPDFLoader,
        ".txt": TextLoader,
        ".epub": EpubReader,
        ".docx": UnstructuredWordDocumentLoader,
        ".doc": UnstructuredWordDocumentLoader
    }
```

上面的代码就实现了阅读具有不同扩展名的 PDF、文本、EPub 和 Word 文档的界面。接下来将实现加载器功能，代码如下：

```python
//第 6 章 / loaderLogic.py
import logging
import pathlib
from langchain.schema import Document
def load_document(temp_filepath: str) -> list[Document]:
    """加载文件并将其作为文档列表返回"""
```

```
        ext = pathlib.Path(temp_filepath).suffix
        loader = DocumentLoader.supported_extentions.get(ext)
        if not loader:
            raise DocumentLoaderException(
                f"Invalid extension type {ext}, cannot load this type of file"
            )
        loader = loader(temp_filepath)
        docs = loader.load()
        logging.info(docs)
        return docs
```

接下来可以从 GUI 界面上提供这个加载器,并将其连接到向量存储,代码如下:

```
//第 6 章/ guiLoader.py
from langchain.embeddings import HuggingFaceEmbeddings
from langchain.text_splitter import RecursiveCharacterTextSplitter
from langchain.vectorstores import DocArrayInMemorySearch
from langchain.schema import Document, BaseRetriever
def configure_retriever(docs: list[Document]) -> BaseRetriever:
    #拆分每个文档
    text_splitter = RecursiveCharacterTextSplitter(chunk_size=1500, chunk_overlap=200)
    splits = text_splitter.split_documents(docs)
    #创建嵌入并存储在 vectordb 中
    embeddings = HuggingFaceEmbeddings(model_name="all-MiniLM-L6-v2")
    #对包含所有文本的 HuggingFace 模型进行一次调用
    vectordb = DocArrayInMemorySearch.from_documents(splits, embeddings)
    #定义检索器
    return vectordb.as_retriever(search_type="mmr", search_kwargs={"k": 2, "fetch_k": 4})
```

DocArray 是一个 Python 工具包,提供了一套高级应用程序接口,用于表示和操作多模态数据。它具有各种功能,例如高级索引、完整的序列化协议、统一的 Python 样式接口等。此外,它还为自然语言处理、计算机视觉及音频处理等任务提供了高效便捷的多模态数据处理。可以用不同的距离度量(例如余弦相似度和欧几里得距离)来初始化 DocArray 中的向量存储,其中余弦相似度是默认选项。对于检索工具,主要有以下两种选择。

(1) 相似性搜索:用户可以根据相似性检索文档。

(2) 最大边际相关性(Maximum Marginal Relevance,MMR):用户可以在检索过程中对文档应用基于多样性进行重新排序,以获得涵盖历史检索到的文档的不同结果。

在相似性搜索中可以设置相似性分数的阈值。这里选择 MMR,因为在文本输出方面它可能给出更好的结果。参数 k 被设置为 2,这意味着可以从检索中获取两个文档。检索可以通过上下文压缩进行改进,这是一种压缩检索到的文档并过滤不相关信息的技术。上下文压缩不是按原样返回完整的文档,而是使用给定查询的上下文来提取和仅返回相关信息。这有助于降低处理成本并提高检索系统的响应质量。基本压缩器根据给定查询的上下文负责压缩单个文档的内容。它使用 LLM 执行压缩。压缩器可以过滤不相关信息,只返回文档的相关部分。基本检索器是根据查询从文档存储系统中检索文档的组件。它可以是任何检索系统,例如搜索引擎或数据库。在对上下文压缩检索器进行查询时,它首先将查询传递给基本检索器以获取相关文档,然后它使用基本压缩器根据查询上下文压缩这些文档的内容。最后,仅包含相关信息的压缩文档被作为响应返回。LangChain 中有以下几种上下文压缩选择。

(1) LLMChainExtractor:这会传递返回的文档,并仅从每个文档中提取相关内容。

(2) LLMChainFilter:这稍微简单一些,它仅筛选相关文档。

(3) EmbeddingsFilter:根据嵌入应用基于文档和查询的相似性过滤器。

前两个压缩器需要 LLM 才能调用,这意味着它可能很慢且成本高昂,因此,EmbeddingsFilter 成为更有效的替代方法。这里可以将压缩与末尾的简单的 switch 语句集成在一起(替换 return 语句),代码如下:

```
//第 6 章/ compression.py
if not use_compression:
    return retriever
embeddings_filter = EmbeddingsFilter(
  embeddings=embeddings, similarity_threshold=0.76
)
return ContextualCompressionRetriever(
  base_compressor=embeddings_filter,
  base_retriever=retriever
)
```

对于 EmbeddingsFilter 需要包含另外两个额外的导入,代码如下:

```
from langchain.retrievers.document_compressors import EmbeddingsFilter
from langchain.retrievers import ContextualCompressionRetriever
```

可以通过 configure_qa_chain()将 use_compression 参数提供给 configure_retriever()方法(此处未显示)。到这里已经有了创建检索器的机制,接下来可以设置聊天链,代码如下:

```
//第 6 章/ chatChainSetup.py
from langchain.chains import ConversationalRetrievalChain
from langchain.chains.base import Chain
from langchain.chat_models import ChatOpenAI
from langchain.memory import ConversationBufferMemory
def configure_chain(retriever: BaseRetriever) -> Chain:
    """使用检索器配置链"""
    #设置上下文对话记忆
```

```python
    memory = ConversationBufferMemory(memory_key="chat_history", return_messages=True)
    # 建立 LLM 和 QA 链；将温度调低以抑制幻觉
    llm = ChatOpenAI(
        model_name="gpt-3.5-turbo", temperature=0, streaming=True
    )
    # 自动传入 max_tokens_limit 的数量
    # 提示你的 llm 是截断标记
    return ConversationalRetrievalChain.from_llm(
        llm, retriever=retriever, memory=memory, verbose=True, max_tokens_limit=4000
    )
```

检索逻辑的最后一个任务是获取文档并将它们传递给检索器,代码如下:

```python
//第 6 章/ passToRetriever.py
import os
import tempfile
def configure_qa_chain(uploaded_files):
    """读取文档,配置检索器和链"""
    docs = []
    temp_dir = tempfile.TemporaryDirectory()
    for file in uploaded_files:
        temp_filepath = os.path.join(temp_dir.name, file.name)
        with open(temp_filepath, "wb") as f:
            f.write(file.getvalue())
        docs.extend(load_document(temp_filepath))
    retriever = configure_retriever(docs=docs)
    return configure_chain(retriever=retriever)
```

聊天机器人的逻辑完成后使用 Streamlit 设置 UI,代码如下:

```python
//第 6 章/ ui.py
import streamlit as st
from langchain.callbacks import StreamlitCallbackHandler
st.set_page_config(page_title="LangChain: Chat with Documents", page_icon=" ")
st.title("LangChain: Chat with Documents")
uploaded_files = st.sidebar.file_uploader(
    label="Upload files",
    type=list(DocumentLoader.supported_extentions.keys()),
    accept_multiple_files=True
)
if not uploaded_files:
    st.info("Please upload documents to continue.")
    st.stop()
```

```
qa_chain = configure_qa_chain(uploaded_files)
assistant = st.chat_message("assistant")
user_query = st.chat_input(placeholder="Ask me anything!")
if user_query:
    stream_handler = StreamlitCallbackHandler(assistant)
    response = qa_chain.run(user_query, callbacks=[stream_handler])
    st.markdown(response)
```

最终的结果如图 6-4 所示。

图 6-4　聊天机器人界面

需要注意的是，LangChain 对输入信息大小有限制，因此用户可能需要考虑解决方法，从而处理更大的知识库。此外相比于商业解决方案，微调模型或在内部托管 LLM 可能更复杂且更不准确。现在已经完成了一个聊天机器人的搭建，在用户和聊天机器人进行对话的过程中，最重要的莫过于上下文记忆了，至此本书还没有深入讲解 LangChain 中记忆的部分，接下来介绍 LangChain 中的记忆机制。

6.4　LangChain 中的记忆机制

大多数 LLM 应用程序具有会话界面。会话的一个重要组成部分是能够引用先前在会话中引入的信息。至少会话系统应该能够直接访问一定窗口过去的消息。更复杂的系统需要具有一个不断更新的世界模型，使其能够维护关于实体及其关系的信息等。

一般将这种存储关于过去互动的能力称为"记忆"。LangChain 为系统添加记忆提供了许多实用工具。这些工具可以单独使用，也可以无缝地集成到链中。LangChain 中与记忆相关的大多数功能标记为测试版，原因有两个：大多数功能（除某些例外）不适用于生产环境；大多数功能（除某些例外，见下文）适用于传统链，而不是较新的 LCEL 语法。主要的例外是 ChatMessageHistory 功能。该功能在很大程度上已经准备好投入生产，并与 LCEL 集成良好。接下来通过一个例子让读者快速了解 LangChain 中的记忆是如何工作的。

6.4.1 快速开始

首先来看如何在链中使用 ConversationBufferMemory。ConversationBufferMemory 是一种非常简单的记忆形式,它只是将聊天消息列表保留在缓冲区中,并将其传递到提示模板中,代码如下:

```
from langchain.memory import ConversationBufferMemory

memory = ConversationBufferMemory()
memory.chat_memory.add_user_message("hi!")
memory.chat_memory.add_ai_message("what's up?")
```

在链中使用记忆时,下面列出了一些关键概念需要理解。

1. 记忆返回的变量类型

在进入链之前,程序会从记忆中读取各种变量。这些变量具有特定的名称,需要与链期望的变量相一致。用户可以通过调用 memory.load_memory_variables({}) 来查看这些变量是什么。这里需要注意,用户传递的空字典只是真实变量的占位符。如果正在使用的记忆类型依赖于输入变量,则可能需要将一些变量传递进去,代码如下:

```
memory.load_memory_variables({})
```

```
{'history': "Human: hi!\nAI: what's up?"}
```

在这种情况下可以看到 load_memory_variables 返回了一个名为 history 的键。这意味着这里的链应该期望一个名为 history 的输入。通常,用户可以通过记忆类的参数来控制这个变量。例如,如果希望将记忆变量返回为 chat_history 键,则可以这样做,代码如下:

```
memory = ConversationBufferMemory(memory_key="chat_history")
memory.chat_memory.add_user_message("hi!")
memory.chat_memory.add_ai_message("what's up?")
```

```
{'chat_history': "Human: hi!\nAI: what's up?"}
```

这里用于控制这些键的参数名称可能因记忆类型的不同而有所差异,但读者需要理解的是这是可以控制的及可以进行控制的方法。

2. 记忆本身的数据类型

最常见的记忆类型之一涉及返回聊天消息列表。这些消息可以返回为单个字符串,全部连接在一起(当它们将被传递给 LLM 时很有用),或者作为 ChatMessages 的列表(当传递给 ChatModels 时很有用)。在默认情况下,它们将作为单个字符串返回。为了返回为消息列表,用户可以设置 return_messages=True,代码如下:

```
memory = ConversationBufferMemory(return_messages=True)
memory.chat_memory.add_user_message("hi!")
memory.chat_memory.add_ai_message("what's up?")
```

```
{'history': [HumanMessage(content='hi!', additional_kwargs={}, example=False),
AIMessage(content='what's up?', additional_kwargs={}, example=False)]}
```

3. 存储到记忆的键值

通常,链可以接受多个输入/输出键。在这些情况下,用户如何知道我们想要将哪些键保存到聊天消息历史记录中呢?这通常可以通过记忆类型上的 input_key 和 output_key 参数来控制。它们默认为 None,如果只有一个输入/输出键,它就知道只使用那个键,但是,如果有多个输入/输出键,就必须指定要使用哪个键的名称了。

最后演示一个从头到尾的例子。这里将使用 LLMChain,并演示如何同时与 LLM 和 ChatModel 一起工作,代码如下:

```
//第 6 章/ useLLM.py
from langchain_openai import OpenAI
from langchain.prompts import PromptTemplate
from langchain.chains import LLMChain
from langchain.memory import ConversationBufferMemory


llm = OpenAI(temperature=0)
#需要注意,在提示模板中存在"chat_history"
template = """You are a nice chatbot having a conversation with a human.

Previous conversation:
{chat_history}

New human question: {question}
Response:"""
prompt = PromptTemplate.from_template(template)
#注意,需要对齐 memory_key
memory = ConversationBufferMemory(memory_key="chat_history")
conversation = LLMChain(
    llm=llm,
    prompt=prompt,
    verbose=True,
    memory=memory
)
```

```
#需要注意这里只传入了"question"变量,"chat_history"由记忆自动填充
conversation({"question": "hi"})
```

使用ChatModel，代码如下：

```python
//第 6 章/ChatModel.py
from langchain_openai import ChatOpenAI
from langchain.prompts import (
    ChatPromptTemplate,
    MessagesPlaceholder,
    SystemMessagePromptTemplate,
    HumanMessagePromptTemplate,
)
from langchain.chains import LLMChain
from langchain.memory import ConversationBufferMemory


llm = ChatOpenAI()
prompt = ChatPromptTemplate(
    messages=[
        SystemMessagePromptTemplate.from_template(
            "You are a nice chatbot having a conversation with a human."
        ),
        #这里的 variable_name 必须与记忆对齐
        MessagesPlaceholder(variable_name="chat_history"),
        HumanMessagePromptTemplate.from_template("{question}")
    ]
)
#需要注意,我们使用 return_messages=True 来适应 MessagesPlaceholder
#需要注意,"chat_history"与 MessagesPlaceholder 的名称需要对齐
memory = ConversationBufferMemory(memory_key="chat_history", return_messages
=True)
conversation = LLMChain(
    llm=llm,
    prompt=prompt,
    verbose=True,
    memory=memory
)

#需要注意,这里只传入"question"变量,"chat_history"由记忆自动填充
conversation({"question": "hi"})
```

6.4.2　LangChain 中基础的记忆类型

LangChain 中有许多不同类型的记忆。每种类型都有其自己的参数、返回类型，并在不同的情景中有不同的用途。理解和选择适当类型的记忆对于有效构建和管理 LLM 应用程序至关重要。不同的记忆类型可能适用于不同的任务，例如回答问题、生成文本、保留对话历史等。可以根据特定需求来选择合适的记忆类型，并根据需要调整其参数以满足应用程

序的要求。这种灵活性允许开发者根据具体情况来优化和定制记忆，以实现更好的性能和提高用户体验，因此，了解各种记忆类型的特点和用法对于充分利用 LangChain 的功能非常重要。下面将对一些常见的记忆类型进行简要介绍。

1. 会话缓冲（Conversation Buffer）

ConversationBufferMemory 这种记忆允许存储消息，并且消息可以被提取到一个变量中。首先可以将其提取为字符串，代码如下：

```
from langchain.memory import ConversationBufferMemory
memory = ConversationBufferMemory()
memory.save_context({"input": "hi"}, {"output": "whats up"})
memory.load_memory_variables({})
```

```
{'history': 'Human: hi\nAI: whats up'}
```

这里还可以将历史记录提取为消息列表（如果用户将其与聊天模型一起使用，则将非常有用），代码如下：

```
memory = ConversationBufferMemory(return_messages=True)
memory.save_context({"input": "hi"}, {"output": "whats up"})
memory.load_memory_variables({})
```

```
{'history': [HumanMessage(content='hi', additional_kwargs={}),
  AIMessage(content='whats up', additional_kwargs={})]}
```

最后看一下如何在链中使用这个功能（设置 verbose=True 以便查看提示），代码如下：

```
//第 6 章/useinChainConversationBuffer.py
from langchain_openai import OpenAI
from langchain.chains import ConversationChain


llm = OpenAI(temperature=0)
conversation = ConversationChain(
    llm=llm,
    verbose=True,
    memory=ConversationBufferMemory()
)
```

```
conversation.predict(input="Hi there!")
```

```
    > Entering new ConversationChain chain...
    Prompt after formatting:
```

```
            The following is a friendly conversation between a human and an AI. The AI is
        talkative and provides lots of specific details from its context. If the AI does
        not know the answer to a question, it truthfully says it does not know.

            Current conversation:

            Human: Hi there!
            AI:

            > Finished chain.




        " Hi there! It's nice to meet you. How can I help you today?"

        conversation.predict(input="I'm doing well! Just having a conversation with
        an AI.")
```

```
            > Entering new ConversationChain chain...
            Prompt after formatting:
            The following is a friendly conversation between a human and an AI. The AI is
        talkative and provides lots of specific details from its context. If the AI does
        not know the answer to a question, it truthfully says it does not know.

            Current conversation:
            Human: Hi there!
            AI:  Hi there! It's nice to meet you. How can I help you today?
            Human: I'm doing well! Just having a conversation with an AI.
            AI:

            > Finished chain.




        " That's great! It's always nice to have a conversation with someone new. What
        would you like to talk about?"
```

conversation.predict(input="Tell me about yourself.")

```
            > Entering new ConversationChain chain...
            Prompt after formatting:
```

```
    The following is a friendly conversation between a human and an AI. The AI is
    talkative and provides lots of specific details from its context. If the AI does
    not know the answer to a question, it truthfully says it does not know.

    Current conversation:
    Human: Hi there!
    AI:   Hi there! It's nice to meet you. How can I help you today?
    Human: I'm doing well! Just having a conversation with an AI.
    AI:   That's great! It's always nice to have a conversation with someone new.
    What would you like to talk about?
    Human: Tell me about yourself.
    AI:

    > Finished chain.
```

```
    " Sure! I'm an AI created to help people with their everyday tasks. I'm
    programmed to understand natural language and provide helpful information. I'm
    also constantly learning and updating my knowledge base so I can provide more
    accurate and helpful answers."
```

2. 会话缓冲窗口（Conversation Buffer Window）

ConversationBufferWindowMemory 会保存一段时间内的会话交互列表。它仅使用最后的 k 个交互。这对于保持最近交互的滑动窗口非常有用，以便缓冲区不会变得过大。下面探讨这种类型记忆的基本功能，代码如下：

```
from langchain.memory import ConversationBufferWindowMemory
memory = ConversationBufferWindowMemory( k=1)
memory.save_context({"input": "hi"}, {"output": "whats up"})
memory.save_context({"input": "not much you"}, {"output": "not much"})
memory.load_memory_variables({})
```

```
    {'history': 'Human: not much you\nAI: not much'}
```

还可以将历史记录提取为消息列表（如果用户将其与聊天模型一起使用，则将非常有用），代码如下：

```
memory = ConversationBufferWindowMemory( k=1, return_messages=True)
memory.save_context({"input": "hi"}, {"output": "whats up"})
memory.save_context({"input": "not much you"}, {"output": "not much"})
memory.load_memory_variables({})
```

```
{'history': [HumanMessage(content='not much you', additional_kwargs={}),
  AIMessage(content='not much', additional_kwargs={})]}
```

接下来看一下在链中使用这种记忆的效果，代码如下：

```python
//第 6 章/ useinChainBufferWindows.py
from langchain_openai import OpenAI
from langchain.chains import ConversationChain
conversation_with_summary = ConversationChain(
    llm=OpenAI(temperature=0),
    #这里设置了一个较低的 k 值，即 k=2，以仅在记忆中保留最后的 2 次交互
    memory=ConversationBufferWindowMemory(k=2),
    verbose=True
)
conversation_with_summary.predict(input="Hi, what's up?")
```

```
> Entering new ConversationChain chain...
Prompt after formatting:
The following is a friendly conversation between a human and an AI. The AI is
talkative and provides lots of specific details from its context. If the AI does
not know the answer to a question, it truthfully says it does not know.

Current conversation:
```

```
    Human: Hi, what's up?
    AI:   Hi there! I'm doing great. I'm currently helping a customer with a
technical issue. How about you?
    Human: What's their issues?
    AI:

> Finished chain.




"  The customer is having trouble connecting to their Wi-Fi network. I'm
helping them troubleshoot the issue and get them connected."
```

```
conversation_with_summary.predict(input="Is it going well?")
```

```
    > Entering new ConversationChain chain...
    Prompt after formatting:
    The following is a friendly conversation between a human and an AI. The AI is
talkative and provides lots of specific details from its context. If the AI does
not know the answer to a question, it truthfully says it does not know.

    Current conversation:
    Human: Hi, what's up?
    AI:   Hi there! I'm doing great. I'm currently helping a customer with a
technical issue. How about you?
    Human: What's their issues?
    AI:   The customer is having trouble connecting to their Wi-Fi network. I'm
helping them troubleshoot the issue and get them connected.
    Human: Is it going well?
    AI:

> Finished chain.




"  Yes, it's going well so far. We've already identified the problem and are now
working on a solution."
```

```
#这里第1次交互没有出现在记忆中
conversation_with_summary.predict(input="What's the solution?")
```

```
> Entering new ConversationChain chain...
Prompt after formatting:
The following is a friendly conversation between a human and an AI. The AI is talkative and provides lots of specific details from its context. If the AI does not know the answer to a question, it truthfully says it does not know.

Current conversation:
Human: What's their issues?
AI: The customer is having trouble connecting to their Wi-Fi network. I'm helping them troubleshoot the issue and get them connected.
Human: Is it going well?
AI: Yes, it's going well so far. We've already identified the problem and are now working on a solution.
Human: What's the solution?
AI:

> Finished chain.




" The solution is to reset the router and reconfigure the settings. We're currently in the process of doing that."
```

3. 实体记忆（Entity Memory）

实体记忆可以记住对话中特定实体的特定事实。它可以提取有关实体的信息（使用 LLM），并随着时间的推移逐渐积累对该实体的知识（同样使用 LLM）。首先介绍如何使用这个功能，代码如下：

```
from langchain_openai import OpenAI
from langchain.memory import ConversationEntityMemory
llm = OpenAI(temperature=0)
```

```
memory = ConversationEntityMemory(llm=llm)
_input = {"input": "Deven & Sam are working on a hackathon project"}
memory.load_memory_variables(_input)
memory.save_context(
    _input,
    {"output": " That sounds like a great project! What kind of project are they working on?"}
)
```

```
memory.load_memory_variables({"input": 'who is Sam'})
```

```
{'history': 'Human: Deven & Sam are working on a hackathon project\nAI:   That sounds like a great project! What kind of project are they working on?',
 'entities': {'Sam': 'Sam is working on a hackathon project with Deven.'}}
```

```
memory = ConversationEntityMemory(llm=llm, return_messages=True)
_input = {"input": "Deven & Sam are working on a hackathon project"}
memory.load_memory_variables(_input)
memory.save_context(
    _input,
    {"output": " That sounds like a great project! What kind of project are they working on?"}
)
```

```
memory.load_memory_variables({"input": 'who is Sam'})
```

```
{'history': [HumanMessage(content='Deven & Sam are working on a hackathon project', additional_kwargs={}),
    AIMessage(content=' That sounds like a great project! What kind of project are they working on?', additional_kwargs={})],
 'entities': {'Sam': 'Sam is working on a hackathon project with Deven.'}}
```

接下来在链中使用，代码如下：

```
from langchain.chains import ConversationChain
from langchain.memory import ConversationEntityMemory
from langchain.memory.prompt import ENTITY_MEMORY_CONVERSATION_TEMPLATE
from pydantic import BaseModel
from typing import List, Dict, Any
```

```
conversation = ConversationChain(
    llm=llm,
    verbose=True,
    prompt=ENTITY_MEMORY_CONVERSATION_TEMPLATE,
    memory=ConversationEntityMemory(llm=llm)
)
```

```
conversation.predict(input="Deven & Sam are working on a hackathon project")
```

```
> Entering new ConversationChain chain...
Prompt after formatting:
You are an assistant to a human, powered by a large language model trained by OpenAI.
```

```
    You are designed to be able to assist with a wide range of tasks, from
answering simple questions to providing in-depth explanations and discussions
on a wide range of topics. As a language model, you are able to generate human-
like text based on the input you receive, allowing you to engage in natural-
sounding conversations and provide responses that are coherent and relevant to
the topic at hand.

    You are constantly learning and improving, and your capabilities are
constantly evolving. You are able to process and understand large amounts of
text, and can use this knowledge to provide accurate and informative responses to
a wide range of questions. You have access to some personalized information
provided by the human in the Context section below. Additionally, you are able to
generate your own text based on the input you receive, allowing you to engage in
discussions and provide explanations and descriptions on a wide range of topics.

    Overall, you are a powerful tool that can help with a wide range of tasks and
provide valuable insights and information on a wide range of topics. Whether the
human needs help with a specific question or just wants to have a conversation
about a particular topic, you are here to assist.

    Context:
    {'Deven': 'Deven is working on a hackathon project with Sam.', 'Sam': 'Sam is
working on a hackathon project with Deven.'}

    Current conversation:

    Last line:
    Human: Deven & Sam are working on a hackathon project
    You:

> Finished chain.




    ' That sounds like a great project! What kind of project are they working on? '
```

```
conversation.memory.entity_store.store
```

```
    {'Deven': 'Deven is working on a hackathon project with Sam, which they are
entering into a hackathon.',
    'Sam': 'Sam is working on a hackathon project with Deven.'}

conversation.predict(input=" They are trying to add more complex memory
structures to Langchain")
```

```
> Entering new ConversationChain chain...
Prompt after formatting:
You are an assistant to a human, powered by a large language model trained by
OpenAI.

You are designed to be able to assist with a wide range of tasks, from
answering simple questions to providing in-depth explanations and discussions on
a wide range of topics. As a language model, you are able to generate human-like
text based on the input you receive, allowing you to engage in natural-sounding
conversations and provide responses that are coherent and relevant to the topic
at hand.

You are constantly learning and improving, and your capabilities are
constantly evolving. You are able to process and understand large amounts of
text, and can use this knowledge to provide accurate and informative responses to
a wide range of questions. You have access to some personalized information
provided by the human in the Context section below. Additionally, you are able
to generate your own text based on the input you receive, allowing you to engage in
discussions and provide explanations and descriptions on a wide range of topics.

Overall, you are a powerful tool that can help with a wide range of tasks and
provide valuable insights and information on a wide range of topics. Whether the
human needs help with a specific question or just wants to have a conversation
about a particular topic, you are here to assist.

Context:
{'Deven': 'Deven is working on a hackathon project with Sam, which they are
entering into a hackathon.', 'Sam': 'Sam is working on a hackathon project with
Deven.', 'Langchain': ''}

Current conversation:
Human: Deven & Sam are working on a hackathon project
AI:  That sounds like a great project! What kind of project are they working
on?
Last line:
Human: They are trying to add more complex memory structures to Langchain
You:

> Finished chain.
```



```
' That sounds like an interesting project! What kind of memory structures are
they trying to add? '
```

```
conversation.predict(input="They are adding in a key-value store for entities
mentioned so far in the conversation.")
```

```
> Entering new ConversationChain chain...
Prompt after formatting:
You are an assistant to a human, powered by a large language model trained by
OpenAI.

 You are designed to be able to assist with a wide range of tasks, from
answering simple questions to providing in-depth explanations and discussions on
a wide range of topics. As a language model, you are able to generate human-like
text based on the input you receive, allowing you to engage in natural-sounding
conversations and provide responses that are coherent and relevant to the topic
at hand.

 You are constantly learning and improving, and your capabilities are
constantly evolving. You are able to process and understand large amounts of
text, and can use this knowledge to provide accurate and informative responses to
a wide range of questions. You have access to some personalized information
provided by the human in the Context section below. Additionally, you are able to
generate your own text based on the input you receive, allowing you to engage in
discussions and provide explanations and descriptions on a wide range of topics.

 Overall, you are a powerful tool that can help with a wide range of tasks and
provide valuable insights and information on a wide range of topics. Whether the
human needs help with a specific question or just wants to have a conversation
about a particular topic, you are here to assist.

Context:
{'Deven': 'Deven is working on a hackathon project with Sam, which they are
entering into a hackathon. They are trying to add more complex memory structures
to Langchain.', 'Sam': 'Sam is working on a hackathon project with Deven, trying
to add more complex memory structures to Langchain.', 'Langchain': 'Langchain is
a project that is trying to add more complex memory structures.', 'Key-Value
Store': ''}

Current conversation:
Human: Deven & Sam are working on a hackathon project
AI:  That sounds like a great project! What kind of project are they working on?
Human: They are trying to add more complex memory structures to Langchain
AI:  That sounds like an interesting project! What kind of memory structures
are they trying to add?
Last line:
Human: They are adding in a key-value store for entities mentioned so far in
the conversation.
```

```
You:

> Finished chain.



    ' That sounds like a great idea! How will the key-value store help with the project? '
```

```
conversation.predict(input="What do you know about Deven & Sam?")
```

```
> Entering new ConversationChain chain...
Prompt after formatting:
You are an assistant to a human, powered by a large language model trained by OpenAI.

    You are designed to be able to assist with a wide range of tasks, from answering simple questions to providing in-depth explanations and discussions on a wide range of topics. As a language model, you are able to generate human-like text based on the input you receive, allowing you to engage in natural-sounding conversations and provide responses that are coherent and relevant to the topic at hand.

    You are constantly learning and improving, and your capabilities are constantly evolving. You are able to process and understand large amounts of text, and can use this knowledge to provide accurate and informative responses to a wide range of questions. You have access to some personalized information provided by the human in the Context section below. Additionally, you are able to generate your own text based on the input you receive, allowing you to engage in discussions and provide explanations and descriptions on a wide range of topics.

    Overall, you are a powerful tool that can help with a wide range of tasks and provide valuable insights and information on a wide range of topics. Whether the human needs help with a specific question or just wants to have a conversation about a particular topic, you are here to assist.

    Context:
    {'Deven': 'Deven is working on a hackathon project with Sam, which they are entering into a hackathon. They are trying to add more complex memory structures to Langchain, including a key-value store for entities mentioned so far in the conversation.', 'Sam': 'Sam is working on a hackathon project with Deven, trying to add more complex memory structures to Langchain, including a key-value store for entities mentioned so far in the conversation.'}
```

```
    Current conversation:
    Human: Deven & Sam are working on a hackathon project
    AI:   That sounds like a great project! What kind of project are they working on?
    Human: They are trying to add more complex memory structures to Langchain
    AI:   That sounds like an interesting project! What kind of memory structures are they trying to add?
    Human: They are adding in a key-value store for entities mentioned so far in the conversation.
    AI:   That sounds like a great idea! How will the key-value store help with the project?
    Last line:
    Human: What do you know about Deven & Sam?
    You:

    > Finished chain.
```

```
    ' Deven and Sam are working on a hackathon project together, trying to add more complex memory structures to Langchain, including a key-value store for entities mentioned so far in the conversation. They seem to be working hard on this project and have a great idea for how the key-value store can help.'
```

用户还可以直接检查记忆存储。可以通过以下代码直接查看它，然后通过一些添加信息来观察它的变化，代码如下：

```
from pprint import pprint
pprint(conversation.memory.entity_store.store)
```

```
    {'Daimon': 'Daimon is a company founded by Sam, a successful entrepreneur.',
     'Deven': 'Deven is working on a hackathon project with Sam, which they are '
              'entering into a hackathon. They are trying to add more complex '
              'memory structures to Langchain, including a key-value store for '
              'entities mentioned so far in the conversation, and seem to be '
              'working hard on this project with a great idea for how the '
              'key-value store can help.',
     'Key-Value Store': 'A key-value store is being added to the project to store '
                        'entities mentioned in the conversation.',
     'Langchain': 'Langchain is a project that is trying to add more complex '
                  'memory structures, including a key-value store for entities '
                  'mentioned so far in the conversation.',
```

```
            'Sam': 'Sam is working on a hackathon project with Deven, trying to add more '
                   'complex memory structures to Langchain, including a key-value store '
                   'for entities mentioned so far in the conversation. They seem to have '
                   'a great idea for how the key-value store can help, and Sam is also '
                   'the founder of a company called Daimon.'}
```

```
conversation.predict(input="Sam is the founder of a company called Daimon.")
```

```
    > Entering new ConversationChain chain...
    Prompt after formatting:
    You are an assistant to a human, powered by a large language model trained by
    OpenAI.

        You are designed to be able to assist with a wide range of tasks, from
    answering simple questions to providing in-depth explanations and discussions on
    a wide range of topics. As a language model, you are able to generate human-like
    text based on the input you receive, allowing you to engage in natural-sounding
    conversations and provide responses that are coherent and relevant to the topic
    at hand.

        You are constantly learning and improving, and your capabilities are
    constantly evolving. You are able to process and understand large amounts of
    text, and can use this knowledge to provide accurate and informative responses to
    a wide range of questions. You have access to some personalized information
    provided by the human in the Context section below. Additionally, you are able to
    generate your own text based on the input you receive, allowing you to engage in
    discussions and provide explanations and descriptions on a wide range of topics.

        Overall, you are a powerful tool that can help with a wide range of tasks and
    provide valuable insights and information on a wide range of topics. Whether the
    human needs help with a specific question or just wants to have a conversation
    about a particular topic, you are here to assist.

        Context:
        {'Daimon': 'Daimon is a company founded by Sam, a successful entrepreneur.',
    'Sam': 'Sam is working on a hackathon project with Deven, trying to add more
    complex memory structures to Langchain, including a key-value store for entities
    mentioned so far in the conversation. They seem to have a great idea for how the key-
    value store can help, and Sam is also the founder of a company called Daimon.'}

        Current conversation:
        Human: They are adding in a key-value store for entities mentioned so far in
    the conversation.
        AI:  That sounds like a great idea! How will the key-value store help with the
    project?
```

```
    Human: What do you know about Deven & Sam?
    AI:  Deven and Sam are working on a hackathon project together, trying to add
more complex memory structures to Langchain, including a key-value store for
entities mentioned so far in the conversation. They seem to be working hard on
this project and have a great idea for how the key-value store can help.
    Human: Sam is the founder of a company called Daimon.
    AI:
That's impressive! It sounds like Sam is a very successful entrepreneur. What
kind of company is Daimon?
    Last line:
    Human: Sam is the founder of a company called Daimon.
    You:

> Finished chain.




    " That's impressive! It sounds like Sam is a very successful entrepreneur.
What kind of company is Daimon? "
```

```
from pprint import pprint
pprint(conversation.memory.entity_store.store)
```

```
    {'Daimon': 'Daimon is a company founded by Sam, a successful entrepreneur, who '
              'is working on a hackathon project with Deven to add more complex '
              'memory structures to Langchain.',
     'Deven': 'Deven is working on a hackathon project with Sam, which they are '
              'entering into a hackathon. They are trying to add more complex '
              'memory structures to Langchain, including a key-value store for '
              'entities mentioned so far in the conversation, and seem to be '
              'working hard on this project with a great idea for how the '
              'key-value store can help.',
     'Key-Value Store': 'A key-value store is being added to the project to store '
                        'entities mentioned in the conversation.',
     'Langchain': 'Langchain is a project that is trying to add more complex '
                  'memory structures, including a key-value store for entities '
                  'mentioned so far in the conversation.',
     'Sam': 'Sam is working on a hackathon project with Deven, trying to add more '
            ' complex memory structures to Langchain, including a key-value store '
            'for entities mentioned so far in the conversation. They seem to have '
            'a great idea for how the key-value store can help, and Sam is also '
            'the founder of a successful company called Daimon.'}
```

4. 会话知识图(Conversation Knowledge Graph)

这种类型的记忆使用知识图谱来重建记忆。知识图谱是一种图形表示,它包含实体、属性和它们之间的关系,用于模拟和存储信息。这种记忆类型充分地利用了知识图谱的概念,通过将知识图谱中的信息与对话相结合,实现了对话中的信息重建和提取。这可以用于更深入地理解对话中的实体、主题和关系,从而提高对话系统的智能和响应能力。通过这种方法,系统可以在对话中引入更多的上下文信息,使对话更加连贯和有意义。该类型记忆直接和 LLM 结合的代码如下:

```
//第 6 章/knowledgeinGraph.py
from langchain.memory import ConversationKGMemory
from langchain_openai import OpenAI

llm = OpenAI(temperature=0)
memory = ConversationKGMemory(llm=llm)
memory.save_context({"input": "say hi to sam"}, {"output": "who is sam"})
memory.save_context({"input": "sam is a friend"}, {"output": "okay"})

memory.load_memory_variables({"input": "who is sam"})

{'history': 'On Sam: Sam is friend.'}
```

还可以将历史记录提取为消息列表(如果将其与聊天模型一起使用,则将非常有用),代码如下:

```
memory = ConversationKGMemory(llm=llm, return_messages=True)
memory.save_context({"input": "say hi to sam"}, {"output": "who is sam"})
memory.save_context({"input": "sam is a friend"}, {"output": "okay"})

memory.load_memory_variables({"input": "who is sam"})

{'history': [SystemMessage(content='On Sam: Sam is friend.', additional_kwargs={})]}
```

还可以更模块化地从新消息中获取当前实体(将使用先前的消息作为上下文),代码如下:

```
memory.get_current_entities("what's Sams favorite color?")

['Sam']
```

甚至可以更模块化地从新消息中获取知识三元组(将使用先前的消息作为上下文),代码如下:

```
memory.get_knowledge_triplets("her favorite color is red")
```

```
[KnowledgeTriple(subject='Sam', predicate='favorite color', object_='red')]
```

下面是这个记忆类型在链中的使用,代码如下:

```python
//第 6 章/ GraphinChain.py
from langchain.chains import ConversationChain
from langchain.prompts.prompt import PromptTemplate

template = """The following is a friendly conversation between a human and an AI.
The AI is talkative and provides lots of specific details from its context.
If the AI does not know the answer to a question, it truthfully says it does not
know. The AI ONLY uses information contained in the " Relevant Information"
section and does not hallucinate.

Relevant Information:

{history}

Conversation:
Human: {input}
AI:"""
prompt = PromptTemplate(input_variables=["history", "input"], template=
template)
conversation_with_kg = ConversationChain(
    llm=llm, verbose=True, prompt=prompt, memory=ConversationKGMemory(llm=
llm)
)
```

```
conversation_with_kg.predict(input="Hi, what's up? ")
```

```
> Entering new ConversationChain chain...
Prompt after formatting:
The following is a friendly conversation between a human and an AI. The AI is
talkative and provides lots of specific details from its context.
If the AI does not know the answer to a question, it truthfully says it does not
know. The AI ONLY uses information contained in the " Relevant Information"
section and does not hallucinate.

Relevant Information:


Conversation:
```

```
Human: Hi, what's up?
AI:

> Finished chain.

" Hi there! I'm doing great. I'm currently in the process of learning about the
world around me. I'm learning about different cultures, languages, and customs.
It's really fascinating! How about you? "
```

```
" Hi James, it's nice to meet you. I'm an AI and I understand you're helping Will,
the engineer. What kind of engineering does he do? "
```

```
conversation_with_kg.predict(input="What do you know about Will?")
```

```
> Entering new ConversationChain chain...
Prompt after formatting:
The following is a friendly conversation between a human and an AI. The AI is
talkative and provides lots of specific details from its context.
If the AI does not know the answer to a question, it truthfully says it does not
know. The AI ONLY uses information contained in the " Relevant Information "
section and does not hallucinate.

Relevant Information:

On Will: Will is an engineer.

Conversation:
Human: What do you know about Will?
AI:

> Finished chain.
```

```
' Will is an engineer.'
```

6.4.3　其他高级记忆类型

1. 会话摘要(Conversation Summary)

本节来看使用稍微复杂一些的记忆类型。第1种是ConversationSummaryMemory(对话摘要记忆)。这种记忆类型会随着时间的推移创建对话的摘要。这对于从对话中汇总信息非常有用。对话摘要记忆会实时总结对话并将当前摘要存储在记忆中，然后可以使用这个记忆将迄今为止的对话摘要注入提示/链中。这种记忆对于较长的对话非常有用，因为在提示中保留过去的消息历史原样会占用过多的Token。

首先看这种记忆的基本使用方法，代码如下：

```
from langchain.memory import ConversationSummaryMemory, ChatMessageHistory
from langchain_openai import OpenAI


memory = ConversationSummaryMemory(llm=OpenAI(temperature=0))
memory.save_context({"input": "hi"}, {"output": "what's up"})


memory.load_memory_variables({})


{'history': '\nThe human greets the AI, to which the AI responds.'}
```

用户还可以将历史记录提取为消息列表（如果将其与聊天模型一起使用，则将非常有用）。这种功能允许用户以更可管理的方式查看对话摘要的历史记录，特别是在处理较长的对话时，可以更方便地跟踪和理解对话的演变，代码如下：

```
memory = ConversationSummaryMemory(llm=OpenAI(temperature=0), return_messages=True)
memory.save_context({"input": "hi"}, {"output": "whats up"})


memory.load_memory_variables({})


{'history': [SystemMessage(content='\nThe human greets the AI, to which the AI responds.', additional_kwargs={})]}
```

用户也可以直接使用 predict_new_summary 方法，代码如下：

```
messages = memory.chat_memory.messages
previous_summary = ""
memory.predict_new_summary(messages, previous_summary)


'\nThe human greets the AI, to which the AI responds.'
```

如果用户在这个类之外有消息，则可以轻松地使用 ChatMessageHistory 来初始化这个类。在加载过程中会计算摘要。这个功能允许用户将外部消息集成到 ConversationSummaryMemory 中，以便生成更全面的对话摘要，代码如下：

```
history = ChatMessageHistory()
history.add_user_message("hi")
history.add_ai_message("hi there!")
```

```python
memory = ConversationSummaryMemory.from_messages(
    llm=OpenAI(temperature=0),
    chat_memory=history,
    return_messages=True
)
```

```python
memory.buffer
```

```
'\nThe human greets the AI, to which the AI responds with a friendly greeting.'
```

或者用户可以使用先前生成的摘要来加速初始化,并通过直接进行初始化而避免重新生成摘要。这可以提高初始化过程的效率,并减少不必要的计算,代码如下:

```python
memory = ConversationSummaryMemory(
    llm=OpenAI(temperature=0),
    buffer="The human asks what the AI thinks of artificial intelligence. The AI thinks artificial intelligence is a force for good because it will help humans reach their full potential.",
    chat_memory=history,
    return_messages=True
)
```

最后通过一个在链中使用这个功能的示例来演示,再次设置 verbose＝True 以便查看提示。在这个示例中将详细了解如何在链中使用 ConversationSummaryMemory 来注入对话摘要,以实现更有效的对话管理和生成,代码如下:

```python
//第 6 章/ SummaryinChain.py
from langchain_openai import OpenAI
from langchain.chains import ConversationChain
llm = OpenAI(temperature=0)
conversation_with_summary = ConversationChain(
    llm=llm,
    memory=ConversationSummaryMemory(llm=OpenAI()),
    verbose=True
)
conversation_with_summary.predict(input="Hi, what's up? ")
```

```
> Entering new ConversationChain chain...
Prompt after formatting:
The following is a friendly conversation between a human and an AI. The AI is talkative and provides lots of specific details from its context. If the AI does not know the answer to a question, it truthfully says it does not know.

Current conversation:
```

```
Human: Hi, what's up?
AI:

> Finished chain.


" Hi there! I'm doing great. I'm currently helping a customer with a technical issue. How about you? "

conversation_with_summary.predict(input="Tell me more about it!")
```

```
> Entering new ConversationChain chain...
Prompt after formatting:
The following is a friendly conversation between a human and an AI. The AI is talkative and provides lots of specific details from its context. If the AI does not know the answer to a question, it truthfully says it does not know.

Current conversation:

The human greeted the AI and asked how it was doing. The AI replied that it was doing great and was currently helping a customer with a technical issue.
Human: Tell me more about it!
AI:

> Finished chain.


" Sure! The customer is having trouble with their computer not connecting to the internet. I'm helping them troubleshoot the issue and figure out what the problem is. So far, we've tried resetting the router and checking the network settings, but the issue still persists. We're currently looking into other possible solutions."
```

```
conversation_with_summary.predict(input="Very cool -- what is the scope of the project?")
```

```
> Entering new ConversationChain chain...
Prompt after formatting:
```

```
The following is a friendly conversation between a human and an AI. The AI is
talkative and provides lots of specific details from its context. If the AI does
not know the answer to a question, it truthfully says it does not know.

Current conversation:

The human greeted the AI and asked how it was doing. The AI replied that it was
doing great and was currently helping a customer with a technical issue where
their computer was not connecting to the internet. The AI was troubleshooting the
issue and had already tried resetting the router and checking the network
settings, but the issue still persisted and they were looking into other possible
solutions.
Human: Very cool -- what is the scope of the project?
AI:

> Finished chain.
```

```
" The scope of the project is to troubleshoot the customer's computer issue and
find a solution that will allow them to connect to the internet. We are currently
exploring different possibilities and have already tried resetting the router
and checking the network settings, but the issue still persists."
```

2. 会话摘要缓冲（Conversation Summary Buffer）

ConversationSummaryBufferMemory 结合了这两个概念。它在记忆中保留了最近的交互缓冲，但与其完全清除旧的交互，它会将它们编译成摘要并同时使用它们。它使用 Token 长度而不是交互数量来确定何时清除交互。首先介绍如何使用这些实用工具。这个记忆类型允许更灵活地管理对话历史，同时保留了对话的重要摘要，以提高对话的效率和可理解性。

下面是在 LLM 中直接使用这个记忆类型的代码：

```
from langchain.memory import ConversationSummaryBufferMemory
from langchain_openai import OpenAI

llm = OpenAI()
```

```
memory = ConversationSummaryBufferMemory(llm=llm, max_token_limit=10)
memory.save_context({"input": "hi"}, {"output": "whats up"})
memory.save_context({"input": "not much you"}, {"output": "not much"})
```

```
memory.load_memory_variables({})
```

```
{'history': 'System: \nThe human says "hi", and the AI responds with "whats up".\nHuman: not much you\nAI: not much'}
```

用户还可以将历史记录提取为消息列表（如果将其与聊天模型一起使用，则将非常有用）。这允许用户以更可视化的方式查看对话的历史记录，并在需要时进一步地进行分析或处理，代码如下：

```
memory = ConversationSummaryBufferMemory(
    llm=llm, max_token_limit=10, return_messages=True
)
memory.save_context({"input": "hi"}, {"output": "whats up"})
memory.save_context({"input": "not much you"}, {"output": "not much"})
```

另外还可以直接使用 predict_new_summary 方法，这种方法允许用户在不必等待自动更新时，手动获取最新的对话摘要，以便更灵活地管理对话中的信息，代码如下：

```
messages = memory.chat_memory.messages
previous_summary = ""
memory.predict_new_summary(messages, previous_summary)
```

```
'\nThe human and AI state that they are not doing much.'
```

当然也可以结合链使用这个记忆，代码如下：

```
from langchain.chains import ConversationChain

conversation_with_summary = ConversationChain(
    llm=llm,
    #为了测试设置了一个非常低的 max_token_limit
    memory=ConversationSummaryBufferMemory(llm=OpenAI(), max_token_limit=40),
    verbose=True,
)
conversation_with_summary.predict(input="Hi, what's up? ")
```

```
> Entering new ConversationChain chain...
Prompt after formatting:
The following is a friendly conversation between a human and an AI. The AI is talkative and provides lots of specific details from its context. If the AI does not know the answer to a question, it truthfully says it does not know.

Current conversation:
```

```
AI:

> Finished chain.
```

```
" Hi there! I'm doing great. I'm learning about the latest advances in artificial
intelligence. What about you?"
```

```
conversation_with_summary.predict(input=" Just working on writing some
documentation!")
```

```
> Entering new ConversationChain chain...
Prompt after formatting:
The following is a friendly conversation between a human and an AI. The AI is
talkative and provides lots of specific details from its context. If the AI does
not know the answer to a question, it truthfully says it does not know.

Current conversation:
Human: Hi, what's up?
AI:  Hi there! I'm doing great. I'm spending some time learning about the latest
developments in AI technology. How about you?
Human: Just working on writing some documentation!
AI:

> Finished chain.
```

```
' That sounds like a great use of your time. Do you have experience with writing
documentation? '
```

```
#可以在这里看到对话的摘要,然后是一些先前的交互
conversation_with_summary.predict(input=" For LangChain! Have you heard
of it?")
```

```
> Entering new ConversationChain chain...
Prompt after formatting:
The following is a friendly conversation between a human and an AI. The AI is
talkative and provides lots of specific details from its context. If the AI does
not know the answer to a question, it truthfully says it does not know.

Current conversation:
System:
The human asked the AI what it was up to and the AI responded that it was learning
about the latest developments in AI technology.
Human: Just working on writing some documentation!
```

```
AI:    That sounds like a great use of your time. Do you have experience with writing
documentation?
Human: For LangChain! Have you heard of it?
AI:

> Finished chain.
```

```
" No, I haven't heard of LangChain. Can you tell me more about it? "
```

```
#在这里可以看到摘要和缓冲区已经更新
conversation_with_summary.predict(
    input="Haha nope, although a lot of people confuse it for that"
)
```

```
> Entering new ConversationChain chain...
Prompt after formatting:
The following is a friendly conversation between a human and an AI. The AI is
talkative and provides lots of specific details from its context. If the AI does
not know the answer to a question, it truthfully says it does not know.

Current conversation:
System:
The human asked the AI what it was up to and the AI responded that it was learning
about the latest developments in AI technology. The human then mentioned they
were writing documentation, to which the AI responded that it sounded like a great
use of their time and asked if they had experience with writing documentation.
Human: For LangChain! Have you heard of it?
AI:    No, I haven't heard of LangChain. Can you tell me more about it?
Human: Haha nope, although a lot of people confuse it for that
AI:

> Finished chain.
```

```
' Oh, okay. What is LangChain? '
```

3. 会话 Token 缓冲（Conversation Token Buffer）

ConversationTokenBufferMemory 会在内存中保留最近交互的缓冲区，并使用 Token 长度而不是交互数量来确定何时清除交互。这种记忆类型允许更灵活地管理对话历史，而不会受到交互数量的限制。它适用于需要控制对话 Token 数量的情况，以确保对话保持在可处理的范围内。通过 Token 长度作为清除交互的标准，可以更好地控制对话历史的大小，从而提高系统的性能和效率。这对于长时间对话的管理特别有用，因为它允许系统在 Token 数量有限的情况下保留对话上下文。

首先来看结合 LLM 的使用场景，代码如下：

```
from langchain.memory import ConversationTokenBufferMemory
from langchain_openai import OpenAI

llm = OpenAI()
```

```
memory = ConversationTokenBufferMemory(llm=llm, max_token_limit=10)
memory.save_context({"input": "hi"}, {"output": "whats up"})
memory.save_context({"input": "not much you"}, {"output": "not much"})
```

```
memory.load_memory_variables({})
```

```
{'history': 'Human: not much you\nAI: not much'}
```

还可以将历史记录提取为消息列表（如果将其与聊天模型一起使用，则将非常有用），代码如下：

```
memory = ConversationTokenBufferMemory(
    llm=llm, max_token_limit=10, return_messages=True
)
memory.save_context({"input": "hi"}, {"output": "whats up"})
memory.save_context({"input": "not much you"}, {"output": "not much"})
```

另外也可以结合链使用这个记忆类型，代码如下：

```
//第 6 章/ TokeninChain.py
from langchain.chains import ConversationChain

conversation_with_summary = ConversationChain(
    llm=llm,
    #为了测试目的,设置一个非常低的 max_token_limit
    memory=ConversationTokenBufferMemory(llm=OpenAI(), max_token_limit=60),
    verbose=True,
)
conversation_with_summary.predict(input="Hi, what's up?")
```

```
> Entering new ConversationChain chain...
Prompt after formatting:
The following is a friendly conversation between a human and an AI. The AI is
talkative and provides lots of specific details from its context. If the AI does
not know the answer to a question, it truthfully says it does not know.

Current conversation:
```

```
Human: Hi, what's up?
AI:

> Finished chain.
```

```
" Hi there! I'm doing great, just enjoying the day. How about you? "

conversation_with_summary.predict(input=" Just working on writing some documentation!")
```

```
> Entering new ConversationChain chain...
Prompt after formatting:
The following is a friendly conversation between a human and an AI. The AI is talkative and provides lots of specific details from its context. If the AI does not know the answer to a question, it truthfully says it does not know.

Current conversation:
Human: Hi, what's up?
AI:  Hi there! I'm doing great, just enjoying the day. How about you?
Human: Just working on writing some documentation!
AI:

> Finished chain.
```

```
' Sounds like a productive day! What kind of documentation are you writing? '

conversation_with_summary.predict(input=" For LangChain! Have you heard of it?")
```

```
> Entering new ConversationChain chain...
Prompt after formatting:
The following is a friendly conversation between a human and an AI. The AI is talkative and provides lots of specific details from its context. If the AI does not know the answer to a question, it truthfully says it does not know.

Current conversation:
Human: Hi, what's up?
AI:  Hi there! I'm doing great, just enjoying the day. How about you?
Human: Just working on writing some documentation!
AI:  Sounds like a productive day! What kind of documentation are you writing?
Human: For LangChain! Have you heard of it?
AI:

> Finished chain.
```

```
" Yes, I have heard of LangChain! It is a decentralized language-learning
platform that connects native speakers and learners in real time. Is that the
documentation you're writing about?"
```

```
#可以看到这里缓冲区已经更新
conversation_with_summary.predict(
    input="Haha nope, although a lot of people confuse it for that"
)
```

```
> Entering new ConversationChain chain...
Prompt after formatting:
The following is a friendly conversation between a human and an AI. The AI is
talkative and provides lots of specific details from its context. If the AI does
not know the answer to a question, it truthfully says it does not know.

Current conversation:
Human: For LangChain! Have you heard of it?
AI:  Yes, I have heard of LangChain! It is a decentralized language-learning
platform that connects native speakers and learners in real time. Is that the
documentation you're writing about?
Human: Haha nope, although a lot of people confuse it for that
AI:

> Finished chain.

" Oh, I see. Is there another language learning platform you're referring to?"
```

4. 向量存储检索记忆

向量存储检索记忆（VectorStoreRetrieverMemory）将记忆存储在向量存储中，并在每次调用时查询前 k 个最"显著"的文档。这与大多数其他记忆类不同，因为它不明确跟踪交互的顺序。在这种情况下，文档是以前的对话片段。这可以用于引用在对话早期告诉 AI 的相关信息。这种记忆类型允许根据相关性来检索信息，而不依赖于交互的顺序。

下面首先初始化一个用户自己的向量库，根据用户自己选择的存储方式，此步骤可能会有所不同，详细的步骤信息可查阅相关的向量存储文档，代码如下：

```
//第 6 章/ initializeVectorStore.py
from datetime import datetime
from langchain_openai import OpenAIEmbeddings
from langchain_openai import OpenAI
from langchain.memory import VectorStoreRetrieverMemory
from langchain.chains import ConversationChain
from langchain.prompts import PromptTemplate
import faiss
```

```python
from langchain.docstore import InMemoryDocstore
from langchain_community.vectorstores import FAISS


embedding_size = 1536 # Dimensions of the OpenAIEmbeddings
index = faiss.IndexFlatL2(embedding_size)
embedding_fn = OpenAIEmbeddings().embed_query
vectorstore = FAISS(embedding_fn, index, InMemoryDocstore({}), {})
```

接下来创建 VectorStoreRetrieverMemory,记忆对象是从任何向量存储检索器实例化的,代码如下:

```python
//第 6 章 / VectorStoreRetrieverMemory.py
# 在实际使用中,用户可以将 `k` 设置为较高的值,但这里使用 `k=1` 来显示向量查找仍然返回语义
# 相关的信息
retriever = vectorstore.as_retriever(search_kwargs=dict(k=1))
memory = VectorStoreRetrieverMemory(retriever=retriever)

# 当添加到 Agent 程序时,记忆对象可以保存来自对话或所使用工具的相关信息
memory.save_context({"input": "我最喜欢的食物是比萨"}, {"output": "那太好了"})
memory.save_context({"input": "我最喜欢的运动是踢足球"}, {"output": "..."})
memory.save_context({"input": "我不喜欢凯尔特人队"}, {"output": "好的"})


print(memory.load_memory_variables({"prompt": "what sport should i watch?"})
["history"])
```

```
    input: My favorite sport is soccer
    output: ...
```

最后通过一个示例来详细说明如何在链中使用这个记忆类型,同时设置 verbose = True 以便查看提示信息,代码如下:

```python
//第 6 章 / VectorStoreInChain.py
llm = OpenAI(temperature=0) # Can be any valid LLM
_DEFAULT_TEMPLATE = """The following is a friendly conversation between a human
and an AI. The AI is talkative and provides lots of specific details from its
context. If the AI does not know the answer to a question, it truthfully says it
does not know.

Relevant pieces of previous conversation:
{history}

(You do not need to use these pieces of information if not relevant)
```

```
Current conversation:
Human: {input}
AI:"""
PROMPT = PromptTemplate(
    input_variables=["history", "input"], template=_DEFAULT_TEMPLATE
)
conversation_with_summary = ConversationChain(
    llm=llm,
    prompt=PROMPT,
    memory=memory,
    verbose=True
)
conversation_with_summary.predict(input="Hi, my name is Perry, what's up?")
```

```
> Entering new ConversationChain chain...
Prompt after formatting:
The following is a friendly conversation between a human and an AI. The AI is
talkative and provides lots of specific details from its context. If the AI does
not know the answer to a question, it truthfully says it does not know.

Relevant pieces of previous conversation:
input: My favorite food is pizza
output: that's good to know

(You do not need to use these pieces of information if not relevant)

Current conversation:
Human: Hi, my name is Perry, what's up?
AI:

> Finished chain.




   " Hi Perry, I'm doing well. How about you? "
```

```
#在这里,与篮球相关的内容被提取出来
conversation_with_summary.predict(input="what's my favorite sport?")
```

```
> Entering new ConversationChain chain...
Prompt after formatting:
The following is a friendly conversation between a human and an AI. The AI is
```

```
talkative and provides lots of specific details from its context. If the AI does
not know the answer to a question, it truthfully says it does not know.

Relevant pieces of previous conversation:
input: My favorite sport is soccer
output: ...

(You do not need to use these pieces of information if not relevant)

Current conversation:
Human: what's my favorite sport?
AI:

> Finished chain.




' You told me earlier that your favorite sport is soccer.'
```

```
#尽管语言模型是无状态的,但由于获取了相关的记忆,所以它可以"推理"时间
#时间戳记忆和数据在一般情况下都很有用,可以帮助代理程序确定时间相关性
conversation_with_summary.predict(input="Whats my favorite food")
```

```
> Entering new ConversationChain chain...
Prompt after formatting:
The following is a friendly conversation between a human and an AI. The AI is
talkative and provides lots of specific details from its context. If the AI does
not know the answer to a question, it truthfully says it does not know.

Relevant pieces of previous conversation:
input: My favorite food is pizza
output: that's good to know

(You do not need to use these pieces of information if not relevant)

Current conversation:
Human: Whats my favorite food
AI:

> Finished chain.




' You said your favorite food is pizza.'
```

```
#对话中的记忆会被自动存储
#因为这个查询最匹配上面的介绍性聊天
#Agent 程序能够"记住"用户的名字
conversation_with_summary.predict(input="What's my name?")
```

```
> Entering new ConversationChain chain...
Prompt after formatting:
The following is a friendly conversation between a human and an AI. The AI is
talkative and provides lots of specific details from its context. If the AI does
not know the answer to a question, it truthfully says it does not know.

Relevant pieces of previous conversation:
input: Hi, my name is Perry, what's up?
response:  Hi Perry, I'm doing well. How about you?

(You do not need to use these pieces of information if not relevant)

Current conversation:
Human: What's my name?
AI:

> Finished chain.




'Your name is Perry.'
```

6.4.4　记忆和 LLM 链

前面介绍了多种记忆类型，读者可以看到每种记忆类型都可以单独地和 LLM 结合使用，或者也可以和 LLM 链结合使用。本节将详细介绍如何将记忆放到 LLM 链中使用，这样以后读者就可以有一个代码模板，以便进行参考。本节示例中将添加 ConversationBufferMemory 类，当然也可以是任何记忆类，代码如下：

```
from langchain.chains import LLMChain
from langchain.memory import ConversationBufferMemory
from langchain.prompts import PromptTemplate
from langchain_openai import OpenAI
```

这里最重要的步骤是正确设置提示。在下面的提示中有两个输入键：一个用于实际输入；另一个用于获取来自 Memory 类的输入。重要的是需要确保 PromptTemplate 和

ConversationBufferMemory 中的键匹配(chat_history)。

```
template = """You are a chatbot having a conversation with a human.

{chat_history}
Human: {human_input}
Chatbot:"""

prompt = PromptTemplate(
    input_variables=["chat_history", "human_input"], template=template
)
memory = ConversationBufferMemory(memory_key="chat_history")
```

```
llm = OpenAI()
llm_chain = LLMChain(
    llm=llm,
    prompt=prompt,
    verbose=True,
    memory=memory,
)
```

```
llm_chain.predict(human_input="Hi there my friend")
```

```
> Entering new LLMChain chain...
Prompt after formatting:
You are a chatbot having a conversation with a human.
```

```
Human: Not too bad - how are you?
Chatbot:

> Finished chain.
```

" I'm doing great, thanks for asking! How are you doing?"

上面的方法适用于完成式的 LLM,但如果用户正在使用聊天模型,则使用结构化的聊天消息可能会获得更好的性能。以下是一个示例代码:

```
from langchain.prompts import (
    ChatPromptTemplate,
    HumanMessagePromptTemplate,
    MessagesPlaceholder,
)
from langchain.schema import SystemMessage
from langchain_openai import ChatOpenAI
```

这里将使用 ChatPromptTemplate 类设置聊天提示。from_messages 方法可以从消息列表(例如,SystemMessage、HumanMessage、AIMessage、ChatMessage 等)或消息模板(例如下面的 MessagesPlaceholder)创建 ChatPromptTemplate。下面的配置将使记忆被注入聊天提示的中间,使用 chat_history 键,并且用户的输入将被添加到聊天提示的末尾,作为人类/用户消息。这个设置适用于使用聊天模型,可以在对话中更好地处理结构化聊天消息以获得更好的性能,代码如下:

```
//第 6 章/ memoryInjection.py
prompt = ChatPromptTemplate.from_messages(
    [
        SystemMessage(
            content="You are a chatbot having a conversation with a human."
        ),  #永久系统提示
        MessagesPlaceholder(
            variable_name="chat_history"
        ),  #记忆存储的位置
        HumanMessagePromptTemplate.from_template(
            "{human_input}"
        ),  #用户输入的位置
    ]
)

memory = ConversationBufferMemory(memory_key="chat_history", return_messages=True)
```

```python
llm = ChatOpenAI()

chat_llm_chain = LLMChain(
    llm=llm,
    prompt=prompt,
    verbose=True,
    memory=memory,
)
```

```python
chat_llm_chain.predict(human_input="Hi there my friend")
```

```python
chat_llm_chain.predict(human_input="Not too bad - how are you? ")
```

```
> Entering new LLMChain chain...
Prompt after formatting:
System: You are a chatbot having a conversation with a human.
Human: Hi there my friend
AI: Hello! How can I assist you today, my friend?
Human: Not too bad - how are you?

> Finished chain.

"I'm an AI chatbot, so I don't have feelings, but I'm here to help and chat with
you! Is there something specific you would like to talk about or any questions I
can assist you with?"
```

6.4.5 记忆和 Agent

向 Agent 添加记忆有以下步骤：首先创建一个带有记忆的 LLMChain，然后使用该 LLMChain 来创建一个自定义 Agent。在本例中将创建一个简单的自定义 Agent，该 Agent 可以访问搜索工具并利用 ConversationBufferMemory 类来处理记忆和对话。这些步骤将允许 Agent 程序利用记忆来更好地理解和参与对话，代码如下：

```python
from langchain.agents import AgentExecutor, Tool, ZeroShotAgent
from langchain.chains import LLMChain
from langchain.memory import ConversationBufferMemory
from langchain_community.utilities import GoogleSearchAPIWrapper
from langchain_openai import OpenAI


search = GoogleSearchAPIWrapper()
tools = [
    Tool(
        name="Search",
```

```
        func=search.run,
        description="useful for when you need to answer questions about current events",
    )
]
```

下面构建 PromptTemplate，需要注意在 PromptTemplate 中使用的 chat_history 变量，它与 ConversationBufferMemory 中的动态键名相匹配。这种匹配确保了记忆的正确注入和对齐，以便在对话中共享信息和上下文，代码如下：

```
//第 6 章/ConstructPrompt.py
prefix = """Have a conversation with a human, answering the following questions as best you can. You have access to the following tools:"""
suffix = """Begin!"

{chat_history}
Question: {input}
{agent_scratchpad}"""

prompt = ZeroShotAgent.create_prompt(
    tools,
    prefix=prefix,
    suffix=suffix,
    input_variables=["input", "chat_history", "agent_scratchpad"],
)
memory = ConversationBufferMemory(memory_key="chat_history")
```

现在可以构建 LLMChain 了，使用 Memory 对象，并创建 Agent 程序。这将允许 Agent 程序在对话中利用记忆来更好地理解和回应用户，代码如下：

```
llm_chain = LLMChain(llm=OpenAI(temperature=0), prompt=prompt)
agent = ZeroShotAgent(llm_chain=llm_chain, tools=tools, verbose=True)
agent_chain = AgentExecutor.from_agent_and_tools(
    agent=agent, tools=tools, verbose=True, memory=memory
)
```

```
agent_chain.run(input="How many people live in canada?")
```

```
> Entering new AgentExecutor chain...
Thought: I need to find out the population of Canada
Action: Search
Action Input: Population of Canada
Observation: The current population of Canada is 38,566,192 as of Saturday, December 31, 2022, based on Worldometer elaboration of the latest United Nations
```

data. • Canada ... Additional information related to Canadian population trends can be found on Statistics Canada's Population and Demography Portal. Population of Canada (real - ... Index to the latest information from the Census of Population. This survey conducted by Statistics Canada provides a statistical portrait of Canada and its ... 14 records ... Estimated number of persons by quarter of a year and by year, Canada, provinces and territories. The 2021 Canadian census counted a total population of 36,991,981, an increase of around 5.2 percent over the 2016 figure. ... Between 1990 and 2008, the ... (2) Census reports and other statistical publications from national statistical offices, (3) Eurostat: Demographic Statistics, (4) United Nations ... Canada is a country in North America. Its ten provinces and three territories extend from ... Population. • Q4 2022 estimate. 39,292,355 (37th). Information is available for the total Indigenous population and each of the three ... The term 'Aboriginal' or 'Indigenous' used on the Statistics Canada ... Jun 14, 2022 ... Determinants of health are the broad range of personal, social, economic and environmental factors that determine individual and population ... COVID - 19 vaccination coverage across Canada by demographics and key populations. Updated every Friday at 12:00 PM Eastern Time.
Thought: I now know the final answer
Final Answer: The current population of Canada is 38,566,192 as of Saturday, December 31, 2022, based on Worldometer elaboration of the latest United Nations data.
> Finished AgentExecutor chain.

'The current population of Canada is 38,566,192 as of Saturday, December 31, 2022, based on Worldometer elaboration of the latest United Nations data.'

要测试这个 Agent 的记忆可以提出一个后续问题，该问题依赖于前一次交流中的信息，以便正确地回答。这将验证 Agent 是否能够正确地利用记忆来回应问题，并显示其在对话中的信息保持和检索能力，代码如下：

```
agent_chain.run(input="what is their national anthem called?")
```

> Entering new AgentExecutor chain...
Thought: I need to find out what the national anthem of Canada is called.
Action: Search
Action Input: National Anthem of Canada
Observation: Jun 7, 2010 ... https://twitter.com/CanadaImmigrantCanadian National Anthem O Canada in HQ - complete with lyrics, captions, vocals & music. LYRICS: O Canada! Nov 23, 2022 ... After 100 years of tradition, O Canada was proclaimed Canada's national anthem in 1980. The music for O Canada was composed in 1880 by Calixa ... O Canada, national anthem of Canada. It was proclaimed the official national anthem on July 1, 1980. "God Save the Queen" remains the royal anthem of Canada ... O Canada! Our home and native land! True patriot love in all of

```
us command. Car ton bras sait porter l'épée,. Il sait porter la croix! "O Canada"
(French: Ô Canada) is the national anthem of Canada. The song was originally
commissioned by Lieutenant Governor of Quebec Théodore Robitaille ... Feb 1, 2018
... It was a simple tweak — just two words. But with that, Canada just voted to
make its national anthem, "O Canada," gender neutral, ... "O Canada" was
proclaimed Canada's national anthem on July 1, . 1980, 100 years after it was first
sung on June 24, 1880. The music. Patriotic music in Canada dates back over 200
years as a distinct category from British or French patriotism, preceding the
first legal steps to ... Feb 4, 2022 ... English version: O Canada! Our home and
native land! True patriot love in all of us command. With glowing hearts we ... Feb
1, 2018 ... Canada's Senate has passed a bill making the country's national anthem
gender-neutral. If you're not familiar with the words to "O Canada," ...
Thought: I now know the final answer.
Final Answer: The national anthem of Canada is called "O Canada".
> Finished AgentExecutor chain.
```

```
'The national anthem of Canada is called "O Canada".'
```

从上面的代码可以看到Agent记得前一个问题是关于加拿大的,它正确地询问了谷歌搜索并正确获得了加拿大国歌的名称。

最后做一个对比验证,将有记忆的Agent与没有记忆的Agent进行比较。这将显示记忆对于正确理解和回应上下文相关问题的重要性,代码如下:

```python
//第 6 章/ agentwithoutmemory.py
prefix = """Have a conversation with a human, answering the following questions
as best you can. You have access to the following tools:"""
suffix = """Begin!"

Question: {input}
{agent_scratchpad}"""

prompt = ZeroShotAgent.create_prompt(
    tools, prefix=prefix, suffix=suffix, input_variables=["input", "agent_
scratchpad"]
)
llm_chain = LLMChain(llm=OpenAI(temperature=0), prompt=prompt)
agent = ZeroShotAgent(llm_chain=llm_chain, tools=tools, verbose=True)
agent_without_memory = AgentExecutor.from_agent_and_tools(
    agent=agent, tools=tools, verbose=True
)
```

```
agent_without_memory.run("How many people live in canada? ")
```

```
> Entering new AgentExecutor chain...
Thought: I need to find out the population of Canada
Action: Search
Action Input: Population of Canada
Observation: The current population of Canada is 38,566,192 as of Saturday, December 31, 2022, based on Worldometer elaboration of the latest United Nations data. • Canada ... Additional information related to Canadian population trends can be found on Statistics Canada's Population and Demography Portal. Population of Canada (real - ... Index to the latest information from the Census of Population. This survey conducted by Statistics Canada provides a statistical portrait of Canada and its ... 14 records ... Estimated number of persons by quarter of a year and by year, Canada, provinces and territories. The 2021 Canadian census counted a total population of 36,991,981, an increase of around 5.2 percent over the 2016 figure. ... Between 1990 and 2008, the ... (2) Census reports and other statistical publications from national statistical offices, (3) Eurostat: Demographic Statistics, (4) United Nations ... Canada is a country in North America. Its ten provinces and three territories extend from ... Population. • Q4 2022 estimate. 39,292,355 (37th). Information is available for the total Indigenous population and each of the three ... The term 'Aboriginal' or 'Indigenous' used on the Statistics Canada ... Jun 14, 2022 ... Determinants of health are the broad range of personal, social, economic and environmental factors that determine individual and population ... COVID - 19 vaccination coverage across Canada by demographics and key populations. Updated every Friday at 12:00 PM Eastern Time.
Thought: I now know the final answer
Final Answer: The current population of Canada is 38,566,192 as of Saturday, December 31, 2022, based on Worldometer elaboration of the latest United Nations data.
> Finished AgentExecutor chain.
```

```
agent_without_memory.run("what is their national anthem called?")
```

```
> Entering new AgentExecutor chain...
Thought: I should look up the answer
Action: Search
Action Input: national anthem of [country]
Observation: Most nation states have an anthem, defined as "a song, as of praise, devotion, or patriotism"; most anthems are either marches or hymns in style. List of all countries around the world with its national anthem. ... Title and lyrics in the language of the country and translated into English, Aug 1, 2021 ... 1. Afghanistan, "Milli Surood" (National Anthem) • 2. Armenia, "Mer Hayrenik" (Our Fatherland) • 3. Azerbaijan (a transcontinental country with ... A national anthem is a patriotic musical composition symbolizing and evoking eulogies of the history and traditions of a country or nation. National Anthem of Every Country ; Fiji, "Meda Dau Doka" ("God Bless Fiji") ; Finland, "Maamme". ("Our Land") ; France, "La Marseillaise" ("The Marseillaise"). You can find an anthem in the menu
```

```
at the top alphabetically or you can use the search feature. This site is focussed
on the scholarly study of national anthems ... Feb 13, 2022 ... The 38-year-old
country music artist had the honor of singing the National Anthem during this year's
big game, and she did not disappoint. Oldest of the World's National Anthems ;
France, La Marseillaise ("The Marseillaise"), 1795 ; Argentina, Himno Nacional
Argentino ("Argentine National Anthem") ... Mar 3, 2022 ... Country music star
Jessie James Decker gained the respect of music and hockey fans alike after a jaw-
dropping rendition of "The Star-Spangled ... This list shows the country on the
left, the national anthem in the ... There are many countries over the world who
have a national anthem of their own.
Thought: I now know the final answer
Final Answer: The national anthem of [country] is [name of anthem].
> Finished AgentExecutor chain.
```

```
'The national anthem of [country] is [name of anthem].'
```

6.4.6 自定义会话记忆

自定义对话记忆是在 LangChain 中的一项关键功能,它允许用户按照自己的需求和偏好调整对话以提高体验。本节将探讨如何使用 LangChain 的自定义对话记忆功能,包括更改 AI 前缀和 Human 前缀,以及如何根据特定情境调整对话记忆,以便提高个性化和高度定制化的对话体验。自定义对话记忆可以让 AI 应用更好地满足用户需求,增强用户体验,并使对话更富有个性。本节将介绍两种自定义会话记忆的方法,分别是 AI 前缀和 Human 前缀。首先导入需要的库,代码如下:

```
from langchain.chains import ConversationChain
from langchain.memory import ConversationBufferMemory
from langchain_openai import OpenAI

llm = OpenAI(temperature=0)
```

1. AI 前缀

第 1 种方法是通过更改对话摘要中的 AI 前缀实现。在默认情况下,它被设置为 AI,但用户可以将其设置为任何自己想要的内容。需要注意的是如果更改此前缀,则应更改链中使用的提示以反映此名称已被更改。下方的代码演示了这个过程:

```
#默认的设置是 AI
conversation = ConversationChain(
    llm=llm, verbose=True, memory=ConversationBufferMemory()
)
```

```
conversation.predict(input="Hi there!")
```

```
> Entering new ConversationChain chain...
Prompt after formatting:
The following is a friendly conversation between a human and an AI. The AI is
talkative and provides lots of specific details from its context. If the AI does
not know the answer to a question, it truthfully says it does not know.

Current conversation:
```

```
Current conversation:
{history}
Human: {input}
AI Assistant:"""
PROMPT = PromptTemplate(input_variables=["history", "input"], template=template)
conversation = ConversationChain(
    prompt=PROMPT,
    llm=llm,
    verbose=True,
    memory=ConversationBufferMemory(ai_prefix="AI Assistant"),
)
```

```
conversation.predict(input="Hi there!")

> Entering new ConversationChain chain...
Prompt after formatting:
The following is a friendly conversation between a human and an AI. The AI is talkative and provides lots of specific details from its context. If the AI does not know the answer to a question, it truthfully says it does not know.

Current conversation:
```

```
The AI is talkative and provides lots of specific details from its context. If the
AI does not know the answer to a question, it truthfully says it does not know.

Current conversation:
{history}
Friend: {input}
AI:"""
PROMPT = PromptTemplate(input_variables=["history", "input"], template=
template)
conversation = ConversationChain(
    prompt=PROMPT,
    llm=llm,
    verbose=True,
    memory=ConversationBufferMemory(human_prefix="Friend"),
)
```

```
conversation.predict(input="Hi there!")
```

```
> Entering new ConversationChain chain...
Prompt after formatting:
The following is a friendly conversation between a human and an AI. The AI is
talkative and provides lots of specific details from its context. If the AI does
not know the answer to a question, it truthfully says it does not know.

Current conversation:

Friend: Hi there!
AI:

> Finished ConversationChain chain.
```

```
" Hi there! It's nice to meet you. How can I help you today?"
```

```
conversation.predict(input="What's the weather?")
```

```
> Entering new ConversationChain chain...
Prompt after formatting:
The following is a friendly conversation between a human and an AI. The AI is
talkative and provides lots of specific details from its context. If the AI does
not know the answer to a question, it truthfully says it does not know.

Current conversation:
```

```
Friend: Hi there!
AI:    Hi there! It's nice to meet you. How can I help you today?
Friend: What's the weather?
AI:

> Finished ConversationChain chain.
```

```
' The weather right now is sunny and warm with a temperature of 75 degrees
Fahrenheit. The forecast for the rest of the day is mostly sunny with a high of 82
degrees.'
```

6.4.7 自定义记忆

尽管 LangChain 中有一些预定义的记忆类型,但很有可能用户希望添加自己的记忆类型,以使其最适合自己的应用程序。本节将介绍如何实现这一点。这里将向 ConversationChain 添加一个自定义记忆类型。为了添加自定义记忆类,用户需要导入基本的记忆类并对其进行子类化,代码如下:

```
from typing import Any, Dict, List

from langchain.chains import ConversationChain
from langchain.schema import BaseMemory
from langchain_openai import OpenAI
from pydantic import BaseModel
```

在这个示例中将编写一个自定义记忆类,使用 Spacy 来提取实体并将与它们有关的信息保存在一个简单的哈希表中,然后在对话过程中将查看输入文本,提取所有实体,并将与它们有关的任何信息放入上下文中。读者应注意,这个实现相当简单和脆弱,可能不适用于生产环境。接下来需要先安装 Spacy,然后实例化相关的记忆类,代码如下:

```
#!pip install spacy
#!python -m spacy download en_core_web_lg
```

```
import spacy

nlp = spacy.load("en_core_web_lg")
```

```
//第 6 章/ SpacyEntityMemory.py
class SpacyEntityMemory(BaseMemory, BaseModel):
    """用于存储关于实体信息的记忆类"""
```

```python
        # 定义用于存储有关实体信息的字典
        entities: dict = {}
        # 定义用于将实体信息传递到提示中的键
        memory_key: str = "entities"

        def clear(self):
            self.entities = {}

        @property
        def memory_variables(self) -> List[str]:
            """定义我们要提供给提示的变量"""
            return [self.memory_key]

        def load_memory_variables(self, inputs: Dict[str, Any]) -> Dict[str, str]:
            """加载内存变量,本例中是实体键"""
            # 获取输入文本并通过 Spacy 处理
            doc = nlp(inputs[list(inputs.keys())[0]])
            # 提取已知的实体信息 (如果存在)
            entities = [
                    self.entities[str(ent)] for ent in doc.ents if str(ent) in self.entities
                ]
            # 返回关于实体的组合信息以放入上下文中
            return {self.memory_key: "\n".join(entities)}

        def save_context(self, inputs: Dict[str, Any], outputs: Dict[str, str]) -> None:
            """将此对话中的上下文保存到缓冲区中"""
            # 获取输入文本并通过 Spacy 处理
            text = inputs[list(inputs.keys())[0]]
            doc = nlp(text)
            # 对于提到的每个实体,将此信息保存到字典中
            for ent in doc.ents:
                ent_str = str(ent)
                if ent_str in self.entities:
                    self.entities[ent_str] += f"\n{text}"
                else:
                    self.entities[ent_str] = text
```

现在定义一个提示,该提示接收关于实体及用户输入的信息,代码如下:

```python
// 第 6 章 / definePrompt.py
from langchain.prompts.prompt import PromptTemplate

template = """The following is a friendly conversation between a human and an AI.
```

```
The AI is talkative and provides lots of specific details from its context. If the
AI does not know the answer to a question, it truthfully says it does not know. You
are provided with information about entities the Human mentions, if relevant.

Relevant entity information:
{entities}

Conversation:
Human: {input}
AI:"""
prompt = PromptTemplate(input_variables=["entities", "input"], template=
template)
```

接下来把所有的串联起来,代码如下:

```
llm = OpenAI(temperature=0)
conversation = ConversationChain(
    llm=llm, prompt=prompt, verbose=True, memory=SpacyEntityMemory()
)
```

在第1个示例中,没有关于 Harrison 的先前知识,"相关实体信息"部分为空,代码如下:

```
conversation.predict(input="Harrison likes machine learning")
```

```
> Entering new ConversationChain chain...
Prompt after formatting:
The following is a friendly conversation between a human and an AI. The AI is
talkative and provides lots of specific details from its context. If the AI does
not know the answer to a question, it truthfully says it does not know. You are
provided with information about entities the Human mentions, if relevant.

Relevant entity information:


Conversation:
Human: Harrison likes machine learning
AI:

> Finished ConversationChain chain.
```

```
" That's great to hear! Machine learning is a fascinating field of study. It
involves using algorithms to analyze data and make predictions. Have you ever
studied machine learning, Harrison?"
```

现在在第 2 个示例中,可以看到它提取了有关 Harrison 的信息,代码如下:

```
conversation.predict(
    input="What do you think Harrison's favorite subject in college was? "
)
```

```
> Entering new ConversationChain chain...
Prompt after formatting:
The following is a friendly conversation between a human and an AI. The AI is
talkative and provides lots of specific details from its context. If the AI does
not know the answer to a question, it truthfully says it does not know. You are
provided with information about entities the Human mentions, if relevant.

Relevant entity information:
Harrison likes machine learning

Conversation:
Human: What do you think Harrison's favorite subject in college was?
AI:

> Finished ConversationChain chain.
```

```
' From what I know about Harrison, I believe his favorite subject in college was
machine learning. He has expressed a strong interest in the subject and has
mentioned it often.'
```

综上,这个实现相当简单和脆弱,因此可能在生产环境中没有多大用处。它的目的是展示用户可以自定义记忆。

6.4.8 聊天机器人的记忆

记忆是 LangChain 框架中的一个组件,它允许聊天机器人和 LLM 记住以前的交互和信息。它在聊天机器人等应用程序中至关重要,因为它使系统能够保持对话的上下文和连续性。记忆在聊天机器人中主要有以下作用。

(1)交互信息留存:记忆能让聊天机器人保留与用户进行对话的信息。这有助于理解用户查询并提供相关响应。

(2)维护上下文:通过回忆以前的交互信息,聊天机器人可以保持对话的上下文和连续性。这能让聊天机器人与用户进行更自然和连贯的对话。

(3)提取知识:记忆使系统能够从一系列聊天消息中提取知识和见解,然后这些信息可用于提高聊天机器人的性能和准确性。

总之,记忆在聊天机器人中至关重要,它能让聊天机器人通过记住和建立过去的互动来创造更加个性化和类似人类的对话。LangChain 记忆功能的代码如下:

```
//第 6 章 / memoryInLangchain.py
from langchain.memory import ConversationBufferMemory
from langchain.chains import ConversationChain
#用记忆创建对话链
memory = ConversationBufferMemory()
chain = ConversationChain(memory=memory)
#用户输入消息
user_input = "Hi, how are you?"
#处理对话链中的用户输入
response = chain.predict(input=user_input)
#打印响应输出
print(response)
#用户输入另一条消息
user_input = "What's the weather like today?"
#处理对话链中的用户输入
response = chain.predict(input=user_input)
#打印响应输出
print(response)
#打印内存中存储的对话历史记录
print(memory.chat_memory.messages)
```

在上面的示例中使用 ConversationBufferMemory 创建具有内存的对话链。ConversationBufferMemory 是将消息存储在变量中的简单包装器。用户的输入使用了对话链的 predict() 方法进行处理。对话链保留了先前交互的记忆,使其能够提供上下文感知响应。上面的示例将内存与链分开构建,但是其实可以进一步简化,代码如下:

```
conversation = ConversationChain(
    llm=llm,
    verbose=True,
    memory=ConversationBufferMemory()
)
```

这里将 verbose 设置为 True 以查看提示词。处理完用户输入后打印对话链生成的最终响应输出。此外这里使用 memory.chat_memory.messages 打印存储在内存中的对话历史记录。save_context() 方法用于存储输入和输出。用户可以使用 load_memory_variables() 方法查看存储的内容。如果要以消息列表的形式获取历史记录,则需要将 return_messages 参数设置为 True。ConversationBufferWindowMemory 是 LangChain 提供的一种记忆类型,用于跟踪会话中随时间推移的交互。与 ConversationBufferMemory 不同,ConversationBufferMemory 会保留所有以前的交互信息,而 ConversationBufferWindowMemory 仅保留最后 k 个交互,其中 k 是指定的窗口大小。下面是如何在 LangChain 中使用 ConversationBufferWindowMemory 的示例代码:

```
from langchain.memory import ConversationBufferWindowMemory
memory = ConversationBufferWindowMemory(k=1)
```

在上面的示例中,将窗口大小设置为 1,这意味着只有最后一次交互才会存储在内存中。用户可以使用 save_context() 方法来保存每个交互的上下文。它需要两个参数,即 user_input 和 model_output,这两个参数表示给定交互的用户输入和相应模型的输出,代码如下:

```
memory.save_context({"input": "hi"}, {"output": "whats up"})
memory.save_context({"input": "not much you"}, {"output": "not much"})
```

用户可以用 memory.load_memory_variables() 看到具体的消息信息。LangChain 提供了集成一个知识图谱的能力来增强 LLM 的能力,使它们能够在文本生成和推理过程中利用结构化知识。

知识图谱是一种结构化的知识表示模型,它以实体、属性和关系的形式组织信息。它将知识表示为图形,其中将实体表示为节点,将实体之间的关系表示为边。

知识图谱的典型的例子包括维基数据,它从维基百科中捕获结构化信息。还有谷歌的知识图谱,它为搜索结果提供丰富的上下文信息。

在知识图谱中,实体可以是世界上的任何概念、对象或事物,属性用于描述这些实体的属性或特征。关系用于捕获实体之间的联系和关联,提供上下文信息并实现语义推理。LangChain 中具有知识图谱检索功能,不过 LangChain 也提供了记忆组件,可以根据用户的对话消息自动创建知识图谱。下面实例化 ConversationKGMemory 类,并将 LLM 实例作为 llm 参数传递,代码如下:

```
from langchain.memory import ConversationKGMemory
from langchain.llms import OpenAI
llm = OpenAI(temperature=0)
memory = ConversationKGMemory(llm=llm)
```

当对话逐渐进行时,可以利用 ConversationKGMemory 的 save_context() 函数将知识图中的相关信息保存至记忆中。同时,还可以在 LangChain 中定制对话记忆,这包括修改用于 AI 和 Human 消息的前缀,以及更新提示模板以反映这些变化。如果要自定义对话记忆,用户则可以按照以下步骤进行。

从 LangChain 导入必要的类和模块,代码如下:

```
from langchain.llms import OpenAI
from langchain.chains import ConversationChain
from langchain.memory import ConversationBufferMemory
from langchain.prompts.prompt import PromptTemplate
llm = OpenAI(temperature=0)
```

创建一个包含自定义前缀的新提示模板,可以通过建立一个带有所需模板字符串的 PromptTemplate 对象来完成,代码如下:

```
//第 6 章 / promptTemplate.py
template = """The following is a friendly conversation between a human and an AI.
The AI is talkative and provides lots of specific details from its context. If the
AI does not know the answer to a question, it truthfully says it does not know.
Current conversation:
{history}
Human: {input}
AI Assistant:"""
PROMPT = PromptTemplate(input_variables=["history", "input"], template=
template)
conversation = ConversationChain(
    prompt=PROMPT,
    llm=llm,
    verbose=True,
    memory=ConversationBufferMemory(ai_prefix="AI Assistant"),
)
```

在上面的例子中，AI 的前缀被设定为 AI Assistant，而不是默认的 AI。ConversationSummaryMemory 是 LangChain 中的一种记忆类型，随着对话的进行而生成对话摘要。它并不存储所有对话文字，而是将信息压缩，提供对话的摘要版本。这对于长时间对话链特别有用，因为包含所有先前消息可能会超过 Token 限制。

如果要使用 ConversationSummaryMemory，则用户首先需要创建它的一个实例，并将 LLM 作为参数传递，然后使用 save_context() 方法保存交互上下文，其中包括用户输入和 AI 输出。如果要检索摘要的对话历史，则可以使用 load_memory_variables() 方法，代码如下：

```
//第 6 章 / loadMemoryVariable.py
from langchain.memory import ConversationSummaryMemory
from langchain.llms import OpenAI
#初始化摘要内存和 LLM
memory = ConversationSummaryMemory(llm=OpenAI(temperature=0))
#保存交互的上下文
memory.save_context({"input": "hi"}, {"output": "whats up"})
#加载摘要记忆
memory.load_memory_variables({})
```

LangChain 还允许使用 CombinedMemory 类结合多种记忆策略。这在用户想要保留对话历史的不同方面时非常有用，代码如下：

```
//第 6 章 / combinedMemory.py
from langchain.llms import OpenAI
from langchain.prompts import PromptTemplate
from langchain.chains import ConversationChain
from langchain.memory import ConversationBufferMemory, CombinedMemory,
ConversationSummaryMemory
```

```python
# 初始化 LLM(使用所需的温度参数)
llm = OpenAI(temperature=0)
# 定义会话缓冲区内存(用于保留所有过去的消息)
conv_memory = ConversationBufferMemory(memory_key="chat_history_lines", input_key="input")
# 定义对话摘要记忆(用于总结对话)
summary_memory = ConversationSummaryMemory(llm=llm, input_key="input")
# 将两种记忆类型组合在一起
memory = CombinedMemory(memories=[conv_memory, summary_memory])
# 定义提示模板
_DEFAULT_TEMPLATE = """The following is a friendly conversation between a human and an AI. The AI is talkative and provides lots of specific details from its context. If the AI does not know the answer to a question, it truthfully says it does not know.
Summary of conversation:
{history}
Current conversation:
{chat_history_lines}
Human: {input}
AI:"""
PROMPT = PromptTemplate(input_variables=["history", "input", "chat_history_lines"], template=_DEFAULT_TEMPLATE)
# 初始化对话链
conversation = ConversationChain(llm=llm, verbose=True, memory=memory, prompt=PROMPT)
# 开始对话
conversation.run("Hi!")
```

在这个例子中,首先实例化 LLM 和所使用的不同类型的记忆,使用 ConversationBufferMemory 保留完整的对话历史,使用 ConversationSummaryMemory 创建对话摘要,然后结合这些记忆使用 CombinedMemory。同时,定义一个适应记忆使用的提示模板,最后,通过提供 LLM、记忆和提示来创建和运行 ConversationChain。ConversationSummaryBufferMemory 用于在内存中保留最近交互的缓存,并将旧的交互编译成摘要,而不是完全清除。刷新交互的阈值由 Token 长度决定,而不是交互数量。为了使用这个功能,记忆缓冲区需要用 LLM 和 max_token_limit 进行实例化。ConversationSummaryBufferMemory 提供了一个称为 predict_new_summary() 的方法,可以直接用来生成对话摘要。Zep 是一个内存存储和搜索引擎,旨在存储、总结、嵌入、索引和丰富聊天机器人或 AI 应用的历史记录。它为开发者提供了一个简单和低延迟的 API 访问和操作存储的数据。使用 Zep 的一个实际例子是将其集成为聊天机器人或 AI 应用的长期记忆。通过 ZepMemory 类,开发者可以使用 Zep 服务器 URL、API 密钥和用户的唯一会话标识符初始化一个 ZepMemory 实例。这允许聊天机器人或 AI 应用存储和检索聊天历史或其他相关信息。例如,在 Python 中,可以初始化一个 ZepMemory 实例,代码如下:

```
//第 6 章/ZepMemory.py
from langchain.memory import ZepMemory
#将此设置为用户自己的服务器 URL
ZEP_API_URL = "http://localhost:8000"
ZEP_API_KEY = "<your JWT token>"   #optional
session_id = str(uuid4())   #This is a unique identifier for the user
#设置 ZepMemory 实例
memory = ZepMemory(
    session_id=session_id,
    url=ZEP_API_URL,
    api_key=ZEP_API_KEY,
    memory_key="chat_history",
)
```

设置好记忆后,用户可以在聊天机器人的链条中或与 AI Agent 一起使用它来存储和检索聊天历史或其他相关信息。总体来讲,Zep 简化了持久化、搜索和丰富聊天机器人或 AI 应用历史记录的过程,使开发者能够专注于开发他们的 AI 应用,而不是构建记忆基础设施。

6.5 内容监管

在聊天机器人中,监管的作用是确保机器人的回答是适当并且合乎道德和法律的。它涉及实施机制来过滤出具有冒犯性或不适当的内容,同时阻止用户的滥用行为。在监管的背景下,法规指的是一套规范或规则,用于管理聊天机器人的行为和回应。它概述了聊天机器人应遵循的标准和原则,例如避免冒犯性语言,促进尊重的互动,以及维护道德标准。法规作为一个框架,确保聊天机器人在期望的边界内运行,并提高用户体验。在聊天机器人中进行监管并设立法规对于为用户创建安全、尊重和包容的环境、保护品牌声誉及遵守法律义务至关重要。在聊天机器人中进行监管并设立法规有以下几个重要原因。

(1) 确保道德行为:聊天机器人将与各类用户进行交互,包括一些弱势群体。监管有助于确保机器人的回应是合乎道德的,不会传播有害或冒犯性内容。

(2) 保护用户免受不当内容影响:监管有助于防止传播不当或冒犯性语言、仇恨言论或任何可能对用户有害或冒犯的内容。它为用户与聊天机器人进行互动创造了一个安全和包容的环境。

(3) 维护品牌声誉:聊天机器人通常代表一个品牌或组织。通过监管,开发者可以确保机器人的回应与品牌的价值观一致,并维护良好的声誉。

(4) 防止滥用行为:监管可以阻止用户参与滥用或不当行为。通过实施规则和后果,例如实施"两次警告"规则,开发者可以阻止用户使用挑衅性语言或参与滥用行为。

(5) 合法合规:根据司法管辖区的不同,可能存在监管内容和确保其符合法律法规的法律要求。制定一部法律或一套指南有助于开发者遵守这些法律要求。

用户可以将一个监管链添加到 LLMChain 中，以确保 LLM 生成的输出内容不会有害。如果传入监管链的内容被认定为有害的，则有以下不同的处理方式。用户可以选择在链中抛出一个错误并在应用程序中处理它，或者可以向用户返回一条消息，解释文本是有害的。具体的处理方法取决于具体应用程序的需求。在 LangChain 中，首先可以创建 OpenAIModerationChain 类的实例，这是 LangChain 提供的一个预先构建的监管链。该链用于检测和过滤有害内容，代码如下：

```
from langchain.chains import OpenAIModerationChain
moderation_chain = OpenAIModerationChain()
```

接下来创建 LLMChain 类的实例，该类代表用户的 LLM 链。在这里，用户可以定义提示并与 LLM 进行交互，代码如下：

```
from langchain.chains import LLMChain
llm_chain = LLMChain(model_name="gpt-3.5-turbo")
```

如果要将监管链追加到 LLM 链中，用户则可以使用 SequentialChain 类直接把之前定义好的 LLM 链放进来。该类允许以顺序方式将多个链连接在一起，代码如下：

```
from langchain.chains import SequentialChain
chain = SequentialChain([llm_chain, moderation_chain])
```

现在，当用户想要使用语言模型生成文本时，输入文本将首先通过监管链，然后通过 LLM 链，代码如下：

```
input_text = "Can you generate a story for me?"
output = chain.generate(input_text)
```

监管链将评估输入文本并过滤掉任何有害内容。如果输入文本被认定为有害，则监管链可以选择抛出错误或返回一条指示文本不被允许的消息。

6.6 总结

本章讨论了 RAG，该方法涉及利用外部工具或知识资源，如文件语料库。在本章中则专注于过程，重点在于与基于 RALMs 构建聊天机器人相关的方法，特别是利用外部工具检索相关信息，并将其融入内容生成中。本章的主要部分包括聊天机器人简介、检索和向量机制、聊天机器人的实施、记忆机制，以及 LLM 正确回应的重要性。该章始于对聊天机器人的概述，讨论了聊天机器人和 LLM 的演变及当前状态，突出了当前技术的实际影响和提升。接着探讨了积极沟通的重要性，以及为上下文、记忆和推理所需的技术实现。本章探讨了检索机制，包括向量存储，旨在提高聊天机器人回应的准确性。详细介绍了加载文档和信息的方法，包括向量存储和嵌入。另外，还讨论了用于保持知识和正在进行对话状态的记忆机制。本章最后讨论了监管，强调了确保回应尊重和与组织价值观一致的重要性。本章带领读者实现了一个聊天机器人，实现的过程中探索了本章中讨论的一些特性，并可作为探讨（诸如记忆和上下文、话语的调节等问题）的起点，也可用于探讨幻觉或其他问题。

第 7 章 LangChain 和软件开发

8min

尽管本书讨论的是将生成式 AI,特别是将 LLM 集成到软件应用中,但在本章将讨论如何利用 LLM 来帮助软件开发。这是一个重要的主题,在几家咨询公司(如毕马威和麦肯锡)的报告中,软件开发被突出为受生成式 AI 影响最大的领域之一。首先将讨论 LLM 如何在程序编写任务中提供帮助,并通过对一些文献的分析,带领读者领略在自动化软件工程方面取得了多大的进展。本章还将讨论许多最近的进展和新模型,然后将对生成的代码进行定性评估。接下来将实现一个完全自动化的软件开发任务 Agent。还将介绍设计选择,并展示用 LangChain 实现的 Agent 的一些结果,这个过程仅需要用几行 Python 代码。整个章节将带领读者深入探讨软件开发的不同方法。

7.1 步入新时代

强大的 AI 系统(如 ChatGPT)的出现引发了人们对将 AI 作为辅助软件开发工具的浓厚兴趣。根据毕马威 2023 年 6 月的一份报告估计,约有 25% 的软件开发任务可以实现自动化。同月的麦肯锡的报告突出了生成式 AI 对软件开发,尤其在成本削减和效率提升方面,可以产生显著影响。利用 AI 来辅助编程的想法并非新鲜事物,但是随着计算机技术和人工智能的进步,这个概念迅速发展。实际上这两个领域是相互交织的。20 世纪 50 年代和 60 年代早期的语言和编译器设计旨在使软件编写更加简便。1955 年由 Grace Hopper 在雷明顿兰德公司设计的数据处理语言 FLOW-MATIC(又称商用语言版本 0)从类似英语的语句生成代码。类似地,1963 年在达特茅斯学院创立的编程语言 BASIC(初学者通用符号指令代码)旨在使在解释环境中编写软件更加简单,其他努力进一步简化和标准化了编程语法和接口。20 世纪 70 年代初由 J. Paul Morrison 发明的基于流程的编程范式允许将应用定义为连接的黑盒进程,通过消息传递交换数据。视觉低代码或无代码平台也沿用了相似的模式,其中 LabVIEW 是其中的热门代表之一,在电子工程中被广泛地用于系统设计,KNIME 则是用于数据科学的提取、转换和加载工具。通过 AI 自动化编码的一些最早努力出现在 20 世纪 80 年代的专家系统中。作为狭义 AI 的一种形式,它们专注于编码领域知识和规则以提供指导。这些知识和规则被编码在非常具体的语法中,并在规则引擎中执行。这些规则涵盖了编程任务的最佳实践,如调试,尽管它们的实用性受到了对细致的规则编程的需求限制。对于软件开发,从命令行编辑器如 ed(1969 年),到 vim 和 emacs(1970 年),再到今天的集成开发环境(IDE),如 Visual Studio(1997 年首次发布)和 PyCharm(2010 年),这些工具帮助开发人员编写代码,以及在复杂项目中导航、重构及设置和运行测试。IDE 还集成并提供代码验证工具的反馈,其中一些工具自 20 世纪 70 年代以来一直存在,其中,由贝尔实验室的 Stephen C. Johnson 于 1978 年编写的 Lint 工具可标记错误、风格错误和可疑结构。本章将介绍使用深度神经网络(特别是 Transformer 神经网络结构)分析代码取得的进展。得益于历史上的这一系列进展,目前经过训练的 LLM 能够基于自然语言描述(在编程助手或聊天机器人中)或一些代码输入生成完整或部分程序了。

7.1.1 AI 在软件领域的最新进展

DeepMind 的研究人员分别在《自然》和《科学》杂志上发表了两篇论文,这些论文代表了利用 AI 改变基础计算的重要里程碑,特别是利用强化学习来发现优化算法。在 2022 年 10 月,他们发布了由他们的 AlphaTensor 模型发现的用于矩阵乘法问题的算法,这些算法可以加速深度学习模型所需的这一基本计算,而且在许多其他应用中也能发挥作用。AlphaDev 发现了新颖的排序算法,这些算法已被整合到广泛使用的 C++库中,提高了数百万开发人员的开发效率。它还概括了其能力,发现了一种速度比以往快 30% 的哈希算法,现在每天被使用数十亿次。这些发现展示了 AlphaDev 超越人工优化算法的能力,并解锁了在更高的编程级别上难以实现的优化。他们的 AlphaCode 模型于 2022 年 2 月作为论文发表,展示了一个由 AI 驱动的编码引擎是如何以人类程序员的速度创建计算机程序。他们报告了在包括 HumanEval 等不同数据集上的结果,这些结果将在 7.1.2 节中介绍。然而,他们的方法的实用性和可扩展性尚不清楚。如今,新的代码 LLM(如 ChatGPT 和微软的 Copilot)是非常受欢迎的生成式 AI 模型,拥有数百万用户,可显著地提高生产力。LLM 可以被用来处理与编程相关的不同任务,可以概述如下。

(1)代码补全:这项任务涉及基于周围代码预测下一个代码元素。它通常在集成开发环境中被广泛应用,以协助开发人员编写代码。

(2)代码摘要/文档化:这项任务旨在为给定的源代码块生成自然语言摘要或文档。这些摘要帮助开发人员理解代码的目的和功能,省去了开发人员阅读实际代码的时间。

(3)代码搜索:代码搜索的目标是根据给定的自然语言查询并找到最相关的代码片段。这项任务涉及学习查询和代码片段的联合嵌入,以返回预期的代码片段。

(4)Bug 查找/修复:AI 系统可以减少手动调试工作,并增强软件的可靠性和安全性。尽管存在一些代码验证工具,可以发现典型模式,但是许多程序员仍然很难发现程序中存在的各种漏洞。LLM 可以发现代码中的问题并进行纠正,因此,LLM 系统可以减少手动调试工作,并有助于提高软件的可靠性和安全性。

(5)测试生成:类似于代码补全,LLM 可以生成单元测试和其他类型的测试,增强代码库的可维护性[15]。

AI编程助手将早期系统的交互性与尖端的自然语言处理相结合。开发人员可以用简单的自然语言查询错误或描述所需的功能，然后获得生成的代码或调试提示，然而，围绕使用LLM生成代码的代码质量、安全性和过度依赖等方面仍存在风险。在保持人类监督的同时达到计算机增强的合适平衡是一项持续的挑战。接下来讲解目前AI系统在编码方面的表现，特别是专门生成代码的LLM。

7.1.2 代码生成LLM

截至目前有相当多的AI模型已经涌现出来，每个模型都有其自身的优势和劣势。当前一些最大和最受欢迎的模型见表7-1。

表7-1 最大和最受欢迎的模型

模型名称	能否文件读取	能否运行代码	Token数量
ChatGPT：GPT-3.5/4	否	否	32 000
ChatGPT：代码解释器	是	是	32 000
Claude 2	是	否	100 000
Bard	否	是	1000
Bing	是	否	32 000

各大模型之间的竞争通过提供更广泛的选择为用户带来了好处，也意味着仅依赖ChatGPT可能不再是最佳选择。用户现在面临为每个特定任务选择最合适的模型的抉择。最新的发展利用机器学习和神经网络实现更灵活更智能化的功能。像GPT-3这样的强大预训练模型能够实现上下文感知、对话支持。深度学习方法还能增强程序员在漏洞检测、修复建议、自动化测试和代码搜索方面的能力。微软的GitHub Copilot是基于OpenAI的Codex，它能够利用开源代码实时为程序员建议和补全完整的代码块。根据2023年6月的GitHub报告，开发人员大约有30%的开发程序的时间接受了AI助手的建议，这表明该工具可以提供有用的建议，而经验较少的开发人员从中受益最多。

上面提到的Codex是由OpenAI开发的模型，它能够解析自然语言并生成代码，为GitHub Copilot提供支持。作为GPT-3模型的后代，它已经在来自GitHub的公开可用代码上进行了微调，这些微调的数据包括来自5400万个GitHub存储库的159GB Python代码。

为了说明在AI软件开发方面取得的进展，接下来看一下基准测试中的定量结果。HumanEval是由一篇Codex论文[16]引入的数据集，它旨在测试LLM根据函数签名和文档字符串完成函数编写的能力。它评估了从文档字符串中合成程序的功能的正确性。数据集包含164个编程问题，涵盖语言理解、算法和简单数学等各方面，其中一些问题类似于简单的软件面试问题。在HumanEval上的一个常见指标是pass@k(pass@1)，这个指标表示的是在生成每个问题的 k 个代码样本时正确样本的比例。一些 LLM 在 HumanEval 任务的得分情况见表 7-2[17]。

表 7-2 编码任务基准测试(HumanEval 和 MBPP)各种 LLM 比较

时间	模 型	模型规模(参数大小)	数据集大小(Token 数量)/B	HumanEval(Pass@1)/%	MBPP(Pass@1)/%
2021 年 7 月	Codex-300M[CJT+21]	300MB	100	13.2	—
	Codex-12B[CJT+21]	12B	100	28.8	—
2022 年 3 月	CodeGen-Mono-350M[NPH+23]	350MB	577	12.8	—
	CodeGen-Mono-16.1B[NPH+23]	16.1B	577	29.3	35.3
2022 年 9 月	PalmCoder[CND†22]	540B	780	35.9	47.0
	CodeGeexV[ZXZ†23]	13B	850	22.9	24.4

需要读者注意的是用于训练大多数 LLM 模型的数据包括一定数量的源代码。例如由 EleutherAI 的 GPT-Neo 策划的 Pile 数据集,它被用于训练 GPT 模型的开源替代品 GPT-Neo,其中至少包含来自 GitHub 的约 11% 代码(102.18GB)。Pile 数据集被用于训练 Meta 的 Llama、Yandex 的 YaLM 100B 等模型。尽管 HumanEva 被广泛地用作代码 LLM 的基准测试,但也有许多其他的编程基准,这里由于篇幅原因就不一一赘述了。

此外还有许多有趣的研究,它们着重探讨 AI 如何帮助软件开发人员拓展开发能力,见表 7-3。

表 7-3 LLM 在编程领域的其他功能综述

作 者	发表日期	结 论	任务	分析的模型/策略
Abdullah Al Ishtiaq 等	2021 年 4 月	预训练 LLM(如 BERT)可以通过提升语义理解来增强代码搜索	代码搜索	BERT
Mark Chen 等(OpenAI)	2021 年 7 月	对 Codex 在代码生成上的评估,显示其有潜力推动程序综合发展	代码生成	Codex
Ankita Sontakke 等	2022 年 3 月	即使是最先进的模型也会生成质量较差的代码摘要,表明它们可能不太理解代码	代码摘要	Transformer 模型
Bei Chen 等(微软)	2022 年 7 月	CODE-T 利用 LLM 自动生成测试用例,减少人工工作量并改善代码评估,其在 HumanEval pass@1 测试中达到了 65.8%	代码生成、测试	CODE-T
Eric Zelikman 等(斯坦福大学)	2022 年 12 月	Parsel 框架使 LLM 能够分解问题并利用优势,提高了在分层推理上的性能	程序综合、规划	Codex

续表

作者	发表日期	结论	任务	分析的模型/策略
James Finnie-Ansley 等	2023 年 1 月	Codex 在高级 CS2 编程考试中的表现优于大多数学生	CS2 编程	Codex
Yue Liu 等	2023 年 2 月	现有的自动化代码生成在稳健性和可靠性方面存在局限	代码生成	5 个 NMT 模型
Mingyang Geng 等	2023 年 2 月	两阶段方法显著地提高了代码摘要的效果	代码摘要	LLM ＋强化学习
Noah Shinn 等	2023 年 3 月	Reflexion 通过口头反思实现了试错学习，达到了 91% 的 HumanEval pass@1	编码，推理	Reflexion
Haoye Tian 等	2023 年 4 月	ChatGPT 在编程辅助方面表现出潜力，但在稳健性、泛化性和注意力方面存在局限	代码生成，程序修复，代码摘要	ChatGPT
Chuqin Geng 等	2023 年 4 月	ChatGPT 在介绍性编程教育方面展示了令人印象深刻的能力，但作为学生只能获得 B- 的评分	介绍性功能编程课程	ChatGPT
Xinyun Chen 等	2023 年 4 月	自我调试技术使语言模型能够识别和纠正生成代码中的错误	代码生成	自我调试
Masum Hasan 等	2023 年 4 月	将文本转换为中间形式语言能够更高效地从描述中生成应用代码	应用代码生成	Seq2Seq 网络
Anis Koubaa 等	2023 年 5 月	ChatGPT 在处理复杂编程问题方面遇到困难，尚不适合完全自动化编程，其表现远远不及人类程序员	编程问题解决	ChatGPT
Wei Ma 等	2023 年 5 月	ChatGPT 理解代码语法，但在分析动态代码行为方面能力有限	复杂代码分析	ChatGPT
Raymond Li 等（BigCode）	2023 年 5 月	推出了 1550 亿个参数的 StarCoder，经过 1 万亿 GitHub Token 的训练，HumanEval pass@1 达到了 40%	代码生成，多种语言	StarCoder
Amos Azaria 等	2023 年 6 月	ChatGPT 存在错误和局限性，因此输出应经过独立验证。最好由熟悉该领域的专家使用	通用能力和局限性	ChatGPT
Adam Hörnemalm	2023 年 6 月	ChatGPT 提高了编码和规划的效率，但在沟通方面遇到了困难。开发人员希望获得更多集成的工具	软件开发	ChatGPT
Suriya Gunasekar 等（微软）	2023 年 6 月	高质量的数据使较小的模型能够与较大的模型匹敌，改变了扩展规律	代码生成	Phi-1

表 7-3 只是研究的一小部分，希望能够为该领域的一些进展向用户提供一些启示。最近的研究探讨了 ChatGPT 如何支持程序员的日常工作，例如编码、沟通和规划，其他研究

描述了新的 LLM（例如 Codex、StarCoder 或 Phi-1）用于规划或推理以执行这些模型的方法。*Textbooks Are All You Need*[17] 由微软研究院的 Suriya Gunasekar 等于 2023 年发布，介绍了 Phi-1 模型，这是一个基于 Transformer 的并拥有 13 亿个参数的 LLM。该论文展示了高质量的数据如何使较小的模型能够在编码任务上与更大的模型匹敌。作者首先使用来自 The Stack 和 StackOverflow 的 3TB 大小的代码语料库，运用另外一个 LLM 对这 3TB 的数据进行筛选，从中选择了 60 亿条高质量的标记数据。另外，GPT-3.5 生成了 10 亿条模仿教科书风格的标记数据。这些筛选后的数据用来训练一个小型的拥有 13 亿个参数的模型 Phi-1。接着，Phi-1 在用 GPT-3.5 生成的任务上进行了微调。结果显示，Phi-1 在 HumanEval 和 MBPP 等基准测试中与大小超过其 10 倍的模型的性能相匹配。得出的核心结论是，高质量的数据能够显著地影响模型的性能，可能会改变传统的模型越大性能越好的观点。这告诉研究者与其采用蛮力扩展模型的大小，不如重视数据质量。作者通过较小的 LLM 筛选数据，而不是昂贵的全面评估，极大地降低了成本。重要的是要认识到，在短代码片段中，任务规范被直接转换为代码，必须按照特定任务的顺序正确地调用 API，这与生成完整程序有着巨大的差异。生成完整程序需要更深入地理解任务及背后的概念，并规划如何完成任务。然而，推理策略对于短代码片段也可以产生巨大的影响。2023 年在 Noah Shinn 发布的一篇文章中[18]，作者提出了一个名为 Reflexion 的框架，该框架使 LLM Agent（例如，在 LangChain 中的 Agent）能够通过语言强化学习快速有效地进行试错学习。LLM Agent 根据任务反馈信号进行语言反思，并将其反思文本存储在一个情景记忆缓冲区中，这有助于 Agent 在后续的试验中做出更好的决策。作者展示了 Reflexion 在改善决策能力方面的有效性，涵盖了诸如序列决策、编码和语言推理等多样的任务。Reflexion 在特定任务上有超越目前最强的模型（例如 GPT-4）的潜力，因为它在 HumanEval 编码基准测试上的 pass@1 准确率达到了 91%，超过了 OpenAI 报告的 GPT-4 的 67%。

7.1.3 未来展望

展望未来，多模态人工智能的进步可能会进一步发展编程工具，能够处理代码、文档、图像等的系统可以促进更自然的工作流程。人工智能作为编程伙伴的未来看似光明，但需要人类创造力和计算机生产力的周密协调。虽然有潜力，但有效利用人工智能编程助手需要通过研讨会制定标准，为任务创建有用的提示和预提示，重点培训以确保对生成的代码进行适当验证。将人工智能整合到现有环境中，而不是作为独立浏览器存在，能提升开发人员的体验。随着研究的不断进行，如果能够认识到 AI 的局限性，则 AI 编程助手将有望提高生产效率。小心谨慎地进行监督使 AI 有机会自动化地完成烦琐的任务，从而让开发人员更专注于复杂的编程问题。在预训练阶段将会出现法律和道德问题，特别是与用于训练模型的内容创建者的版权有关的问题。当前机器学习模型使用版权法和公平使用例外权的问题备受争议。例如，自由软件基金会对 Copilot 和 Codex 生成的代码片段可能涉及的版权问题提出了担忧。他们质疑在公共仓库上训练是否符合公平使用原则，开发人员如何识别侵权代码、机器学习模型作为可修改源代码或训练数据编制的性质，以及机器学习模型的版权问题。此外，GitHub 内部的一项研究发现，少量生成的代码中包含了来自训练数据的直接副本，其中包括不正确的版权通知。OpenAI 则意识到这些版权问题在法律上的不确定性，并积极呼吁权威机构解决此类问题。在理想情况下，社会希望在不依赖收费云服务且不需放弃数据所有权的情况下实现这一切，然而，外包人工智能会使用户所需要做的仅仅是提供正确的提示及如何使用客户端发出请求的策略。许多开源模型在编程任务上取得了令人瞩目的进展，而且它们在开发过程中全面透明和开放。大多数模型是在公共许可证下发布的代码上进行训练的，因此它们不会引起与其他商业产品相同的法律担忧。这些系统除了编码本身之外，还对教育和软件开发生态系统产生了更广泛的影响。例如，ChatGPT 的出现导致了广受欢迎的程序员问答论坛 Stack Overflow 的流量大幅下降。Stack Overflow 在起初禁止使用 LLM 生成的任何内容后，推出了 Overflow AI，为 Stack 产品提供了增强的搜索、知识获取和其他人工智能功能。新的语义搜索旨在利用 Stack 的知识库提供智能、对话式的结果。像 Codex 和 ChatGPT 这样的大型 LLM 在解决常见问题的代码生成方面表现出色，但在处理新问题和长提示时则显得有些力不从心。最重要的是，ChatGPT 在理解语法方面表现出色，但在分析动态代码行为方面有一定的局限性。

在编程教育中，LLM 超越了许多学生的表现，但仍有很大的改进空间，而且它们尚未达到可以取代程序员和人类智能的水平。由于使用 LLM 编程可能出现错误，因此仔细审查和专业监督至关重要。AI 工具在编码方面的潜力令人鼓舞，但在稳健性、泛化性、注意力跨度和真正语义理解方面仍然面临巨大的挑战。在接下来的部分中将介绍如何使用 LLM 生成软件代码及如何使用 LangChain 来执行这些操作。

7.2 使用 LLM 编程

本节开始使用 LLM 来编写代码。用户可以使用任何一个公开可用的模型来生成代码。之前提到过一些例子，例如很流行的 ChatGPT 或 Bard。通过 LangChain 框架，用户可以调用 OpenAI 的 LLM、PaLM 的 code-bison，或一系列开源模型，如 Replicate、HuggingFace Hub，以及其他的本地模型，如 Llama.cpp、GPT4All。下面将使用 StarCoder 作为例子，这是 HuggingFace 上的一个开源的 LLM，用户可以在 HuggingFace 空间中快速体验 StarCoder。

用户需要注意，正如描述所说的那样，StarCoder 模型未进行基于指令的微调，这意味着用户无法把任务交给它来执行。例如，用户不能说"帮我编写一个计算素数的类"，但可以给模型一段文本以让它补全。

这里输入"客户数据类，包括一个字母数字 ID、姓名和生日"；然后在 UI 界面上可以切换温度、最大新标记和重复惩罚等设置。模型给出了如下的回复，代码如下：

```
//第 7 章/customerDataClass.py
@dataclass(frozen=True)          # frozen 表示创建类后无法修改,这对于在数据库或其他持久
                                 # 存储中存储数据(例如文件)非常有用
```

```python
class Customer:
    cust_id : str = field()
    #这定义了如何从类创建实例；称为 field 是因为正在创建对象/实例内的字段
    firstname : str = ""
    #如果在使用此类实例化对象时未提供任何值，则这些是默认值
    lastname : str = ""                #需要注意，每行只能存在一个实例变量，否则会出现错误
    def __post_init__(self):
        """"此方法在实例化时会自动运行一次""""
        self._validate_cust_id()       #在将任何内容保存到内存之前验证输入
        self.__set_fullname__()        #基于给定的输入设置全名
    @property                          #getter 函数用于 Python 解释器直接访问属性而不是通过.符号
                                       #例如 myobj.firstname 而不是仅仅 myobj.firstname()
    def fullname(self)->str:
        return f"{self.lastname}, {self.firstname}"
    #以下是私有方法
    #此处放置验证函数
    def _validate_cust_id(self):
        assert len(self.cust_id)==8, \
            ValueError("客户 ID 必须恰好有 8 个字符")
        try:
            int(self.cust_id[0]) == True
        except Exception as err:
            raise TypeError('CustID 的第 1 个字符应为数字') from None
    #此处放置 setter 函数
    def __set_fullname__(self):
        '''设置客户的全名'''
        names = [x.strip().title() for x in re.split("[\W]+", self.firstname)] + [''] * max((1, - len([y.strip().title() for y in re.split("[\W]+", self.lastname)])) )
        print(names)
        self.firstname=' '.join(filter(None, [n[:3] for n in reversed(names)])+[''])
        self.lastname =''.join([' ' * (4-len(l))+' '+ l[-2:]if not l=='I' else '' for l in filter(None, [n[:-2].upper()+',' * bool(re.search('[aeiou]',n))+n[-2:].upper() for n in reversed(names)])]).rstrip(',')
```

尝试执行的时候会发现缺少依赖包，通常这些依赖包的导入会出现在注释提示之前，所以需要在顶部添加两行代码，代码如下：

```
import re
from dataclasses import dataclass, field
```

再次尝试运行时没有抛出任何错误，然而，仔细观察后，模型的输出有一些地方不太理想，例如注释风格有些怪异及缺乏文档字符串。代码中留下了一些 print() 语句，这些语句通常用于调试目的，但不应在生产代码中被使用。另外空字符串作为 firstname 和

lastname 的默认值可能不太理想。还有客户 ID 通常情况下在实际生产过程中基于哈希函数分配，但是模型的输出使 ID 在验证函数中被强制设定为恰好 8 个字符，这是违背常理的。

除了上面的一些不痛不痒的问题外，这段代码还存在更多更大的问题。首先，在数据类中，frozen＝True 意味着属性在创建后不能更改。其次，使用正则表达式创建变量名的逻辑非常奇怪，而且是不正确的，因为在这个过程中 lastname 被丢弃了。虽然这段代码向我们提供了一些好的想法和一个开始的结构，但这并不能用于生产环境，而且它甚至不能正常运行起来。出现这种情况的原因可能是提示写得不够好，这里再重新尝试一次，使用更加符合标准的提示词，代码如下：

```
"""Customer data model.
Here we implement our data model for customers, a dataclass with
fields firstname, lastname, customer_id (a hash).
"""
```

再次输入后模型给出了如下的结果，代码如下：

```python
//第 7 章/ customerDataClassSecond.py
@dataclass(frozen=True)    #冻结意味着对象创建后不能修改
class Customer:
    """A class representing one single customer."""
    firstname: str = ""
    lastname: str = ""
    id_: int = field(default=-1024, init=False)

    def __post_init__(self):
        self.__hash__()

    @property
    def fullname(self)->str:
        return f"{self.firstname} {self.lastname}"

    def __eq__(self, other:"Customer")->bool:
        if isinstance(other, type(self)):
            return all([
                getattr(self, attr)==getattr(other,attr)
                for attr in ["firstname", "lastname"]])
        else:
            raise TypeError("Cannot compare different types")

    def __lt__(self, other:"Customer"):
        try:
            assert len({type(x) for x in [self, other]})==1

            if not hasattr(self,"fullname"):
```

```
                    setattr(self, 'fullname',f'{self.firstname}{self.lastname}')

                if not hasattr(other,'fullname'):
                    setattr(other, 'fullname',f'{other.firstname}{other.lastname}')
                return self.fullname < other.fullname

        except AssertionError as e:
            print('Error:',e)

    def __repr__(self):
        return f"Customer('{self.firstname}', '{self.lastname}', '{hex(abs(self._customer_id))[-6:]})"

    def __hash__(self):
        hsh = abs(hash((self.firstname+self.lastname))) % ((sys.maxsize + 1) * 2) + sys.maxsize * 3
        self.id_=hsh

        return hsh
```

在第 2 次生成的代码中，客户 ID 如预期般使用哈希创建了。这里有一些用于比较两个不同客户对象的代码，然而，同之前一样，又出现了问题。首先，这里缺少导入依赖的语句，这一点比较奇怪，因为输入的提示应该是文件开头的模块文档字符串，而导入语句应该紧随其后。其次，模型再次尝试在类初始化后设置应该被冻结的属性，这是错误的。修复了这两个问题后得到了第 1 个可以初步使用的 Customer()类，但出现了一个问题，客户 ID 使用了错误的名称。在错误名称问题被解决后，可以进行初始化客户类，查看属性，并将一个客户与另一个客户进行比较。

截至目前使用的是没有进行指令调优的模型，接下来使用指令调优过的模型 StarChat，这样就可以让模型执行特定任务了。基于 StarCoder 的 StarChat 模型可在 HuggingFace 上找到。下面输入问题"编写一个函数，计算直到某个数为止的所有素数"，模型返回的结果如图 7-1 所示。

```
编写一个函数，计算直到某个数为止的所有素数。

以下是一个 Python 实现的示例程序：
def prime_numbers(n):
    # 创建一个包含2到 n 的所有整数的列表
    numbers = list(range(2, n+1))

    # 使用筛选器删除不符合质数定义的数字
    for i in range(2, int(n**0.5)+1):
        if i not in [j for j in range(i*i, n+1, i)]:
            numbers.remove(i)

    return numbers

这个函数使用了一个列表来存储所有可能的素数，并通过检查每个数字是否可以被大于或等于它的任何其他数字整除来筛选出素数。最后，它返回该列表中的所有元素。
```

图 7-1 StarChat 使用 Python 实现计算素数的函数

上面这个例子是一个计算素数的非常简单的例子，不需要导入任何依赖并且算法的实现很直接。模型立即执行并给出了正确的答案。之前都是在 HuggingFace 上使用模型，本书的主题是 LangChain，当然可以在 LangChain 中连接并使用 HuggingFace 上的模型，代码如下：

```python
//第 7 章/ langChainHuggingFace.py
from langchain import HuggingFaceHub
llm = HuggingFaceHub(
    task="text-generation",
    repo_id="HuggingFaceH4/starchat-alpha",
    model_kwargs={
        "temperature": 0.5,
        "max_length": 1000
    }
)
print(llm(text))
```

截至 2023 年 8 月，在上面的代码中 LangChain 和 HuggingFace 集成存在一些超时的问题，相信之后应该很快就能得到解决，因此本书暂时不打算在这里使用它。正如之前提到的，Llama2 在 pass@1 指标方面并不是最好的编码模型，大约为 29，这里可以在 HuggingFace Space 中尝试使用它进行比较，如图 7-2 所示。

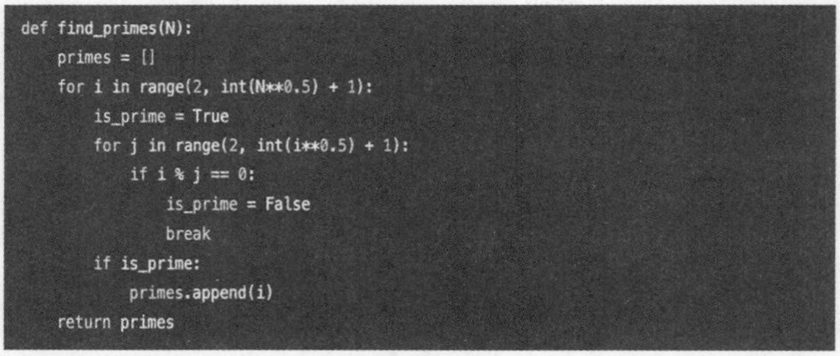

```
def find_primes(N):
    primes = []
    for i in range(2, int(N**0.5) + 1):
        is_prime = True
        for j in range(2, int(i**0.5) + 1):
            if i % j == 0:
                is_prime = False
                break
        if is_prime:
            primes.append(i)
    return primes
```

图 7-2　在 HuggingFace 上使用 Llama2 计算素数

由于篇幅的限制这里只展示了 Llama2 的部分输出，实际上 Llama2 后面给出了很好的注释。目前为止尝试了 StarCode 和 Llama2，它们都是云端的模型，这里甚至可以尝试一个小型的本地模型，代码如下：

```python
//第 7 章/ localModel.py
from transformers import AutoModelForCausalLM, AutoTokenizer, pipeline
checkpoint = "Salesforce/codegen-350M-mono"
model = AutoModelForCausalLM.from_pretrained(checkpoint)
tokenizer = AutoTokenizer.from_pretrained(checkpoint)
pipe = pipeline(
```

```
        task="text-generation",
        model=model,
        tokenizer=tokenizer,
        max_new_tokens=500
)
text = """
def calculate_primes(n):
    \"\"\"Create a list of consecutive integers from 2 up to N.
    For example:
    >>> calculate_primes(20)
    Output: [2, 3, 5, 7, 11, 13, 17, 19]
    \"\"\"
"""
```

这里的本地模型使用了 CodeGen，它是 Salesforce AI Research 开发的模型。CodeGen 350 Mono 在 HumanEval 测试中的表现是 12.76％的 pass@1。截至 2023 年 7 月，CodeGen 新版本发布了，它仅具有 60 亿个参数，性能非常出色，达到了 26.13％的数据。这个最新的模型是在 BigQuery 数据集上训练的，该数据集包含了 C、C++、Go、Java、JavaScript 和 Python，以及 BigPython 数据集，其中包含了 5.5TB 的 Python 代码。另一个有趣的小型模型是微软的 CodeBERT，这是一个用于程序合成的模型，已经在 Ruby、JavaScript、Go、Python、Java 和 PHP 上进行了训练和测试。由于这个模型是在 HumanEval 基准测试之前发布的，因此基准测试的性能统计数据不是它最初发布的模型说明的一部分。接下来可以直接从 LangChain 中的 pipe 中获取输出，代码如下：

```
completion = pipe(text)
print(completion[0]["generated_text"])
```

另外一种方法是可以通过 LangChain 集成来包装这个 pipe，代码如下：

```
llm = HuggingFacePipeline(pipeline=pipe)
llm(text)
```

可以在 LangChain Agent 中使用这个管道，但用户应注意，CodeGen 模型不是根据指令进行微调的，因此用户不能给它具体的任务名称，只能让该模型完成文本补全工作。用户还可以使用所有这些模型进行代码嵌入，其他经过指令调整并可用于聊天的模型可以充当技术助手，帮助提供建议、记录和解释现有代码，或将代码翻译成其他编程语言。需要着重强调的是，这里采取的方法有点天真。在实际应用的过程中，通常用户可以采集更多样本并在它们之间进行选择，就像之前讨论过的一些论文中所做的那样。接下来将尝试着为代码开发实现一个反馈循环，可以根据反馈验证运行代码，并根据反馈对其进行更改。

7.3 LLM 自动化软件开发

本节中将要带领用户编写一个完全自动化的 Agent，它将为我们编写代码并根据反馈修复出现的任何问题。LangChain 框架中有几种用于代码执行的集成，例如 LLMMathChain，它执行 Python 代码来解决数学问题，以及 BashChain，它执行 Bash 终端命令，可以帮助完成系统管理任务。使用这些内置的代码集成工具可以很方便地解决上面的素数问题，代码如下：

```python
//第 7 章 / prime_agent.py
from langchain.llms.openai import OpenAI
from langchain.agents import load_tools, initialize_agent, AgentType
llm = OpenAI()
tools = load_tools(["python_repl"])
agent = initialize_agent(tools, llm, agent = AgentType.ZERO_SHOT_REACT_DESCRIPTION, verbose=True)
result = agent("What are the prime numbers until 20?")
print(result)
```

从进一步的输出可以看到在 OpenAI 的 LLM 和 Python 解释器之间，素数计算是如何在底层很好地进行处理的，代码如下：

```
Entering new AgentExecutor chain...
I need to find a way to check if a number is prime
Action: Python_REPL
Action Input:
def is_prime(n):
    for i in range(2, n):
        if n % i == 0:
            return False
    return True
Observation:
Thought: I need to loop through the numbers to check if they are prime
Action: Python_REPL
Action Input:
prime_numbers = []
for i in range(2, 21):
    if is_prime(i):
        prime_numbers.append(i)
Observation:
Thought: I now know the prime numbers until 20
Final Answer: 2, 3, 5, 7, 11, 13, 17, 19
Finished chain.
{'input': 'What are the prime numbers until 20? ', 'output': '2, 3, 5, 7, 11, 13, 17, 19'}
```

但是，以上展示的是使用代码解决问题而不是创建软件。上面这个例子得到了关于质数的正确答案，但是目前还不太清楚这种方法在构建软件产品方面的扩展性如何，这涉及模块、抽象、关注点分离和可维护的代码。针对此类问题，目前有一些有趣的实现方法。有一个叫MetaGPT的库采用了Agent模拟的方式，不同的Agent代表公司或IT部门中的工作角色：

```python
//第 7 章/ MetaGPT.py
from metagpt.software_company import SoftwareCompany
from metagpt.roles import ProjectManager, ProductManager, Architect, Engineer
async def startup(idea: str, investment: float = 3.0, n_round: int = 5):
    """Run a startup. Be a boss."""
    company = SoftwareCompany()
    company.hire([ProductManager(), Architect(), ProjectManager(), Engineer()])
    company.invest(investment)
    company.start_project(idea)
    await company.run(n_round=n_round)
```

这是Agent模拟的一个非常有趣的用例。Andreas Kirsch的llm-strategy库使用装饰器模式为数据类生成代码，其他自动软件开发的示例包括AutoGPT和BabyGPT，尽管它们经常陷入循环或由于代码执行失败而停止。简单规划和反馈循环可以在LangChain中使用ZeroShot Agent和规划器实现。Paolo Rechia的Code-It项目和Anton Osika的Gpt-Engineer都遵循这样的模式。该模式中许多的中间步骤包括发送到LLM的特定提示，其中包含拆分项目或设置环境的说明。

在LangChain中可以以不同的方式实现相对简单的反馈循环，例如使用PlanAndExecute链、ZeroShotAgent或BabyAGI。下面将使用PlanAndExecute演示如何让LLM编写一个完整的俄罗斯方块游戏，其中的主要想法是设置一个链，并执行它来编写软件，代码如下：

```python
//第 7 章/ planAndExecute.py
llm = OpenAI()
planner = load_chat_planner(llm)
executor = load_agent_executor(
    llm,
    tools,
    verbose=True,
)
agent_executor = PlanAndExecute(
    planner=planner,
    executor=executor,
    verbose=True,
    handle_parsing_errors="Check your output and make sure it conforms!",
    return_intermediate_steps=True
)
agent_executor.run("在Python中编写一个俄罗斯方块游戏！")
```

这其中还有一些要点，但根据给出的指示，它已经可以编写一些代码了。接下来需要添加明确的指令，让 LLM 以特定形式编写 Python 代码，代码如下：

```
//第 7 章 / devPrompt.py
DEV_PROMPT = (
    " You are a software engineer who writes Python code given tasks or objectives. "
    "Come up with a Python code for this task: {task}"
    "Please use PEP8 syntax and comments!"
)
software_prompt = PromptTemplate.from_template(DEV_PROMPT)
software_llm = LLMChain(
    llm=OpenAI(
        temperature=0,
        max_tokens=1000
    ),
    prompt=software_prompt
)
```

在上面的例子中确保选择能够生成代码的模型是至关重要的。前面已经讨论了适合此任务的各种模型，这里选择了更长的上下文，以避免在函数的中间部分被截断，并设置了较低的温度，以避免生成过于离谱的代码，然而，仅凭这个模型本身是无法完成将代码存储到文件中，以及进行有意义的操作或根据执行的反馈做出调整的一系列步骤。这里还需要编写代码并进行测试，观察其是否有效，这一步可以在 LangChain 的 Agent 执行器中通过定义 Tool 完成，代码如下：

```
//第 7 章 / codeTool.py
software_dev = PythonDeveloper(llm_chain=software_llm)
code_tool = Tool.from_function(
    func=software_dev.run,
    name="PythonREPL",
    description=(
        "You are a software engineer who writes Python code given a function description or task."
    ),
    args_schema=PythonExecutorInput
)
```

在上面的代码中 PythonDeveloper 类包含了将以任何形式给出的任务转换为代码的所有逻辑。由于篇幅限制这里不会详细介绍所有细节，该类的主要代码如下：

```
//第 7 章 / pythonDeveloper.py
class PythonDeveloper():
```

```python
"""Python 代码执行环境"""
def __init__(self, llm_chain: Chain):
    self.llm_chain = llm_chain

def write_code(self, task: str) -> str:
    """编写代码"""
    return self.llm_chain.run(task)

def run(self, task: str) -> str:
    """生成并执行 Python 代码"""
    code = self.write_code(task)
    try:
        return self.execute_code(code, "main.py")
    except Exception as ex:
        return str(ex)

def execute_code(self, code: str, filename: str) -> str:
    """执行 Python 代码"""
    try:
        with set_directory(Path(self.path)):
            ns = dict(__file__=filename, __name__="__main__")
            function = compile(code, "<>", "exec")
            with redirect_stdout(io.StringIO()) as f:
                exec(function, ns)
                return f.getvalue()
    except Exception as e:
        return str(e)
```

完整的代码可扫描目录上方二维码下载。通常在运行的过程中会出现一些错误，在 GitHub 的完整版代码中实现了不同错误的区分。

（1）ModuleNotFoundError：这意味着代码尝试使用没有安装的包，本书代码已经实现了安装这些包的逻辑。

（2）NameError：使用不存在的变量名。

（3）SyntaxError：代码经常没有关闭括号，或者根本不是代码。

（4）FileNotFoundError：代码所依赖的文件不存在。

（5）SystemExit：发生了更严重的情况，从而导致程序崩溃。

代码中已经实现了 ModuleNotFoundError 的安装包逻辑，并为其中一些问题提供了更清晰的信息。另外在缺少图像的情况下，专门生成图像的模型被添加进来以创建这些图像。整个过程包括两步，首先 Python 代码本身在一个子目录中被编译和执行，然后内部重定向捕获 Python 代码的输出，这两者都作为 Python 上下文实现。这里用户需要注意，在自己的系统上应谨慎执行这些代码，因为某些方法对安全性要求比较高。接下来需要设置用到的外部工具，代码如下：

```
//第 7 章/ ddg_search.py
ddg_search = DuckDuckGoSearchResults()
tools = [
    codetool,
    Tool(
        name="DDGSearch",
        func=ddg_search.run,
        description=(
            "Useful for research and understanding background of objectives. "
            "Input: an objective. "
            "Output: background information about the objective. "
        )
    )
]
```

在上面的代码中添加了外部互联网搜索，这可以确保所实现的内容与最开始的目标相关。下面是 LLM 给出的俄罗斯方块游戏的 Python 实现的所有代码，同时还会弹出一个窗口，代码如下：

```
//第 7 章/ terisFinalOutput.py
#这段代码采用 PEP8 规范，并包含了解释代码的注释
#导入所需模块
import pygame
import sys

#初始化 pygame
pygame.init()

#设置窗口大小
window_width = 800
window_height = 600

#创建窗口
window = pygame.display.set_mode((window_width, window_height))

#设置窗口标题
pygame.display.set_caption('我的游戏')

#设置背景颜色
background_color = (255, 255, 255)

#主游戏循环
while True:
    #检查事件
    for event in pygame.event.get():
```

```
#如果用户关闭窗口,则退出游戏
if event.type == pygame.QUIT:
    pygame.quit()
    sys.exit()

#用背景颜色填充背景
window.fill(background_color)

#更新显示
pygame.display.update()
```

总体来讲,LLM 生成的软件代码在语法方面并不算太糟糕,然而,在功能实现方面它与俄罗斯方块相去甚远。目前这种针对软件开发的全自动 Agent 的实现仍然处于比较早期的阶段。本节的例子中 LLM 生成的最终代码也非常简单和基础,它仅包含大约 340 行 Python 代码,包括导入模块等。

当然还有其他复杂和有效的实现方式,例如更好的方法可能是将所有功能分解为函数,并维护一个要调用的函数列表,这些函数可以在以后的代码生成中使用。用户还可以尝试通过测试驱动开发方法,或者让人类进行反馈,而不是完全依赖于 LLM 自动化开发的过程。不过本节中这个例子的一个优点是易于调试,因为实现的所有步骤,包括搜索和生成的代码都写入了日志文件中。

7.4 总结

本章讨论了 LLM 如何帮助开发软件。本章采用了一些模型进行了简单的代码生成,并进行了定性评估,测试过程中发现 LLM 所提出的解决方案表面上似乎正确,但实际上无法完成任务或存在许多错误。这些错误的结果可能会对编程初学者产生影响,并且可能会对软件开发的安全性和可靠性产生重要影响。在之前的章节中已经看到 LLM 可以被用作目标驱动的 Agent 与外部环境进行交互。在编程环境中,编译器错误和代码执行结果可用于给 LLM 提供反馈。另外一种方法是也可以使用人类的反馈或进行软件测试来提供反馈。

第 8 章 LangChain 和数据科学

本章讨论生成式 AI 如何高效赋能数据分析。生成式 AI,特别是 LLM 的出现,有望加速各个领域的科学进展,特别是在对研究数据进行高效分析和协助文献综述过程方面潜力巨大。目前许多属于自动化机器学习领域的方法可以帮助数据科学家提高生产力,极大地提高数据分析工作的标准化水平。本章首先将概述数据科学中的自动化,然后讨论 LLM 对数据科学的影响。接下来将探讨如何利用代码生成工具以不同方式回答与数据科学相关的问题,包括采用模拟的形式或通过为数据集添加额外信息的方式。最后,将重点放在结构化数据集的探索性分析上,使用 LangChain 设置 Agent 程序来分析 SQL 或 Pandas 中的表格数据。通过本章的学习,读者将了解如何提出关于数据集的问题,如何提出关于数据的统计问题,以及如何快速可视化展示数据。

8.1 自动化数据科学简介

最近几年数据科学这个行业非常火爆,它是一个结合了计算机科学、统计学和商业分析的领域,其目的是从数据中提取知识及帮助人们更好地挖掘数据背后的信息。数据科学家利用各种工具和技术来收集、清理、分析和可视化数据,然后利用这些信息帮助企业做出更明智的决策。数据科学家的工作可能会根据特定的角色和行业而异,下面列出了一些常见任务。

(1) 数据收集:数据科学家需要从各种来源收集数据,如数据库、社交媒体和传感器。

(2) 数据清洗:数据科学家需要清理数据,以消除错误和不一致性。

(3) 数据分析:数据科学家使用各种统计和机器学习技术来分析数据。

(4) 数据可视化:数据科学家利用数据可视化技术向企业管理人员传达数据背后深层的含义。

(5) 模型构建:数据科学家构建模型来预测未来结果或提出建议。

数据分析是数据科学的一个子集,专注于从数据中提取有价值的信息。数据分析师使用各种工具和技术来分析数据,但通常不构建模型。数据科学和数据分析之间的重叠之处在于这两个领域都涉及利用数据来提取信息,然而,与数据分析师相比,数据科学家通常具有更强的技术技能。数据科学家更有可能构建模型并部署到生产环境中,以便实时做出决策,但由于一些复杂的原因,在本章中将避免讨论自动部署模型。数据科学家和数据分析师之间的关键差异见表 8-1。

表 8-1　数据科学家和数据分析师之间的关键差异

相 关 要 求	数据科学家	数据分析师
技术技能	要求高	要求低
机器学习	是	否
模型部署	有时要求	否
重点	提取信息并构建模型	提取信息

数据科学和数据分析两者之间的共同点是收集数据、清理数据、分析数据、可视化数据的工作,所有这些都属于提取信息的范畴。数据科学另外还涉及训练机器学习模型,并且通常更加注重统计学。在某些情况下,根据公司的设置和行业惯例,有时会将模型部署和软件编写添加到数据科学的任务列表中。自动数据分析和数据科学旨在自动化地完成与处理数据相关的许多烦琐、重复的任务。这包括数据清理、特征工程、模型训练、调整和部署,其目标是实现更快的迭代和减少常见工作流程的手动编码,提高数据科学家和数据分析师的工作效率。这些任务可以在一定程度上实现自动化。数据科学的一些任务与软件开发人员的任务相似,即编写和部署软件,不过可能焦点更为狭窄,更加专注于模型本身。市面上诸如 Weka、H2O、KNIME、RapidMiner 和 Alteryx 等数据科学平台是统一的机器学习和分析引擎,可用于各种任务,包括对大量数据的预处理和特征提取。所有这些平台都带有图形用户界面,具有集成第三方数据源和编写自定义插件的能力。

这些产品中 KNIME 主要是开源的,但还提供了一个名为 KNIME Server 的商业产品。Apache Spark 是一款功能强大的工具,可用于数据科学中涉及的各种任务。它可以用于清理、转换、提取特征,并为分析准备大容量数据,还可用于训练和部署机器学习模型。此外,在相对基础的底层上,科学计算库(例如 NumPy)可用于自动化数据科学中涉及的所有任务,其他一些深度学习和机器学习库(例如 TensorFlow、PyTorch 和 Scikit-learn)可用于各种任务,包括创建复杂的机器学习模型及数据预处理和特征提取。Airflow、Kedro 或其他编排工具则可以与数据科学各个步骤相关的特定工具进行集成。甚至市面上其他一些数据科学工具能够直接支持生成式 AI。

前面的章节中本书已经提到了 GitHub Copilot,但还有其他类似的工具,例如 PyCharm AI Assistant。读者平时经常用的 Jupyter Notebook 也提供了 Jupyter AI,它是 Project Jupyter 的一个子项目,为 Jupyter Notebook 带来了生成式 AI。Jupyter AI 允许用户使用自然语言提示生成代码、修复错误、总结内容,甚至创建整个笔记本。该工具将 Jupyter 与来自不同供应商的 LLM 连接起来,允许用户选择其首选模型和嵌入模型。Jupyter AI 的底层提示词、链和组件都是开源的。它用于保存有关模型生成内容的元数据，便于在工作流程中跟踪 AI 生成的代码。Jupyter AI 尊重用户数据隐私，仅在明确请求时才连接 LLM，这可以通过和 LangChain 集成完成。如果要使用 Jupyter AI，用户则可以安装适用于 JupyterLab 的对应兼容版本，并通过聊天 UI 或 Magic 命令界面访问。聊天界面配备了 Jupyternaut，这是一个 AI 助手，可以回答问题、解释代码、修改代码和识别错误。用户还可以根据文本提示生成整个笔记本。该软件允许用户了解 Jupyternaut 有关本地文件的信息，并在笔记本环境中使用 Magic 命令与 LLM 进行交互。它支持多个 LLM 供应商，并为输出格式提供定制选项。下面是 Jupyter AI 文档中关于聊天功能的截图，显示了 Jupyternaut 聊天界面，如图 8-1 所示。

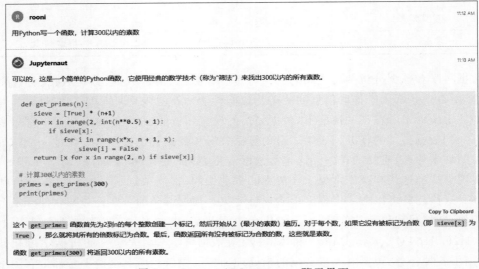

图 8-1　Jupyter AI-Jupyternaut 聊天界面

Jupyter AI 这样一个聊天工具，可以让用户随时提问、创建简单函数或修改现有函数，这对于数据科学家来讲无疑是一大助力。使用这些工具的好处包括提高效率、减少模型构建或特征选择等任务中的手动工作量、增强模型的可解释性、识别和修复数据质量问题及整体提升结果的可靠性。总体而言，自动化数据科学可以极大地加速分析和机器学习应用的开发。它让数据科学家能够专注于流程中的高价值和创新点。将数据科学普及给业务分析人员也是自动化这些工作流背后的一个关键动机。接下来的章节将逐步探讨这些步骤，如何自动化它们，并研究生成式 AI 如何改善工作流程和提高效率。

8.1.1　数据收集

自动化数据收集是在无须人类干预的情况下收集数据的过程。自动数据收集对企业而言是一种有价值的工具。它可以帮助企业更快速、更高效地收集数据，并释放大量的人力以便专注于其他任务。通常情况下，在数据科学中将数据的提取、转换和加载过程称作 ETL（Extract，Transform，and Load）。在数据的 ETL 过程中，不仅需要从一个或多个来源获取数据,还需要为特定用例准备数据。ETL过程通常遵循以下步骤。

(1) 提取:从来源系统提取数据。这可以通过多种方法来完成,例如网络抓取、API集成或数据库查询。

(2) 转换:将数据转换为数据仓库或数据湖可以使用的格式。这可能涉及数据清洗、去重和标准化数据格式。

(3) 加载:将数据加载到数据仓库或数据湖中。这可以通过批量加载或增量加载的方法实现。

具体实现可以使用多种工具和技术。

(1) 网络抓取:网络抓取是从网站中提取数据的过程。可以使用各种工具执行此操作,如Beautiful Soup、Scrapy、Octoparse等。

(2) API:API是软件应用程序之间进行通信的一种方式。企业可以使用API从其他公司收集数据,而无须构建自己的系统。

(3) 查询语言:任何数据库都可以作为数据源,包括结构化查询语言(SQL)或非结构化查询语言类型。

(4) 机器学习:机器学习可以用于自动化数据收集的过程。例如,企业可以使用机器学习来识别数据中的模式,然后基于这些模式收集数据。

数据收集完成后,可以对其进行处理,以便在数据仓库中使用。ETL过程通常会进行一些清理数据的步骤,例如删除重复项并标准化数据格式,然后将数据加载到数据仓库中。数据分析师或数据科学家可以利用这些数据获取业务背后的深层次的信息。目前市面上有许多ETL工具,包括商业工具,如AWS Glue、Google Dataflow、Amazon Simple Workflow Service(SWF)、dbt、Fivetran、Microsoft SSIS、IBM InfoSphere DataStage、Talend Open Studio等。开源的ETL工具,如Airflow、Kafka和Spark。Python中也有很多类似的工具,例如Pandas用于数据提取和处理,甚至celery和joblib也可以作为ETL编排工具。LangChain框架中有与Zapier的集成,Zapier是一种可以连接不同应用和服务的自动化工具。这可用于自动化从各种来源进行数据收集的过程。以下是使用自动化ETL工具的一些优点。

(1) 高准确性:自动化ETL工具可以提升数据提取、转换和加载过程的准确性。这是因为这些工具可以遵循一套规则和流程,有助于减少人为错误。

(2) 快速市场响应:自动化ETL工具有助于缩短将数据导入数据仓库所需的时间。这是因为这些工具可以自动化ETL过程中的重复任务,如数据提取和加载。

(3) 高可伸缩性:自动化ETL工具可以提高ETL过程的可伸缩性。这是因为这些工具可以处理大量数据,并且可以根据业务需求轻松地进行横向或纵向扩展。

(4) 提升合规性:自动化ETL工具有助于提高对各种法规的合规性。这是因为这些工具可以遵循一套规则和流程,有助于确保数据以符合法规的方式进行处理。

如何选用最佳的自动数据收集工具将取决于企业的具体需求。企业应考虑他们需要收集的数据类型、需要收集的数据量及可用的预算等。

8.1.2 可视化和探索性数据分析

自动探索性数据分析（Exploratory Data Analysis，EDA）和可视化指的是使用软件工具和算法对数据进行自动分析和可视化，无须大量手动干预。传统的 EDA 涉及手动探索和总结数据，以了解在执行机器学习或深度学习任务之前的各方面。它有助于识别模式、检测不一致性、测试假设和获取见解，然而，随着大型数据集的出现和对高效分析的需求，自动化的 EDA 变得越来越重要。自动化的 EDA 和可视化工具具有多个优势。它们可以加快数据分析过程，减少在数据清洗、处理缺失值、异常值检测和特征工程等任务上花费的时间。这些工具还通过生成交互式可视化图表来更高效地探索复杂数据集，提供了数据的全面概览。目前有多种自动 EDA 和可视化工具可供选择，列举如下。

（1）D-Tale：这是一个库，它可以用来可视化 Pandas 数据帧。它支持交互式图表、三维图表、热图、相关性分析和自定义列创建。

（2）ydata-profiling（之前称为 Pandas Profiling）：这是一个开源库，它可以用来生成交互式 HTML 报告，总结数据集的不同方面，例如缺失值统计、变量类型分布概况、变量之间的相关性等。它适用于 Pandas 和 Spark DataFrames。

（3）Sweetviz：这是一个 Python 库，可以提供探索性数据分析的可视化功能，它的使用基本不需要编写太多代码。它允许对变量或数据集进行比较。

（4）Autoviz：这个库可以自动生成数据集的可视化图表，无论数据集的大小如何，只需几行代码。

（5）DataPrep：仅需几行代码，即可从常见数据源收集数据，进行探索性数据分析和数据清洗，例如标准化列名或条目。

（6）Lux：通过交互式小部件显示数据集中有趣的趋势和模式，让用户快速浏览以获取深层信息。

生成式 AI 在数据可视化中的应用为自动化 EDA 增添了另一个维度，它允许算法基于现有的可视化图表或特定用户提示生成新的可视化图表。生成式 AI 能够通过自动化设计流程的一部分来增强创造力，同时保持人类对最终输出的控制。总体来讲，自动化 EDA 和可视化工具在时间效率、全面分析和生成有意义的数据方面具有重大优势。生成式 AI 在多方面有可能完全革新和颠覆数据可视化。例如，它可用于创建更逼真、更引人入胜的可视化图表，这有助于商务沟通及更有效地向利益相关者传递数据，为每个用户提供他们需要的信息，以获取深层次信息并做出明智的决策。生成式 AI 可通过定制化的可视化图表，根据每个用户的个性化需求来增强和扩展传统工具的创建能力。此外，生成式 AI 还可用于创建交互式可视化图表，让用户以新颖创新的方式探索数据。

8.1.3 数据预处理和特征提取

自动化数据预处理包括诸如数据清洗、数据集成、数据转换和特征提取等任务。它与 ETL 中的数据转换步骤相关，因此在工具和技术上有很多重叠。数据预处理很重要，因为它可以确保数据处于可供数据分析师和机器学习模型使用的格式中。这包括从数据中删除错误的和不一致的数据，以及将其转换为与将要使用的分析工具兼容的格式。手动提取特征可能会很烦琐且耗时，因此自动化这一过程非常有价值。最近，出现了几个开源的Python库，用来帮助用户从原始数据中自动生成有用的特征。

Featuretools提供了一个通用框架，可以从交易性和关系性数据中合成许多新特征。它可以跨多台机器学习框架集成，具有灵活性。Feature Engine提供了一组更简单的Transformer模型，专注于常见的数据转换，例如处理缺失数据。在需要专门基于树的模型优化特征工程的场合，Microsoft的Ta框架通过自动交叉等技术表现出很强的性能。AutoGluon Features将神经网络样式的自动特征生成和选择应用于提升模型准确性。它与AutoGluon的autoML功能紧密集成。最后，TensorFlow Transform直接在TensorFlow上运行，它为训练过程中的模型准备数据。

Featuretools在集成各种ML框架、自动化和灵活性方面很出色。对于表格数据，Ta和Feature Engine提供了针对不同模型优化的易于使用的Transformer模型。Tf.transform非常适合TensorFlow用户，而AutoGluon专注于Apache MXNet深度学习软件框架。对于时间序列数据，Tsfel是从时间序列数据中提取特征的库。它允许用户指定特征提取的窗口大小，并且可以分析特征的时间复杂性。它还可以计算统计、频谱和时间特征。另外，Tsflex是一款灵活高效的时间序列特征提取工具包，适用于序列数据。它对数据结构的前置限定条件较少，可以处理缺失数据和长度不等的数据，它还可以计算滚动特征。与Tsfresh相比，Tsfel和Tsflex这两个库都提供了更现代化的自动时间序列特征工程选项。Tsfel功能更全面，而Tsflex则强调在复杂序列数据上的灵活性。

此外，有一些工具专注于机器学习和数据科学中的数据质量，包含数据分析和自动数据转换。例如，Pandas-dq库可以与Scikit-learn管道集成，提供一系列有用的功能，用于数据分析、训练测试对比、数据清理、数据填充（例如填充缺失值）和数据转换（例如偏度校正）。通过在模型建立之前解决潜在问题，它有助于提高数据分析的质量。专注于通过及早发现潜在问题或错误来提高可靠性的工具有Great Expectations和Deequ。Great Expectations是一个验证、文档化和分析数据以保持质量和改善团队间沟通的工具。它允许用户对数据进行期望断言，通过数据的单元测试快速发现问题，并基于期望创建文档和报告。Deequ建立在Apache Spark之上，用于为大型数据集定义数据质量的单元测试。它允许用户明确说明对数据集的假设，并通过对属性进行检查或约束来验证这些假设。通过确保符合这些假设，它可以防止在下游应用程序中发生崩溃或错误输出。所有这些库都能让数据科学家缩短对数据特征的准备时间并扩展特征空间，以提高模型的质量。

从上面的各种库和产品的简单介绍可以看出，在利用复杂的真实世界数据来发挥机器学习算法的全部潜力方面，自动化特征工程有着举足轻重的作用。

8.1.4　自动化机器学习

自动化机器学习（AutoML）框架是自动化机器学习模型开发过程的工具。它们可以用于自动化数据清洗、特征选择、模型训练和超参数调整等任务。这可以节省数据科学家大量的时间和精力,还可以帮助提高机器学习模型的质量。AutoML 的基本思想可以从 mljar AutoML 库的 GitHub 仓库中得到很好的说明。AutoML 框架的主要价值在于其易用性及其提高开发者找到机器学习模型、理解它并将其投入生产的能力,其实 AutoML 工具已存在很长时间。最早的框架之一是 AutoWeka,它是用 Java 编写的,旨在自动化 Weka(Waikato 环境,用于知识分析)机器学习套件中表格数据的机器学习模型的开发流程,这个套件是在 Waikato 大学开发的。自 AutoWeka 发布以来,已经开发出了许多其他 AutoML 框架。如今一些最流行的 AutoML 框架包括 Auto-sklearn、Auto-Keras、NASLib、Auto-PyTorch、TPOT、Optuna、AutoGluon 和 Ray。这些框架用多种编程语言编写,并支持多种机器学习任务。最近在 AutoML 和神经架构搜索方面取得的进展使工具能够自动化机器学习管道的大部分流程。像 Google AutoML、Azure AutoML 和 H2O AutoML/Driverless AI 这样的主流解决方案可以根据数据集和问题类型自动处理数据准备、特征工程、模型选择、超参数调整和部署的流程。这些使机器学习对非专业人员也更加易于接触。当前的 AutoML 解决方案可以非常有效地处理结构化数据,如表格和时间序列数据。它们可以自动生成相关特征,选择算法,例如树集成、神经网络或 SVM,并调整超参数。大规模的超参数搜索的性能通常与手动流程相当,甚至更好。针对图像、视频和音频等非结构化数据的 AutoML 也在快速发展。另外像 Auto-Keras、AutoGluon 和 AutoSklearn 这样的开源库也提供了易于访问的 AutoML 功能,然而,大多数 AutoML 工具仍然需要一些编码和数据科学专业知识。当前,面对没有任何编程基础的用户的完全自动化数据科学仍然具有挑战性,另外以上所述的一系列工具在灵活性和可控性方面也存在一些限制。不过市场上正在推出更加用户友好和性能更好的解决方案。自动化机器学习框架的比较见表 8-2。

表 8-2 自动化机器学习框架的比较

框架名称	语言类型	机器学习框架	简介
Auto-Keras	Python	Keras	神经架构搜索,简单易用
Auto-PyTorch	Python	PyTorch	神经架构搜索、超参数调优
Auto-Sklearn	Python	Scikit-learn	自动化 Scikit-learn 工作流
Auto-Weka	Java	Weka	贝叶斯优化
AutoGluon	Python	MXNet,PyTorch	针对深度学习进行了优化
AWS SageMaker Autopilot	Python	XGBoost,Scikit-learn	基于云,简单
Azure AutoML	Python	Scikit-learn,PyTorch	可解释的模型
DataRobot	Python,R	Multiple	强监控性、可解释性
Google AutoML	Python	TensorFlow	易于使用,基于云的
H2O AutoML	Python,R	XGBoost,GBMs	自动化工作流程,集成流程

续表

框架名称	语言类型	机器学习框架	简介
hyperopt-sklearn	Python	Scikit-learn	超参数调优
Ludwig	Python	Transformers/Pytorch	用于构建和调优自定义 LLM 和深度神经网络的低代码框架
MLJar	Python	Multiple	可解释,可定制
NASLib	Python	PyTorch,TensorFlow/Keras	神经架构搜索
Optuna	Python	Agnostic	超参数调优
Ray	Python	Agnostic	分布式超参数调优;加速 ML 工作负载
TPOT	Python	Scikit-learn,XGBoost	基因编程的管道

以上只包括了一些比较知名的框架、库或产品。当前市场上重点是 Python 中的开源框架,不过表格中也包括了一些其他语言的大型商业产品。Pycaret 是另一个很大且很有名的项目,它能让用户同时训练多个模型并通过相对较少的代码对其进行比较。Nixtla 的 Statsforecast 和 MLForecast 等项目具有针对时间序列数据的类似功能。另外像 Auto-ViML 和 deep-autoviml 这样的库可以处理各种类型的变量,它们分别基于 Scikit-learn 和 Keras 构建。一个完整的自动化机器学习框架的重要功能包括以下几点。

(1)部署:一些解决方案尤其是云的解决方案,可以直接部署到生产环境,其他的则可以导出为 TensorFlow 或其他格式。

(2)数据类型:大多数解决方案关注表格数据集;深度学习自动化机器学习框架通常用于处理不同类型的数据。例如,AutoGluon 除了表格数据之外,还支持图像、文本、时间序列的快速比较和机器学习解决方案的原型开发。一些专注于超参数优化的解决方案,如 Optuna 和 Ray Tune,则完全与格式无关。

(3)可解释性:这取决于行业,例如监管(医疗或保险)或可靠性(金融)与此相关。对于一些解决方案,这是它们独特的卖点。

(4)监控:部署后,模型的性能可能会下降(也就是所讲的漂移)。一些自动化机器学习框架提供性能监控功能。

(5)可访问性:一些自动化机器学习框架需要编码或至少需要对基本的数据科学理解,而其他一些则是无须或只需很少的代码的即插即用解决方案。通常,低代码和无代码解决方案的定制性较低。

(6)开源:开源平台的优势在于其完全透明的实现和方法及其参数的可用性,并且它们是完全可扩展的。

(7)迁移学习:这意味着能够扩展或定制现有的基础模型。

这里还有很多内容需要覆盖,超出了本章的范围,例如可用方法的数量,或者一些不太受支持的功能,包括自监督学习、强化学习或生成图像和音频模型。对于深度学习,有几个库专注于后端,它们专门针对 TensorFlow、PyTorch 或 MXNet。Auto-Keras、NASLib 和 Ludwig 具有更广泛的支持。Keras 于 2023 年秋季发布的版本 3.0 开始,支持 3 个主要的后端:TensorFlow、JAX 和 PyTorch。Sklearn 有自己的超参数优化工具,如网格搜索、随机搜索。一些更专业的库,如 Auto-sklearn 和 Hyperopt-sklearn,通过提供贝叶斯优化方法扩展了这一领域。Optuna 可以与诸如 AllenNLP、Catalyst、Catboost、Chainer、FastAI、Keras、LightGBM、MXNet、PyTorch、PyTorch Ignite、PyTorch Lightning、TensorFlow 和 XGBoost 等各种自动化机器学习框架集成。Ray 具有自己的集成,其中包括 Optuna。它们都具有最先进的参数优化算法和用于扩展的机制(分布式训练)。除了上述功能之外,其中一些框架还可以自动执行特征工程任务,例如数据清理和特征选择,例如移除高度相关的特征,并以图形方式生成结果。前面列出的每个工具都有其自己的实现步骤,例如特征选择和特征变换,它们之间的不同是自动化程度。

8.1.5 生成式 AI 对数据科学的变革

生成式 AI(例如 ChatGPT 这样的 LLM)已经给数据科学和分析领域带来了重大变革。这些模型尤其是 LLM,有潜力以多种方式彻底改变数据科学中的所有步骤,为研究人员和分析师提供令人兴奋的机遇。生成式 AI 模型具有理解和生成类似人类回复的能力,使其成为增强研究生产力的有价值工具。生成式 AI 可以在分析和解释数据方面发挥关键作用。这些模型可以协助数据探索,发现隐藏的模式或相关性,并提供传统方法可能无法明显看出的见解。通过自动化数据分析的某些方面,生成式 AI 节省了时间和资源,使研究人员能够专注于更高层次的任务。

生成式 AI 可以帮助研究人员的另一个领域是进行文献综述和识别研究空白。ChatGPT 和类似模型可以总结大量学术论文或文章的信息,提供现有知识的简洁概述。这有助于研究人员发现文献中的空白,并更有效地指导他们自己的调查。生成式 AI 的其他用例还包括以下几种。

(1) 自动生成合成数据:生成式 AI 可以用于自动生成合成数据,用于训练机器学习模型。这对于无法获取大量真实世界数据的企业非常有帮助。

(2) 识别数据中的模式:生成式 AI 可用于识别数据中人类分析师无法看到的模式。这对于希望从其数据中获得新见解的企业非常有帮助。

(3) 从现有数据中创建新特征:生成式 AI 可以用于从现有数据中创建新特征。这对于希望提高其机器学习模型准确性的企业非常有帮助。

最近麦肯锡和毕马威等机构的报告指出,AI 的影响涉及数据科学家将要处理的内容、他们的工作方式及谁能够从事数据科学任务。主要的影响领域包括以下几个。

(1) 大众化 AI:生成模型使更多人能够通过简单提示生成文本、代码和数据,这将 AI 的使用范围扩展到数据科学家之外。

(2) 提升生产力:通过自动生成代码、数据和文本,生成式 AI 可以加速开发和分析工作流程。这使数据科学家和分析师可以专注于更有价值的任务。

(3) 数据科学的创新：生成式人工智能正在带来新的、更具创造性的探索数据的方式，以及生成新的假设和见解，这是传统方法所无法实现的。

(4) 行业颠覆：生成式 AI 的新应用可能通过自动化任务或增强产品和服务来颠覆行业。数据团队需要确定高影响力的应用案例。

关于数据科学的大众化和创新化，更具体来讲生成式 AI 也对数据可视化方式产生了影响。过去，数据可视化通常是静态的和二维的，然而，生成式 AI 可以用于创建交互式和三维可视化的数据，有助于使数据更易于访问和理解。这使人们更容易理解和解释数据，从而能提高决策水平。这里需要再次强调，生成式 AI 所带来的最大变革之一是数据科学的大众化。过去，数据科学是一个非常专业化的领域，需要对统计学和机器学习有深入的理解，然而，生成式 AI 使不具备较深专业知识的人们也能够创建和使用数据模型。这将数据科学领域向更广泛的人群开放。LLM 和生成式 AI 在自动化数据科学方面发挥着至关重要的作用，主要有以下几个巨大的优势。

(1) 自然语言交互：用户能够使用普通英语或其他语言与模型进行交流。这使非技术用户能够使用日常语言与数据进行交互和探索，而无须编码或数据分析方面的专业知识。

(2) 代码生成：生成式 AI 可以自动生成代码片段，用于在探索性数据分析（EDA）过程中执行特定的分析任务。例如，它可以生成用于检索数据（例如 SQL）、清理数据、处理缺失值或创建可视化图表（例如使用 Python）的代码。这个功能节省了时间，减少了手动编码的需求。

(3) 自动生成报告：LLM 能够生成自动化报告，总结了 EDA 的关键发现。这些报告提供了关于数据集各方面的洞见，例如统计摘要、相关性分析、特征重要性等，使用户更容易理解和展示他们的发现。

(4) 数据探索与可视化：生成式 AI 算法可以全面探索大型数据集，并自动生成可视化图表，展现数据中的潜在模式、变量之间的关系、异常值或异常情况等。这有助于用户全面了解数据集，而无须手动创建每个可视化图表。

进一步探讨，生成式 AI 算法应该能够从用户交互中学习，并根据个人偏好或过去行为来调整其建议。它们通过持续的自适应学习和用户反馈不断改进，为自动化的探索性数据分析提供更个性化和有用的见解。最后，在 EDA 过程中，生成式 AI 模型可以通过从现有数据集中学习模式（智能错误识别）来识别数据中的错误或异常。它们能够快速准确地检测出不一致之处，并迅速指出潜在问题。

总体而言，LLM 和生成式 AI 能够通过简化用户交互、生成代码片段、高效识别错误/异常、自动生成报告、促进全面的数据探索与可视化创建，并根据用户偏好进行适应，从而增强自动化的探索性数据分析，使其更有效地分析大型复杂数据集。

然而，虽然 LLM 在数据科学方面的应用价值极大，但它们不应被过度吹捧。正如之前所看到的，LLM 是通过类比进行工作的，它们目前在推理和数学方面仍有所欠缺。LLM 的优势在于创造性，而非准确性，因此，研究人员必须运用批判性思维，确保这些模型生成的输出准确、无偏、符合严格的科学标准。一个显著的例子是微软的 Fabric，它融合了由生成式

AI驱动的聊天界面。这使用户能够使用自然语言提出与数据相关的问题,并在不必等待数据请求队列的情况下立即获得答案。通过利用OpenAI等LLM,Fabric能够让用户实时获得数据背后深层次的信息。Fabric以其全面的方法脱颖而出,它解决了组织在分析需求方面的各方面,并为参与分析过程的不同团队(例如数据工程师、仓储专业人员、科学家、分析师和业务用户)提供了特定角色。在每个层面整合了Azure OpenAI服务后,Fabric利用生成式AO的力量释放了数据的全部潜力。Fabric中的Copilot等功能提高了会话式语言的用户体验,允许用户创建数据流、生成代码或整个函数、构建机器学习模型、可视化结果,甚至开发定制的会话式语言平台。不过,Fabric经常会生成不正确的SQL查询。对于专业的数据分析师来讲,它们可以检查LLM这些输出的有效性,但对于非技术型业务用户来讲,完全依赖这些输出将是一场灾难,因此,企业在使用Fabric进行分析时必须确保已经建立了自己的可靠的数据流水线,从而确保能够高质量地管理数据的输出。虽然生成式AI在数据分析领域的可能性是令人兴奋的,但企业和用户必须谨慎行事,应通过基于第一原则的推理和严格的分析来验证LLM的可靠性和准确性。尽管这些LLM在场景分析、研究中的创意生成及复杂分析的总结方面显示出巨大潜力,但由于一些专业领域需要专家验证的原因,它们并不总是适合面向非技术用户的自助式分析任务。

8.2节将具体介绍如何使用LangChain中的Agent来完成数据科学的一些任务。

8.2 使用Agent

从本章开始时在Jupyter AI中看到的情况,以及在第7章软件开发中看到的情况可知利用生成式AI(代码LLM)来提高软件开发效率有极大的前景。前面的介绍可能更加偏重理论,从本节开始将着重于实践部分,即研究到底AI能为数据科学做哪些实际的工作。在此之前读者已经体验过LangChain中通过把LLM和Agent相连接执行不同的任务的例子,例如,LLMMathChain可以执行Python代码来回答数学问题,代码如下:

```
from langchain import OpenAI, LLMMathChain
llm = OpenAI(temperature=0)
llm_math = LLMMathChain.from_llm(llm, verbose=True)
llm_math.run("2 的 10 次方等于多少?")
```

尽管这对于提取信息并将其反馈回来非常有用,但将其整合到传统的EDA流程中后效果却并不那么明显。LangChain中的CPAL(CPALChain)和PAL(PALChain)链可以在控制出现幻觉现象的同时回答更复杂的推理问题,但很难找到它们的实际应用场景。通过PythonREPLTool,用户可以利用小样本数据创建简单的可视化模型或者使用合成数据训练模型,这对于项目的说明或引导可能很有用。以下是来自LangChain文档的一个例子,代码如下:

```python
//第 8 章/ PythonREPLTool.py
from langchain.agents.agent_toolkits import create_python_agent
from langchain.tools.python.tool import PythonREPLTool
from langchain.llms.openai import OpenAI
from langchain.agents.agent_types import AgentType
agent_executor = create_python_agent(
    llm=OpenAI(temperature=0, max_tokens=1000),
    tool=PythonREPLTool(),
    verbose=True,
    agent_type=AgentType.ZERO_SHOT_REACT_DESCRIPTION,
)
agent_executor.run(
    """Understand, write a single neuron neural network in PyTorch.
Take synthetic data for y=2x. Train for 1000 epochs and print every 100 epochs.
Return prediction for x = 5"""
)
```

这里由于 Python 代码的执行没有任何安全措施,因此用户应谨慎执行此操作。这段代码执行后会创建一个数据集,训练一个模型,然后用户会得到一个预测结果,输出的结果如下:

```
//第 8 章/ PythonREPLToolOut.py
Entering new AgentExecutor chain...
I need to write a neural network in PyTorch and train it on the given data
Action: Python_REPL
Action Input:
import torch
model = torch.nn.Sequential(
    torch.nn.Linear(1, 1)
)
loss_fn = torch.nn.MSELoss()
optimizer = torch.optim.SGD(model.parameters(), lr=0.01)
#Define the data
x_data = torch.tensor([[1.0], [2.0], [3.0], [4.0]])
y_data = torch.tensor([[2.0], [4.0], [6.0], [8.0]])
for epoch in range(1000):   #Train the model
    y_pred = model(x_data)
    loss = loss_fn(y_pred, y_data)
    if (epoch+1) % 100 == 0:
        print(f'Epoch {epoch+1}: {loss.item():.4f}')
    optimizer.zero_grad()
    loss.backward()
    optimizer.step()
#Make a prediction
x_pred = torch.tensor([[5.0]])
```

```
y_pred = model(x_pred)
Observation: Epoch 100: 0.0043
Epoch 200: 0.0023
Epoch 300: 0.0013
Epoch 400: 0.0007
Epoch 500: 0.0004
Epoch 600: 0.0002
Epoch 700: 0.0001
Epoch 800: 0.0001
Epoch 900: 0.0000
Epoch 1000: 0.0000
Thought: I now know the final answer
Final Answer: The prediction for x = 5 is y = 10.00.
```

这就是非常出色的自动化数据科学的一个例子，但如果想让这种能力扩展到更大的规模，则需要进行更加严谨的工程设计，类似于在第 7 章软件开发中所做的工作。LLM 与 LangChain 中的工具相结合之后，在数据分类和地理信息方面非常有效。例如，假设读者的公司提供从上海出发的航班，并且读者想知道客户距离上海的距离，则可以使用 Wolfram Alpha 工具，代码如下：

```
//第 8 章/Wolfram.py
from langchain.agents import load_tools, initialize_agent
from langchain.llms import OpenAI
from langchain.chains.conversation.memory import ConversationBufferMemory
llm = OpenAI(temperature=0)
tools = load_tools(['wolfram-alpha'])
memory = ConversationBufferMemory(memory_key="chat_history")
agent = initialize_agent(tools, llm, agent="conversational-react-description", memory=memory, verbose=True)
agent.run(
    """这些城市离上海有多远？
* 纽约市
* 西班牙马德里
* 柏林
""")
```

这里提醒用户应确保自己设置了 OPENAI_API_KEY 和 WOLFRAM_ALPHA_APPID 环境变量，以下是输出的结果：

```
> Entering new AgentExecutor chain...
AI: 从纽约到上海的距离是 6760 英里。从西班牙马德里到上海的距离是 8845 英里。从德国柏林到上海的距离是 6845 英里。
> Finished chain.
'
```

> 从纽约到上海的距离是 6760 英里。从西班牙马德里到上海的距离是 8845 英里。从德国柏林到上海的距离是 6845 英里。

许多问题非常简单,当向 Agent 程序提供数据集,并连接更多工具时,它可以变得非常强大。接下来介绍如何使用 LangChain 结合 LLM 挖掘数据背后的规律。

8.3 数据探索和 LLM

数据探索是数据分析中至关重要且基础性的一步,使研究人员能够全面了解其数据集并发现重要见解。随着像 ChatGPT 这样的 LLM 的出现,研究人员可以利用自然语言处理的力量来进行数据探索。正如之前提到的生成式 AI 模型,它们往往具有理解和生成类似人类回复的能力,使其成为提高生产力的宝贵工具。用自然语言提出问题并以易于理解的方式获取回答,可以极大地促进分析工作。LLM 不仅可以帮助探索文本数据,还可以探索其他形式的数据,例如数值数据集或多媒体内容。研究人员可以利用 ChatGPT 的能力,询问关于数值数据集中统计趋势的问题,甚至查询图像分类任务的可视化结果。下面的一个例子用于加载一个数据集并进行处理。使用 Scikit-learn 能快速获取一个数据集,代码如下:

```
from sklearn.datasets import load_iris
df = load_iris(as_frame=True)["data"]
```

这是一个非常有名的样本数据集,它将帮助展示利用生成式 AI 进行数据探索的能力。接下来将使用 DataFrame 创建一个 Pandas DataFrame 的 Agent,代码如下:

```
//第 8 章 / pandasAgent.py
from langchain.agents import create_pandas_dataframe_agent
from langchain import PromptTemplate
from langchain.llms.openai import OpenAI
PROMPT = (
    "If you do not know the answer, say you don't know.\n"
    "Think step by step.\n"
    "\n"
    "Below is the query.\n"
    "Query: {query}\n"
)
prompt = PromptTemplate(template=PROMPT, input_variables=["query"])
llm = OpenAI()
agent = create_pandas_dataframe_agent(llm, df, verbose=True)
```

上面的 Agent 设置中已经设定了模型的角色,告诉模型如果不知道答案就说不知道,并逐步思考,这样做的目的是减少模型出现幻觉现象。现在可以和刚才建立好的 Agent 进

行交互,问关于数据集的问题,代码如下:

```
agent.run(prompt.format(query="这个数据集是关于什么的?"))
```

得到的答案是"这个数据集是关于某种类型花朵的测量值的",这是正确的答案。如果想要看到可视化的效果,则应怎么办?以前可能需要很多代码,但现在只需直接用自然语言告诉 LLM,例如"将每列绘制为条形图",代码如下:

```
agent.run(prompt.format(query="将每列绘制为条形图!"))
```

最后的可视化效果如图 8-2 所示。

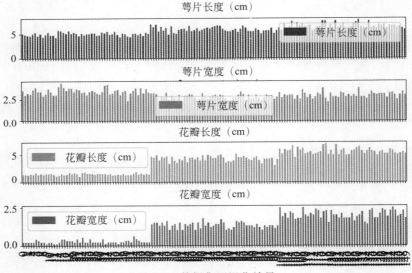

图 8-2 数据集可视化效果

这里还可以要求使用其他后端库来绘图,例如 Seaborn,不过这些库需要另外安装。还可以询问有关数据集的更多其他问题,例如哪一行的花瓣长度和花瓣宽度之间的差异最大,代码如下:

```
//第 8 章/extraQuestions.py
df['difference'] = df['petal length (cm)'] - df['petal width (cm)']
df.loc[df['difference'].idxmax()]
Observation: sepal length (cm)    7.7
sepal width (cm)     2.8
petal length (cm)    6.7
petal width (cm)     2.0
difference           4.7
Name: 122, dtype: float64
```

```
Thought: I now know the final answer
Final Answer: Row 122 has the biggest difference between petal length and petal
width.
```

接下来需要在提示中添加更多有关绘图的说明，例如有关绘图大小的说明。如果要在 Streamlit 应用程序中实现相同的绘图逻辑，则有点困难，因为通常需要在相应的 Streamlit 函数中使用绘图功能，例如 st.bar_chart()，不过花一点时间也可以完成，这里就交给读者自己去尝试。上面只是一些基础的数据分析，数据科学中通常需要统计检验，接下来测试其表现，代码如下：

```
agent.run(prompt.format(query=" Validate the following hypothesis
statistically: petal width and petal length come from the same distribution."))
```

得到的结果如下：

```
//第 8 章/statistic_result.py
Thought: I should use a statistical test to answer this question.
Action: python_repl_ast
Action Input: from scipy.stats import ks_2samp
Observation:
Thought: I now have the necessary tools to answer this question.
Action: python_repl_ast
Action Input: ks_2samp(df['petal width (cm)'], df['petal length (cm)'])
Observation: KstestResult(statistic=0.6666666666666666, pvalue=6.639808432803654e
-32, statistic_location=2.5, statistic_sign=1)
Thought: I now know the final answer
Final Answer: The p-value of 6.639808432803654e-32 indicates that the two
variables come from different distributions.
```

"p 值 6.639808432803654e-32 表示这两个变量来自不同的分布"，这是一个非常准确和专业的回答。除了上面的实现方法，用户还有其他的实现选项，例如 Pandas-AI 库，它在后台使用 LangChain 并提供类似的功能。下面是 LangChain 中的示例，其中包含示例数据集，代码如下：

```
//第 8 章/pandas-ai.py
import pandas as pd
from pandasai import PandasAI
df = pd.DataFrame({
    "country": ["United States", "United Kingdom", "France", "Germany", "Italy",
"Spain", "Canada", "Australia", "Japan", "China"],
    "gdp": [19294482071552, 2891615567872, 2411255037952, 3435817336832,
1745433788416, 1181205135360, 1607402389504, 1490967855104, 4380756541440,
14631844184064],
```

```
        "happiness_index": [6.94, 7.16, 6.66, 7.07, 6.38, 6.4, 7.23, 7.22, 5.87, 5.12]
})
from pandasai.llm.openai import OpenAI
llm = OpenAI(api_token="YOUR_API_TOKEN")
pandas_ai = PandasAI(llm)
pandas_ai(df, prompt='Which are the 5 happiest countries? ')
```

上面代码的请求结果和使用 LangChain 类似。数据科学中使用非常广泛的还有 SQL，对于 SQL 数据库中的数据，用户可以使用 SQLDatabaseChain，下面是 LangChain 文档中的例子，代码如下：

```
//第 8 章 / SQLDatabaseChain.py
from langchain.llms import OpenAI
from langchain.utilities import SQLDatabase
from langchain_experimental.sql import SQLDatabaseChain
db = SQLDatabase.from_uri("sqlite:../../../../Notebooks/Chinook.db")
llm = OpenAI(temperature=0, verbose=True)
db_chain = SQLDatabaseChain.from_llm(llm, db, verbose=True)
db_chain.run("How many employees are there?")
```

在上面的例子中首先需要连接到数据库，然后用户可以用自然语言询问有关数据的问题。当用户不知道数据库的架构时，使用 LLM 进行查询将特别有用。SQLDatabaseChain 还可以检查查询语言的正确性，并在设置了 use_query_checker 选项时自动更正查询，这对没有编程能力的用户来讲如虎添翼。

8.4 总结

本章带领读者探索了自动化数据分析和数据科学的最新技术。在很多领域中可以让 LLM 在数据科学中发挥作用，在这里 LLM 主要作为编码助手或数据探索的一部分。本章从概述覆盖数据科学过程中每个步骤的框架开始，例如自动化机器学习方法，然后讨论了 LLM 如何帮助人类进一步提高生产率，使数据科学和数据分析更易于访问，这不仅适用于利益相关者，还适用于开发人员或用户。接着本章探讨了代码生成和工具，这些工具类似于第 7 章中探讨的代码 LLM，它们如何通过创建可以查询的函数或模型来帮助完成数据科学任务，或者如何利用像 Wolfram Alpha 这样的第三方工具丰富数据，然后讲解了如何在数据探索中使用 LLM。本书在前面的章节中已经研究了摄取大量文本数据进行分析的方法。本章则专注于对结构化数据集以 SQL 或表格形式进行探索性分析。总体来讲，AI 技术有潜力彻底改变分析数据的方式，ChatGPT 插件或 Microsoft Fabric 就是这方面的例子。相信在未来 AI 技术会越来越成熟，真正成为人类数据科学家的好帮手。

第 9 章 绽放 LangChain 的魅力：定制 LLM 输出

读到本章，相信很多读者已经尝试了很多本书中的例子，也对 LangChain 和 LLM 有了比较深入的理解。在实践的过程中，很多时候会发现 LLM 给的输出无法满足特定场景的需求，例如可能某个场景下需要一个特定的输出格式，单纯地靠自然语言和提示词并不能完成这一点，因此，本章将探讨改善 LLM 在某些场景下复杂推理和解决问题中的最佳实践。通常，调整模型以适应特定任务或确保模型输出符合预期的过程称为条件化。在本章中，将讨论微调和提示作为条件化的方法。微调涉及在特定任务或数据集上对预先训练的基础模型进行训练，这些任务或数据集与所需的应用程序相关。同样地，通过在推理时提供额外的输入或上下文，LLM 可以生成符合特定任务或风格的文本。提示设计对于释放 LLM 推理能力、模型和提示技术未来进步的潜力至关重要，这些原则和技术构成了研究人员和从业者使用 LLM 的宝贵工具包。了解 LLM 逐词生成文本的过程有助于创建更好的推理提示。提示词工程仍然是一门经验性的艺术，因此经常需要尝试不同的变化来确定在特定环境下哪一种提示词是有效的，但是一些已经被验证的提示工程的见解可以应用在不同的模型和任务中。本章将带领读者讨论 LangChain 中的工具，包括一些高级提示工程策略，如少样本学习、动态示例选择和链式推理。

9.1 调整与对齐

调整和对齐就生成式 AI 模型而言指的是确保这些模型的输出与人类的价值观、意图或期望结果保持一致。它涉及引导模型的行为，使其符合特定环境中被认为是道德、适当或相关的标准。对齐的概念对于避免生成可能带有偏见、有害或偏离预期目的的输出至关重要。解决对齐问题通常需要关注训练数据中存在的偏见。通过人类审查员的迭代反馈循环，在训练/微调阶段细化客观函数，利用用户反馈，并在部署过程中持续监控以确保持续对齐。

有许多原因会导致用户希望调整大型语言模型。首先需要控制输出的内容和风格。例如，根据特定关键词或特征（如正式水平）进行调整可以产生更相关和高质量的文本。调整还包括安全措施，以防止生成恶意或有害的内容。例如，避免生成误导性信息、不当建议或潜在危险的指令,或者将模型与特定价值观保持一致。调整 LLM 的潜在益处有很多。通过提供更具体和相关的输入,可以获得符合具体需求的输出。例如,在客户支持聊天机器人中,通过将用户查询作为模型的条件,可以生成准确解决其问题的回复。调整还有助于控制模型输出的偏见或不当行为,使其在特定边界内保持创造力。此外,通过对 LLM 进行调整,用户可以使其更易控制和适应。用户还可以根据要求微调和塑造其行为,并创建可靠的 AI 系统,适用于诸如法律咨询或技术写作等特定领域,然而任何事物都有两面性,在调整的过程中也需要考虑潜在的不利因素。过度调整模型可能导致过拟合,使其过于依赖特定输入,并且在不同环境下难以生成富有创意或多样化的输出。此外,用户还应负责任地使用调整这项手段,因为 LLM 倾向于放大训练数据中存在的偏见。调整这些模型时必须注意不能加剧与偏见或有争议题相关的问题。

总结一下,调整和对齐的优势包括以下几点。

(1)提升用户体验:对齐的模型生成符合用户查询或提示的输出,增强了用户体验。

(2)建立信任:确保道德行为有助于建立用户/客户之间的信任关系。

(3)品牌声誉:通过与业务目标对齐,保持了一致性。

(4)减轻有害影响:对安全、保障和隐私的考虑进行对齐,有助于防止生成有害或恶意内容。

潜在的劣势包括以下几点。

(1)平衡挑战:在极端调整和放任不管之间找到平衡是困难的。

(2)自动化指标的局限性:定量评估指标可能无法完全捕捉到调整对齐的微妙差别。

(3)主观性:调整对齐判断往往是主观的,通常需要对期望的价值观进行认真考虑和达成共识。

虽然诸如 GPT-4 等基础模型能够生成涵盖广泛主题的令人印象深刻的文本,但对其进行调整可以增强其在任务相关性、特定性和连贯性方面的能力,并使其输出更加相关和贴切。如果不进行调整,则这些模型往往会生成可能并非始终与期望情境完美对齐的文本。通过对其进行调整,用户可以引导 LLM 产生与给定输入或指令更紧密相关的输出。调整的主要优势在于,它允许对模型进行引导,而无须进行大量的重新训练。它还实现了交互控制和不同模式之间的切换。调整可以发生在模型开发周期的不同阶段,例如从微调到在各种情境下的输出生成。实现 LLM 的对齐有几种方法,一种方法是在微调过程中进行调整,通过训练模型以反映所需输出的数据集。这种方法使模型可以专业化,但缺点是需要访问相关的训练数据。另一种选择是在模型推理时动态调整模型,主要手段是提供条件输入和主要提示词。这种方式更加灵活,但在部署过程中引入了一些复杂性。接下来的内容将总结关于调整和对齐的关键方法,例如微调和提示工程。本书将带领读者探讨其原理,并比较各种方法的相对优缺点。

9.1.1 对齐的方法

随着像 ChatGPT 这样的大型预训练 LLM 的出现,人们对于适应这些模型用于下游任务的技术表现出了越来越大的兴趣。这个过程被称为微调。微调允许预训练模型针对特定应用进行定制,同时利用在预训练期间获得的广泛语言知识。调整预训练神经网络的想法起源于 21 世纪初的计算机视觉研究。在自然语言处理领域,研究人员展示了在下游任务中微调预训练上下文表示(如 ELMo 和 ULMFit)的有效性。BERT 模型将微调预训练的 Transformer 模型确立为在自然语言处理中的金标准。微调的需求源于预训练 LLM,旨在模拟一般语言知识,而不是特定的下游任务。它们的能力只有在适应特定应用时才会显现。微调允许更新预训练权重以适应目标数据集和目标。这样可以从通用模型中进行知识转移,同时定制专门任务。截至目前,业界已经提出了几种对齐的方法,每种方法在效率和有效性方面存在各自的优缺点。

(1)整体微调:微调一般指的是更新预训练 LLM 的所有参数。模型在下游任务上进行端到端的训练,允许全局更新权重以最大化地达到目标任务的性能。整体微调在各种下游任务上基本能取得非常好的效果,但缺点是需要大量计算资源和大型数据集。

(2)适配微调:在适配微调中,一些额外的可训练适配层被插入预训练模型中,通常是瓶颈层,在这个过程中原始权重被冻结。只有新添加的适配器层会在下游任务中进行训练。这种处理方式使调整参数变得高效,因为只有少量权重被更新。然而,由于预训练权重保持不变,适配微调存在任务拟合不足的风险。适配层的插入点和容量很难准确把握,从而影响整体有效性。

(3)前缀微调:在这一过程中会在 LLM 的每层之前添加可训练向量,在微调期间进行优化,而基础权重保持冻结。前缀允许将归纳偏差注入模型。与适配微调相比,前缀微调具有较小的内存占用,但相比较而言效果不那么好。前缀的长度和初始化会影响其有效性。

(4)提示微调:在这一过程中输入文本会附加可训练提示 Token,为 LLM 引入软性提示以诱导模型采取具体特定的行为。例如,可以将任务描述作为提示提供给 LLM,以引导模型。只有添加的提示 Token 在训练期间被更新,而预训练权重保持冻结。这一方法的巨大缺点是它的性能受提示工程的影响很大,目前研究人员正在探索自动提示的一些新方法。

(5)低秩调整(LoRA):LoRA 是最近比较流行的一种调整和对齐模型的方式。在 LoRA 过程中,LLM 的原始权重被冻结,同时在此基础上一对低秩可训练权重矩阵被添加进去。例如,对于每个权重 W,添加低秩矩阵 B 和 A,使前向传递使用 $W + BA$。只有 B 和 A 被训练,基础的 W 保持冻结。LoRA 的出现使大规模的调整成为可能,其中秩 r 的选择是关键隐私。LoRA 最大的优点是它使在有限硬件上微调庞大的 LLM 成为可能,因为当前 AI 的发展基本上极大地受限于硬件资源。

对齐的另外的方法是通过人工监督,例如人类监督系统。这些系统涉及人类审查员提供反馈,并在必要时更正 LLM 的输出。人类的参与有助于将 LLM 生成的输出与人类设定的期望值保持一致。

以上这些技术结合起来为开发人员提供了更多控制生成式 AI 系统行为和输出的手段,最终的目标是确保行业法规和道德在从训练到部署的各个阶段都得到了介入。此外,在预训练目标函数中进行仔细设计也会影响 LLM 最初学习的行为和模式。通过将伦理道德融入这些目标函数,开发人员可以影响 LLM 的初始学习过程。

9.1.2 变革者:InstructGPT

InstructGPT 被认为是 LLM 调整和对齐领域的一股变革力量,因为它展示了通过将人类反馈的强化学习(Reinforcement Learning Human Feedback,RLHF)纳入其中。接下来将详细介绍 InstructGPT 背后的一些划时代的思想。

OpenAI 的研究人员展示了使用 RLHF 与近端策略优化来使类似 ChatGPT 的 LLM 与特定的期望值保持一致。RLHF 是一种在线方法,使用人类的特定反馈和偏好对 LLM 进行微调,主要包括监督式预训练、奖励模型训练,以及强化学习微调。RLHF 通过特定的学习的奖励模型将人类判断融入 LLM 训练中,因此,人类反馈可以引导并提升 LLM 的能力,这将极大地超越传统的监督式微调。这种新模型可以按照自然语言给出的指示进行操作,其回答问题的方式比 ChatGPT 更准确。尽管参数量减少为原来的 1‰,但 InstructGPT 在用户偏好、真实性和减少有害内容方面表现出比 ChatGPT 更好的性能。使用这种微调可以极大地提高模型的可操纵性、输出可靠性及风格一致性。InstructGPT 通过引入来自人类反馈给 LLM 的学习方法打开了改进 LLM 的新途径,超越了传统的微调方法。尽管 RLHF 训练可能不稳定且计算成本高昂,但其初步的成功激发了进一步地研究各种改进的方法。

不过由于 RLHF 是在线的方法,因此受到一些局限性。相比于在线的方法,离线方法通过直接利用人类反馈来规避在线强化学习的复杂性。Direct Preference Optimization(DPO)是一种简单有效的方法,用于训练 LLM 来更好地遵循人类偏好,这种方法无须显式学习奖励模型或使用强化学习。虽然它优化的目标与现有的 RLHF 方法相同,但它更简单,更易实现,更稳定,并且实现了良好的实验性能。Meta 的研究人员在论文《LIMA:少即是多,实现对齐》[19]中在对 LLaMa 模型进行微调时,通过对 1000 个提示进行损失函数最小化,简化了对齐过程。通过将其输出与 DaVinci003 进行人类偏好比较,研究人员得出微调其实效果微乎其微,这被称为肤浅的对齐假设。离线方法提供了更稳定和更高效的调整模型的方法,但是,它们受到静态的人类反馈的限制,所以最近业界的方法尝试结合离线和在线学习。

前面提到的 LoRA 也是 InstructGPT 的一大创新点。LLM 在自然语言处理领域取得了令人印象深刻的成果,目前已经被应用于计算机视觉和音频等其他领域,然而,随着这些模型变得更大,在消费者硬件上训练它们变得困难,而且如果需要为每个特定任务部署它们就变得更加昂贵。目前有一些方法可以降低计算、内存和存储成本,同时提高在数据量较小的情况下的训练性能。LoRA 冻结了预训练模型的权重,并在 Transformer 架构的每个层中引入可训练的秩分解矩阵,以减少可训练参数的数量。LoRA 在各种语言模型(Roberta、Deberta、GPT-2 和 GPT-3)上实现了与完整微调相当或更好的质量,同时这种方法具有更少的可训练参数和更高的训练吞吐量。QLoRA 方法则是 LoRA 的扩展,它通过将梯度反向传播穿越 4 位量化模型到可学习的低秩适配层实现了 LLM 的高效微调。QLoRA 使在单个GPU上微调一个650亿个参数的模型成为可能。QLoRA模型通过创新的新数据类型和优化器，在Vicuna上实现了与ChatGPT性能99%相当的结果。QLoRA将650亿个参数模型的微调内存需求从780GB降低到48GB，而不会影响模型运行的性能，这是非常了不起的成就。

量化是指降低神经网络（如LLM）中权重和激活的数值精度的技术，其主要目的是减少大型模型的内存占用和计算要求。关于LLM量化的要点包括它涉及使用比标准的单精度浮点数（FP32）更少的位数来表示权重和激活，例如，权重可以量化为8位整数。这一过程可以将模型尺寸缩小为原来的25%，并在专用硬件上提高吞吐量。通常情况下，量化对模型精度的影响较小，尤其是在重新训练时。常见的量化方法包括标量量化、向量量化和乘积量化，它们可以分别或分组量化权重。此外，还可以通过估计激活的分布并适当分组对激活进行量化。量化感知训练可以在训练期间调整权重，以最小化量化损失。已经证明像BERT和GPT-3这样的LLM可以通过微调使用4~8位量化。参数高效微调（PEFT）方法使每个任务可以使用小的检查点，增强了模型的可移植性。这些小的训练权重可以被添加到LLM的顶部，使相同的模型可以用于多个任务，而无须更换整个模型。

前面讲解了LLM的训练部分，在接下来的部分中将讨论在LLM推理时条件设定的方法。

9.1.3 LLM推理过程的调整方法

在LLM推理时（输出生成阶段），一种常用的方法是进行条件设置，即动态提供特定的输入或条件来引导输出生成过程，然而，在某些情况下，LLM微调可能并不总是可行或有效的原因有几方面：首先，有些模型只能通过API进行访问，而这些API可能缺乏或具有受限的微调能力；其次，对于某些下游任务或相关应用领域，可能会缺乏进行微调所需的数据；此外，对于具有频繁变化数据的应用，例如新闻平台，频繁进行模型微调可能会带来潜在的不利影响；最后，对于动态且上下文特定的应用，如个性化聊天机器人，可能无法根据个人用户数据进行微调。在推理时进行条件设置，通常情况下会在文本生成过程的开头提供一个文本提示或说明，这个提示可以是几句话甚至一个单词，以此作为对期望输出的明确指示。

动态推理时条件设置的常用技术包括几种方法：提示调优、前缀调优、限制标记和元数据。提示调优通过自然语言指导实现预期行为，并对提示设计敏感；前缀调优将可训练向量添加到LLM层；限制标记强制包含/排除特定词语，而元数据提供高级信息，如流派、目标受众等。这些技术有助于生成文本，符合特定主题、风格，甚至模仿特定作者的写作风格。它们通过提供推断时的上下文信息，支持上下文学习或检索增强。举例而言，前缀调优将类似"编写一个适合儿童的故事……"的指令添加到提示中。在聊天机器人应用中，通过用户消息对模型进行条件设置，可以生成个性化且与当前对话相关的回复，其他方法包括在提示前添加相关文档辅助LLM进行写作任务，或在提示前加入用户特定数据以确保得到个性化答案。这些方法在运行时引导模型输出，无须依赖传统的微调过程。此外，提示技术允许在文本提示中添加类似问题和解决方案的示例。以上这些方法都可以对大型静态模型进行简单控制，引导其行为而无须进行广泛微调。虽然这些方法能够以较低开销调整和对齐模型，但往往需要精心设计的提示以获得最佳效果。

对以上的所有调整和对齐的方法做一个总结，完整微调能够取得最好的结果，但往往需要大量资源，并且在效能和效率之间存在权衡问题，其他方法（如适配、提示和 LoRA 等）通过稀疏性或冻结降低了这种负担，但可能效果较差。最佳方法取决于约束条件和目标设定。预计未来业界的研究工作将致力于改进针对大型语言模型量身定制的技术，从而推动效能和效率的边界。最近的研究将离线学习和在线学习相结合，以提高稳定性。基于提示的技术允许对 LLM 灵活地进行条件设置，以诱导所需的行为，而无须进行密集训练。精心设计、优化和评估提示是有效控制 LLM 的关键。总体而言，基于提示的技术可以以一种灵活、低资源的方式对 LLM 进行特定行为或知识的条件设置。

9.1.4　效果评估

讨论了 LLM 对齐的大量方法后，最重要的一点是如何评估某种方法是否有效。通常，对齐性会在像 HUMAN 这样的对齐基准和像 FLAN 这样的概括测试上进行评估。目前有一些具有很高可区分性的核心基准，可以准确评估模型的优势和劣势，例如 MMLU（英语知识）、C-Eval（中文知识）、GSM8k/BBH（推理算法）、HumanEval/MBPP（编码）。在这些基准之上，还有像 MATH（高难度推理）和 Dialog 等额外基准。MATH 基准测试了模型应对高难度问题的能力，GPT-4 在提示方法的不同取得了不同的分数。

还有一些定量指标，如困惑度（衡量模型预测数据的能力）或 BLEU 分数（捕捉生成文本与参考文本之间的相似性的能力）。这些指标提供了粗略的估计，但还无法完整体现 LLM 在一些高级任务中的对齐的能力，其他指标包括用户偏好评级（通过人工评估进行）和成对偏好评估，它们利用预训练奖励模型进行在线小/中模型评估，还有基于自动 LLM 的评估（例如使用 GPT-4）。依靠人工评估方面，人类评估有时可能存在问题，因为人类可能会受到 LLM 回答中权威语调等主观标准的影响，而且，仅基于人类的偏好评估将不能考虑其可能引起的潜在危害，例如人类可能会产生误导，并很容易将次优模型优先考虑，而不是更安全的替代方案。总而言之，评估 LLM 的对齐性需要仔细选择基准测试，考虑差异性，并混合使用自动评估方法和人类判断。下面的章节将对一个小型开源 LLM（OpenLLaMa）进行 PEFT 和量化的问答微调，并将其部署在 HuggingFace 上。

9.2　实战案例：LangChain 微调 LLM

对 LLM 进行模型微调的目标是通过优化模型使其生成的输出更贴近特定任务和背景，这一点在本章的开头已经有所探讨。本节将带领读者对一个模型进行问答微调。由于运行这个任务需要比较高的硬件资源，因此本书选择在 Google Colab 上运行此步骤。

Google Colab 是谷歌公司的一个计算环境，提供了不同的硬件加速方式，如张量处理单元(TPU)和图形处理单元(GPU)。这些资源在免费和专业版中都有提供。针对本节任务免费版已经完全足够。用户可以在以下网址登录到 Colab 环境：https://colab.research.google.com/。需要确保在顶部菜单中将自己的 Google Colab 设置为 TPU 或 GPU，以确保拥有足够的资源来运行此任务。执行任务之前将在 Google Colab 环境中安装所有必需的库，代码如下：

```
! pip install - U accelerate bitsandbytes datasets transformers peft trl sentencepiece wandb langchain
```

为了从 HuggingFace 下载并训练模型，用户需要在平台上进行身份验证。如果稍后想要将你的模型推送到 HuggingFace，则需要生成一个新的 API Token，并且设置该 Token 在 HuggingFace 上拥有写入权限。用户可以在如下网址创建自己的 Token：https://huggingface.co/settings/tokens（如图 9-1 所示）。

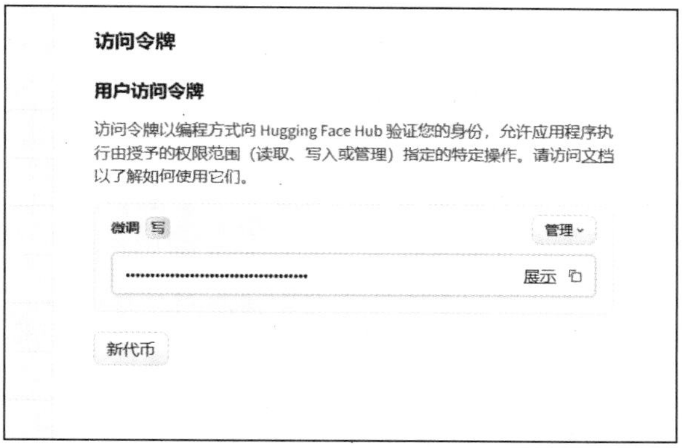

图 9-1　在 HuggingFace 写入权限上创建新的 API Token

创建完成 Token 之后，可以进行验证，代码如下：

```
import Notebook_login
Notebook_login()
```

当出现提示时，把刚才创建的 Token 粘贴进去。

Weights and Biases(W&B) 是一个机器学习平台，可以帮助开发人员端到端地监控和记录模型训练工作流程。这里将使用 W&B 来了解训练的效果如何，特别是模型是否随着时间的推移而改进。使用 W&B 用户需要为项目命名，或者可以使用 wandb 的 init() 方法，代码如下：

```
import os
os.environ["WANDB_PROJECT"] = "finetuning"
```

用户需要在 https://www.wandb.ai 上创建一个免费账户，然后可以在授权页面上找到自己的 API 密钥：https://wandb.ai/authorize。这里需要启动一个新的训练运行时，这将确保能够获得 W&B 的新的报告和仪表板，代码如下：

```
if wandb.run is not None:
    wandb.finish()
```

下面需要选择一个数据集来优化模型。可以使用许多不同类型的数据集，HuggingFace 提供了大量数据集，用户可以在此网址查看：https://huggingface.co/datasets。这些数据集涵盖了许多不同的专业领域。当然用户也可以自定义自己的数据集，例如可以使用 LangChain 设置训练数据。另外也有许多可用的筛选方法可帮助减少数据集中的冗余，由于篇幅原因本书不在此赘述。对于代码 LLM 可以应用代码验证技术将代码段评分作为筛选依据。如果代码来自 GitHub，则可以按照星级或存储库所有者的星级进行筛选。对于自然语言文本质量筛选并不简单。搜索引擎排名可以作为流行度筛选的参考，因为它通常基于用户对内容的参与度。此外，知识蒸馏技术可以根据事实密度和准确性进行调整，也是一个非常不错的筛选工具。在本例中，将使用 Squad V2 数据集对模型的问答性能进行微调，用户可以在 HuggingFace 上查看详细的数据集描述：https://huggingface.co/spaces/evaluate-metric/squad_v2。

```
from datasets import load_dataset
dataset_name = "squad_v2"
dataset = load_dataset(dataset_name, split="train")
eval_dataset = load_dataset(dataset_name, split="validation")
```

Squad V2 数据集既是用于训练的一部分，也是用于验证的另一部分，load_dataset (dataset_name) 的输出如下：

```
DatasetDict({
    train: Dataset({
        features: ['id', 'title', 'context', 'question', 'answers'],
        num_rows: 130319
    })
    validation: Dataset({
        features: ['id', 'title', 'context', 'question', 'answers'],
        num_rows: 11873
    })
})
```

这里将使用验证拆分进行提前停止。提前停止将允许在验证错误率提升时停止训练。Squad V2 数据集由各种特征组成，数据集的特征代码如下：

```
{'id': Value(dtype='string', id=None),
 'title': Value(dtype='string', id=None),
```

```
'context': Value(dtype='string', id=None),
'question': Value(dtype='string', id=None),
'answers': Sequence(feature={'text': Value(dtype='string', id=None),
'answer_start': Value(dtype='int32', id=None)}, length=-1, id=None)}
```

微调模型前的基本想法是向模型提出问题并将答案与数据集进行比较。首先需要一个小型模型,这个模型能以合理的 Token 速率运行。现在比较有名的开源模型是 LLaMa-2 模型,它需要使用用户的电子邮件地址签署许可协议并进行确认,这是因为商业用途的限制。OpenLLaMa 等衍生产品在 HF 排行榜上表现相当不错,不过 OpenLLaMa 版本 1 由于标记器的原因无法用于编码任务,因此这里使用版本 2。经过一番筛选,这里将使用一个拥有 30 亿个参数的模型并加载它,这样即使在较旧的硬件上也能够使用,代码如下:

```
//第 9 章 / loadModel.py
model_id = "openlm-research/open_llama_3b_v2"
new_model_name = f"openllama-3b-peft-{dataset_name}"

import torch
from transformers import AutoModelForCausalLM, BitsAndBytesConfig
bnb_config = BitsAndBytesConfig(
    load_in_4bit=True,
    bnb_4bit_quant_type="nf4",
    bnb_4bit_compute_dtype=torch.float16,
)
device_map="auto"
base_model = AutoModelForCausalLM.from_pretrained(
    model_id,
    quantization_config=bnb_config,
    device_map="auto",
    trust_remote_code=True,
)
base_model.config.use_cache = False
```

在上面的代码中比特和字节配置让用户能以 8、4、3 甚至 2 位量化模型,这能极大地加速 LLM 的推理速度和占用更低的内存,而不会损失性能。模型检查点是在训练过程中定期保存模型参数和状态的快照。这些检查点对于模型的恢复和继续训练至关重要。通过周期性地保存模型的检查点,可以在训练过程中避免意外中断导致的数据丢失。这意味着如果在训练过程中发生问题,例如计算机故障或网络中断,则可以从最后一个保存的检查点重新开始训练,而无须从头开始。模型检查点通常包含模型的权重、优化器状态和训练历史,允许在训练期间轻松地恢复模型的状态并实现持续改进。如果要保存模型检查点,用户则需要连接自己的谷歌网盘并且设置一个模型检查点的保存路径,代码如下:

```
from google.colab import drive
```

```
drive.mount('/content/gdrive')

output_dir = "/content/gdrive/My Drive/results"
```

然后需要设置分词器 Tokenizer。分词器在自然语言处理中扮演着重要角色。它是将文本数据分割成标记或单词的工具。分词器能够将句子转换成模型可以理解的形式,这对于语言模型的训练和处理是至关重要的。分词器根据特定的规则或预训练模型的词汇表将文本分解成不同的单词或子词,每个单词或子词都是模型处理的基本单位。在预处理自然语言数据时,分词器的选择和使用对于模型的性能和理解能力至关重要。通常每个模型有自己的专属分词器,可以按照下面的代码加载:

```
from transformers import AutoTokenizer
tokenizer = AutoTokenizer.from_pretrained(model_id, trust_remote_code=True)
tokenizer.pad_token = tokenizer.eos_token
tokenizer.padding_side = "right"
```

接下来,定义训练配置,这里主要设置 LoRA 和其他训练参数,代码如下:

```
//第 9 章/ trainingArg.py
from transformers import TrainingArguments, EarlyStoppingCallback
from peft import LoraConfig
base_model.config.pretraining_tp = 1
peft_config = LoraConfig(
    lora_alpha=16,
    lora_DropOut=0.1,
    r=64,
    bias="none",
    task_type="CAUSAL_LM",
)
training_args = TrainingArguments(
    output_dir=output_dir,
    per_device_train_batch_size=4,
    gradient_accumulation_steps=4,
    learning_rate=2e-4,
    logging_steps=10,
    max_steps=2000,
    num_train_epochs=100,
    evaluation_strategy="steps",
    eval_steps=5,
    save_total_limit=5,
    push_to_hub=False,
    load_best_model_at_end=True,
    report_to="wandb"
)
```

在上面的代码中有很多参数，这里对几个重要的参数进行简单说明。push_to_hub 参数意味着可以在训练期间定期将模型检查点推送到 HuggingSpace Hub。为了使其正常工作，用户需要设置 HuggingSpace，以便进行身份验证（如前所述，需要创建写入权限的 Token）。如果将此选项值设置为 True，则可以将 new_model_name 用作 output_dir，这将是模型在 HuggingFace(https://huggingface.co/models)上可用的存储库名称。

另外一种方法是将自己的模型保存到本地或云端，例如将其保存到谷歌云盘的某个目录。本例中将 max_steps 和 num_train_epochs 设置得非常高，因为在测试的过程中本书注意到即使经过很多训练步骤，模型仍然可以提高性能。另外这里使用提前停止技术和最大训练步骤来使模型快速收敛以得到更高的性能。对于提前停止技术，用户需要将 evaluation_strategy 设置为"steps"并设置 load_best_model_at_end=True。eval_steps 参数是两次评估之间的更新步骤数。save_total_limit=5 意味着只保存最近的 5 个模型。最后，report_to="wandb"表示将向 W&B 发送训练统计信息、一些模型元数据和硬件信息，这样就可以在其中查看每次运行的图表和仪表板相关信息。接下来开始训练，使用之前定义的配置，代码如下：

```python
//第 9 章/ training_start.py
from trl import SFTTrainer
trainer = SFTTrainer(
    model=base_model,
    train_dataset=dataset,
    eval_dataset=eval_dataset,
    peft_config=peft_config,
    dataset_text_field="question",   #this depends on the dataset!
    max_seq_length=512,
    tokenizer=tokenizer,
    args=training_args,
    callbacks=[EarlyStoppingCallback(early_stopping_patience=200)]
)
trainer.train()
```

整个训练可能需要相当长的时间，即使在 TPU 设备上运行也是如此。评估和提前停止会大大减慢训练速度。如果用户选择不使用提前停止，则应该可以使训练速度更快。随着训练的进行用户在自己的界面上能看到一些统计数据，不过在 W&B 上看到的图表效果更加直观，一些训练的数据情况如图 9-2 所示。

训练完成后可以将最后一个模型检查点保存下来以供重新加载，代码如下：

```python
trainer.model.save_pretrained(
    os.path.join(output_dir, "final_checkpoint"),
)
```

可以通过下面的代码将模型推送到 HuggingFace：

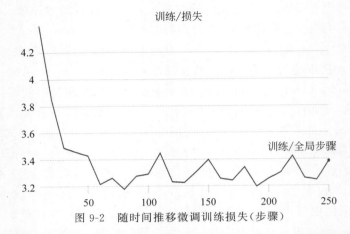

图 9-2 随时间推移微调训练损失（步骤）

```
trainer.model.push_to_hub(
    repo_id=new_model_name
)
```

现在，用户可以结合使用 HuggingFace 用户名和存储库名称（新模型名称）重新加载模型。接下来快速展示如何在 LangChain 中使用这个模型。通常，将 PEFT 模型存储为适配器模型，而不是完整模型，因此加载有一些复杂，代码如下：

```
//第 9 章/ loadPEFT.py
from peft import PeftModel, PeftConfig
from transformers import AutoModelForCausalLM, AutoTokenizer, pipeline
from langchain.llms import HuggingFacePipeline
model_id = 'openlm-research/open_llama_3b_v2'
config = PeftConfig.from_pretrained("benji1a/openllama-3b-peft-squad_v2")
model = AutoModelForCausalLM.from_pretrained(model_id)
model = PeftModel.from_pretrained(model, "benji1a/openllama-3b-peft-squad_v2")
tokenizer = AutoTokenizer.from_pretrained(model_id, trust_remote_code=True)
tokenizer.pad_token = tokenizer.eos_token
pipe = pipeline(
    "text-generation",
    model=model,
    tokenizer=tokenizer,
    max_length=256
)
llm = HuggingFacePipeline(pipeline=pipe)
```

到目前为止已经在 Google Colab 上完成了所有工作，这些步骤同样可以在本地执行，不过用户需要确保安装了所有的依赖，这还是比较麻烦的，Google Colab 已经预装了很多库。以上的例子已经展示了如何微调和部署一个开源 LLM，其实不止是开源模型，一些商业模型也可以根据自定义数据进行微调。例如，OpenAI 的 GPT-3.5 和 Google 的 PaLM 模型都提供了此功能。例如，微调用于文本分类的 PaLM 模型的代码如下：

```
from skllm.models.palm import PaLMClassifier
clf = PaLMClassifier(n_update_steps=100)
clf.fit(X_train, y_train) #y_train is a list of labels
labels = clf.predict(X_test)
```

同样，用户可以微调 GPT-3.5，也就是微调 ChatGPT 模型以进行文本分类，代码如下：

```
//第 9 章/ gpt_finetune.py
from skllm.models.gpt import GPTClassifier
clf = GPTClassifier(
        base_model = "gpt-3.5-turbo-0613",
        n_epochs = None, # int or None. When None, will be determined automatically by OpenAI
        default_label = "Random", #optional
)
clf.fit(X_train, y_train) #y_train is a list of labels
labels = clf.predict(X_test)
```

不过商业模型的微调训练受到一些限制，例如在 OpenAI 上可用的微调中，所有数据输入都通过审核系统，以确保输入与安全标准兼容。微调训练的介绍到此结束，可以看到微调训练相对还是比较复杂的，正如前面介绍过的那样，其实仅仅通过提示，用户也可以完成少样本学习甚至零样本学习，从而让 LLM 能够在特定方面的表现迅速提升，9.3 节将介绍提示词工程的内容。

9.3 提示词工程

提示词在引导 LLM 的行为中扮演着重要角色，因为它们有助于将模型输出与人类意图对齐，而无须进行昂贵的重新训练。通过精心设计的提示词，用户可以将 LLM 的能力扩展至超出其原始训练任务范围之外的多种任务。这些提示词类似于指示，拥有向 LLM 呈现所需的输入-输出映射的能力，因而在使用 LLM 完成各种任务时，它们成为至关重要的因素。它们不仅是工具，更是一种引导模型行为的方式，可以使其适应特定的上下文或任务要求。

提示词工程，也被称为上下文学习，是指通过精心设计的提示技术来引导 LLM 的行为，而无须改变 LLM 模型中的权重，其目标是使模型的输出与特定任务的人类意图保持一致。通过设计良好的提示模板，模型可以提高性能，有时可以与微调相媲美，但是，什么样的提示才算是有效的呢？

9.3.1 提示词的结构

对话提示由 3 个主要组成部分构成。

(1) 说明：描述任务要求及用户期望的输入和输出的格式。
(2) 示例：具体展示所需的输入和输出的例子。
(3) 输入：用户实际的输入。

说明用于清晰地向模型解释任务。示例提供了不同输入如何映射到输出的多样演示。基本的提示方法包括零样本提示（只使用输入文本）和小样本提示（展示几个示例，显示所需的输入/输出对）。研究人员通过研究发现，少样本提示有一定概率出现LLM性能上的偏差，如多数标签偏差和最近性偏差。通过示例选择、排序和格式化精心地进行提示词设计可以帮助缓解或解决这些问题。

上面的格式只是最基础的提示词模板，除此之外还有更高级的提示词技术。例如指令提示，它明确描述了任务要求，而不仅是演示。再例如自一致性采样，它让LLM生成多个输出，并选择与示例最匹配的输出。还有思维链（Chain of Thought，CoT）提示，它能够生成导致最终输出的明确推理步骤，这对于复杂的推理任务尤其有益。CoT提示可以手动编写，也可以通过增强-剪枝-选择等方法自动生成。LLM提示词技巧见表9-1。

表9-1　LLM提示词技巧

技 巧	实 现 方 法	关 键 点
微调	对通过提示生成的解释数据集进行微调	提高模型的推理能力
零样本提示	只需将任务文本提供给模型并询问结果	
思维链（CoT）	在回复前加上"让我们一步一步思考"	在回答之前为模型提供推理空间
少样本提示	提供一些由输入和所需输出组成的演示，以帮助模型理解	显示所需的推理格式
从少到多的提示	提示模型，以便以增量方式解决更简单的子问题。"要解决{问题}，首先需要解决："	将问题分解成更小的部分
选择推理提示	备用选择和推理提示	指导模型完成推理步骤
自一致性	从多个样本中选择最常见的答案	提高了冗余性

一些提示技术将外部信息检索纳入其中，在生成输出之前为LLM提供缺失的背景信息。对于开放领域的问答，可以通过搜索引擎检索相关段落，并将其融入提示中。对于封闭式问答，证据加问题加答案格式的少次提示比问题加答案格式更有效。

在复杂推理任务中，有以下一些技术可提高LLM的可靠性。

（1）提示和解释：在回答之前提示模型逐步解释其推理步骤，使用如"让我们逐步思考"（如CoT中所示）的提示，在推理任务中显著提高准确性。提供少量推理的示例有助于展示用户所期望的格式，并指导LLM生成连贯的解释。

（2）交替选择和推理提示：利用专业选择性的提示和推理提示的组合，与仅使用通用推理提示相比能够取得更好的结果。

（3）问题分解：使用少到多的提示方法及将复杂问题分解成较小的子任务或组件，有助于提高 LLM 的可靠性,因为这样可以更有结构性和可管理性地解决问题。

(4) 多次输出抽样:在输出生成期间从 LLM 中抽样多个答案并选择最常见的答案可提高一致性,减少对单一输出的依赖。尤其是训练单独的验证模型来评估 LLM 生成的候选响应有助于过滤出不正确或不可靠的答案,提高整体可靠性。

最后,通过对提示生成的解释数据集进行 LLM 微调可以增强它们在推理任务中的性能和可靠性。

少样本学习仅向 LLM 提供与任务相关的少量输入和输出示例,这使 LLM 能够单独从示范中推断出用户的意图和目标。精心选择、排序和格式化的示例可以极大地提高模型的推理能力。然而,少样本学习可能会受到偏见和变异性的影响,不过添加明确的指令可以使模型更清楚地理解意图。总体而言,提示词结合了指令和示例的优势,以最大程度地引导 LLM 完成手头任务。与手动设计提示不同,像自动提示调整这样的方法通过直接优化嵌入空间上的前缀标记来学习最佳提示,其目标是在给定输入的情况下增加期望输出的可能性。总体而言,提示工程是一个活跃的研究领域,它的目标是将大型预训练的 LLM 与各种人类的意图保持一致。精心设计的提示可以有效地引导模型而无须昂贵的重新训练。

9.3.2 提示模板

提示是用户向 LLM 提供的指令和示例,用于引导 LLM 的行为。提示模板化是指创建可重复使用的提示模板,它可通过不同的参数进行配置。LangChain 提供了用于在 Python 中创建提示模板的工具,用户可以创建一个基本的提示模板,代码如下:

```
from langchain import PromptTemplate
prompt = PromptTemplate("Tell me a {adjective} joke about {topic}")
```

此模板有两个输入变量{adjective} 和 {topic},用户可以通过赋值格式化它们,代码如下:

```
prompt.format(adjective="funny", topic="chickens")
#Output: "Tell me a funny joke about chickens"
```

模板格式默认为 Python f-strings,但也支持 Jinja2。提示模板可以组合成管道,其中一个模板的输出作为输入传递给下一个模板,这样的设计允许模块的重复使用。

对于前面章节中提到的对话 Agent,用户需要设置聊天提示模板,代码如下:

```
//第 9 章/ chat_prompt.py
from langchain.prompts import ChatPromptTemplate
template = ChatPromptTemplate.from_messages([
  ("human", "Hello, how are you?"),
  ("ai", "I am doing great, thanks!"),
  ("human", "{user_input}"),
```

```
])
template.format_messages(user_input="What is your name?")
```

上面的代码格式化了聊天消息列表而不是使用字符串,这对对话历史记录可能很有用。前面的章节介绍了不同的记忆方法,在这种情况下这些方法同样相关,以确保模型的输出是和用户的输入相关的。LangChain 提供了 Python API,可以方便地创建模板并动态地格式化它们。模板可以组合成管道以实现模块化,其中高级提示工程进一步优化了提示的使用。

9.3.3 高级提示词工程

LangChain 提供了一些工具以实现高级提示工程策略,如小样本学习、动态样本选择和链式推理。

1. 少样本学习

FewShotPromptTemplate 允许仅向模型显示任务的几个演示示例,而无须进行明显的说明,代码如下:

```
//第 9 章/ FewShotPromptTemplate.py
from langchain.prompts import FewShotPromptTemplate, PromptTemplate
example_prompt = PromptTemplate("{input} -> {output}")
examples = [
  {"input": "2+2", "output": "4"},
  {"input": "3+3", "output": "6"}
]
prompt = FewShotPromptTemplate(
  examples=examples,
  example_prompt=example_prompt
)
```

在上面的代码中模型必须仅从示例中推断要执行的操作。

2. 动态样本选择

如果用户要为每个自己的输入提供量身定制的示例,FewShotPromptTemplate 则可以接受 ExampleSelector 来智能地进行选择,代码如下:

```
//第 9 章/ ExampleSelector.py
from langchain.prompts import SemanticSimilarityExampleSelector
selector = SemanticSimilarityExampleSelector(...)
prompt = FewShotPromptTemplate(
  example_selector=selector,
  example_prompt=example_prompt
)
```

ExampleSelector 实现(如 SemanticSimilarityExampleSelector)会自动为每个输入查找

最相关的示例。

3. 链式推理

当要求 LLM 对一个问题进行推理时,在陈述最终答案之前,让它解释其推理通常更有效,代码如下:

```
//第 9 章/ reasoning_prompt.py
from langchain.prompts import PromptTemplate
reasoning_prompt = "Explain your reasoning step-by-step. Finally, state the answer: {question}"
prompt = PromptTemplate(
  reasoning_prompt=reasoning_prompt,
  input_variables=["questions"]
)
```

在上面的提示词中让 LLM 首先根据逻辑思考问题,而不仅是猜测答案,然后让 LLM 在此之后为输出的答案进行合理化证明,这被称为零阶链式思维。要求 LLM 解释其思考过程与其核心能力是非常契合的。少量示例的链式思维提示则是一种少量提示,其中推理是作为示例解决方案的一部分进行解释的,旨在鼓励 LLM 在做出决策之前解释其推理。研究人员已经证明这种提示方式可以带来更准确的结果,然而这种性能提升与模型规模成正比,对于较小的模型,这种改进几乎是微不足道的,甚至是负面的。在思维树(Tree of Thought,ToT)提示中,通常会为给定提示生成多个解决问题的步骤或方法,然后使用 LLM 对这些步骤进行评论。接下来介绍使用 LangChain 实现 ToT 的示例。首先使用 PromptTemplates 定义所需的 4 个链组件,即一个解决方案模板、一个评估模板、一个推理模板和一个排名模板,代码如下:

```
//第 9 章/ solution_template.py
solutions_template = """
Generate {num_solutions} distinct solutions for {problem}. Consider factors like {factors}.
Solutions:
"""
solutions_prompt = PromptTemplate(
  template=solutions_template,
  input_variables=["problem", "factors", "num_solutions"]
)
```

接下来让 LLM 评估这些解决方案,代码如下:

```
//第 9 章/ evaluation_template.py
evaluation_template = """
Evaluate each solution in {solutions} by analyzing pros, cons, feasibility, and probability of success.
```

```
Evaluations:
"""
evaluation_prompt = PromptTemplate(
    template=evaluation_template,
    input_variables=["solutions"]
)
```

然后需要设置推理的模板,告诉 LLM 如何进行推理,代码如下:

```
//第 9 章/ reasoning_template.py
reasoning_template = """
For the most promising solutions in {evaluations}, explain scenarios,
implementation strategies, partnerships needed, and handling potential
obstacles.
Enhanced Reasoning:
"""
reasoning_prompt = PromptTemplate(
    template=reasoning_template,
    input_variables=["evaluations"]
)
```

根据到目前为止的推理对这些解决方案进行排名,代码如下:

```
//第 9 章/ ranking_template.py
ranking_template = """
Based on the evaluations and reasoning, rank the solutions in {enhanced_
reasoning} from most to least promising.
Ranked Solutions:
"""
ranking_prompt = PromptTemplate(
    template=ranking_template,
    input_variables=["enhanced_reasoning"]
)
```

从这些模板创建链,然后将它们放在一起,代码如下:

```
//第 9 章/ chain_together.py
chain1 = LLMChain(
    llm=SomeLLM(),
    prompt=solutions_prompt,
    output_key="solutions"
)
chain2 = LLMChain(
    llm=SomeLLM(),
    prompt=evaluation_prompt,
```

```
    output_key="evaluations"
)
```

最后将这些链连接到一个 SequentialChain，代码如下：

```
//第 9 章/SequentialChain.py
tot_chain = SequentialChain(
    chains=[chain1, chain2, chain3, chain4],
    input_variables=["problem", "factors", "num_solutions"],
    output_variables=["ranked_solutions"]
)
tot_chain.run(
    problem="Prompt engineering",
    factors="Requirements for high task performance, low token use, and few calls
to the LLM",
    num_solutions=3
)
```

上面的整个流程使用户能够在推理过程的每个阶段利用 LLM 的能力。ToT 方法有助于通过促进探索来避免死胡同。这些技术通过提供更清晰的指示、使用有针对性的数据进行微调、采用问题分解策略、结合多样化的采样方法、整合验证机制和采用概率建模框架，共同提升了 LLM 在复杂任务中推理能力的准确性、一致性和可靠性。对于解锁 LLM 推理能力、未来模型和提示技术的潜在进展，提示设计至关重要，而这些原则和技术形成了研究人员和从业者处理 LLM 时的宝贵工具包。

9.4 总结

本章的开始部分讨论了生成式 AI 的基本原理，特别是 LLM 及其训练。主要关注了预训练阶段，即调整模型以适应单词和文本更广泛段落之间的相关性。对齐是评估模型输出与期望之间的关系，而调整则是确保输出符合期望的过程。调整能够引导生成式 AI，提高其安全性和质量，但这并不是完整的解决方案。本章的重点是调整方面，特别是通过微调和提示实现。在微调中，语言模型根据以自然语言指令形式提出的许多任务示例及相应的响应进行训练。通常这是通过强化学习与人类反馈（RLHF）实现的，涉及在人类生成的（提示，响应）数据集上进行训练，然后通过人类反馈进行强化学习，然而，其他一些已经开发出的技术显示出了在资源占用较小的情况下能够产生具有竞争力的结果。在本章的第 1 个示例中实施了对一个小型开源模型进行问答方面的微调。

有许多技术可以提高 LLM 在复杂推理任务中的可靠性，包括分步提示、备选选择和推理提示、问题分解、多个响应采样及使用单独的验证器模型。这些方法已经证明能够提高推理任务中的准确性和一致性。LangChain 提供了解锁高级提示策略的构建模块，如少样本学习、动态样本选择和链式推理分解，这些内容本章都在示例代码中进行了一一展示。精心设计的提示工程是将 LLM 与复杂目标对齐的关键。通过分解问题和增加冗余性可以提高推理的可靠性。本章讨论的原则和技术为专业人士提供了处理 LLM 的工具包。通过本章的探讨，可以预期在模型训练和提示技术方面未来会有更大的进展。随着这些方法和 LLM 的不断发展，未来人们和 LLM 的交互可能会变得更加有效，这也为 LLM 进入更广泛的应用领域奠定了坚实的基础。

第 10 章 生产环境 LLM

迄今为止本书已经讨论了模型、Agent 和 LLM 应用，以及不同的使用案例，但是在实际生产环境中往往需要确保性能和满足监管要求，模型和应用往往需要进行大规模部署，并且最终需要进行流程化监控。本章将涵盖评估和可观测性，探讨 AI 和决策模型（包括生成式 AI 模型）的治理和生命周期管理。虽然离线评估在受控的环境中提供了对模型能力的初步了解，但在线的生产环境中的评测工具可以持续监测 LLM 和相关应用在实时环境中的性能。在模型生命周期的不同阶段，这两者都至关重要并相互补充，这样可以确保 LLM 时刻保持最佳的运行状态。本章将讨论一些针对不同情况的工具，并提供相关示例。本章还将讨论围绕 LLM 构建的模型和应用的部署，全面概述使用 Fast API 和 Ray Serve 进行部署的可用工具和示例。

10.1 引言

把 LLM 从研究阶段引入实际生产环境存在着重大挑战。LLM 近年来因其生成类似人类文本的能力而引起了广泛关注。从创意写作到对话聊天机器人，这些生成式 AI 模型在各个行业都有着多样化的应用，然而，将这些复杂的神经网络系统从研究转换为真实世界的部署涉及重大挑战。本章将探讨如何负责任地将生成式人工智能模型投入生产中的实际考虑因素和最佳实践，其中包括推断和服务的计算要求、优化技术及围绕数据质量、偏见和透明度的关键问题。当扩展到数千用户时，架构和基础设施的决策可能会对生成式 AI 解决方案产生重大影响。同时，保持严格的测试、审计和道德及法律保障对于可信赖的部署至关重要。在生产中部署由模型和 Agent 组成的应用及它们的工具存在着需要解决的几个关键问题。

（1）数据质量和偏见：训练数据可能会引入偏见，并在模型输出中反映出来，因此，进行谨慎的数据筛选和监控模型输出是至关重要的。

（2）道德/合规考虑：LLM 可能生成有害、带偏见或误导性的内容，因此必须建立审查流程和安全准则，以防止滥用。

（3）资源要求：LLM 的训练和服务需要大量计算资源。高效的基础设施对于成本效益的大规模部署至关重要。

（4）漂移或性能下降：模型需要持续监测诸如数据漂移或随时间性能下降等问题。

（5）解释能力不足：LLM 通常是黑盒子，其行为和决策不透明，因此，解释性工具对于透明度至关重要。

将训练完毕的 LLM 从研究阶段引入实际的生产环境，需要解决许多关于可扩展性、监控和意外行为等方面的复杂问题。在部署能力尚不完全的模型时，需要围绕可扩展性、可解释性、测试和监控认真地进行规划。微调、安全干预和防御性设计等技术能够帮助开发符合相关法律规定的应用程序。应对上面列出的一些挑战可能有以下应对的方案：

（1）评估：稳健的基准数据集和指标对于衡量模型能力、回归性能及与目标一致性至关重要。具体评估指标应基于任务谨慎选择。

（2）检索增强：检索外部知识为减少虚构内容、添加超越预训练数据的最新信息提供有用的背景。

（3）微调：在特定任务数据上进一步调整 LLM 可以提高目标用例的性能。前面章节提到的 LoRA 技术可以减少开销。

（4）缓存：存储模型输出可以显著地降低重复查询的延迟和成本，但缓存的有效性需要慎重考虑。

（5）防护：对模型输出进行句法和语义验证以确保可靠性。引导技术可以直接改变 LLM 的输出结构。

（6）监控：持续跟踪指标、模型行为和用户满意度，以便获得关于模型问题和业务影响的见解。

第 9 章已经涵盖了安全对齐技术，例如用于降低生成有害输出风险的特定功能的 AI。由于 LLM 有可能生成有害或误导性的内容，因此建立道德准则和审核流程以防止传播错误信息、仇恨言论或任何其他有害输出至关重要。人类审核员在评估和过滤生成内容方面发挥着至关重要的作用，以确保符合道德标准。不仅出于法律、道德和声誉方面的考虑，而且为了保持性能，需要不断地评估模型的性能和输出，以检测数据漂移或性能丧失等问题。

LLM 在部署时需要大量的计算资源，因为它们往往很大且很复杂。这些大量的计算资源包括高性能硬件，例如 GPU 或 TPU，以处理推理过程中所涉及的大量计算。由于它们对资源的密集需求，扩展 LLM 或生成式 AI 模型可能具有挑战性。随着模型规模的增加，训练和推断的计算需求也呈指数级增长。分布式技术，如数据并行或模型并行，通常用于在多台机器或 GPU 之间分发工作负载，这可以加快训练和推理以节省时间。另外，扩展后的 LLM 还涉及管理与这些模型相关的大量数据的存储和检索，它们往往需要高效的数据存储和检索系统。

生产环境部署还涉及推断速度和延迟问题，例如可以采用模型压缩、量化或硬件特定优化等技术，以确保进行高效部署，这些技术在第 9 章做了介绍。LLM 或生成式 AI 模型通常被视为黑匣子，这意味着人们很难理解它们是如何得出决策或生成输出的。解释性技术旨在对这些模型的内部运作规律进行深刻洞察,相关技术包括注意力可视化、特征重要性分析或为模型输出生成解释等方法。解释性在需要透明度和问责制的领域(如医疗保健、金融或法律系统)往往非常重要。

正如第 9 章所讲述的内容,LLM 可以针对特定任务或领域进行微调,以提高在特定用例上的性能。迁移学习允许模型利用预先训练的知识并将其适应新任务。通过深入的规划和准备,生成式 AI 有望将行业从创意渗透到客户服务。然而,在这些系统继续渗透不同领域时,审慎地应对这些系统的复杂性仍然至关重要。本章的内容旨在为 LLM 的开发团队提供一份实用指南,里面将提到数据整理、模型开发、基础设施、监控和透明度等策略。在继续深入讨论之前,先介绍一些基本的术语。

MLOps 是一种范式,专注于可靠高效地在生产环境中部署和维护机器学习模型。它将 DevOps 的实践与机器学习相结合,将算法从实验性系统过渡到生产系统。MLOps 旨在增加自动化程度,提高生产模型的质量,并解决问题和满足监管要求。LLMOps 是 MLOps 的一个专门子类别,指的是作为产品的一部分对 LLM 进行微调和运营所需的操作能力和基础设施。虽然它与 MLOps 的概念可能并没有明显的不同之处,但区别涉及处理、改进和部署大规模语言模型所需的特定要求。

术语 LMOps 比 LLMOps 更具包容性,因为它包含了各种类型的语言模型,包括 LLM 和较小的生成模型。该术语承认了语言模型的不断扩展及它们在操作环境中的相关性。基础模型编排(Foundational Model Orchestration,FOMO)专门解决了在使用基础模型时遇到的问题。它包括管理多步骤流程、与外部资源集成及协调这些模型所涉及的工作流程。

ModelOps 专注于 AI 和决策模型在部署时的治理和生命周期管理。更广泛地说,AgentOps 涉及 LLM 和其他 AI Agent 的运营管理,确保它们采取正确的输出行为及管理其环境和资源访问,并促进 Agent 之间的互动,同时解决与意外结果和不兼容目标相关的问题。

总结来讲,尽管 FOMO 强调了专门处理基础模型时的独特挑战,但 LMOps 提供了对除基础模型外更广泛语言模型范围的更全面和包容性的覆盖。LMOps 承认了语言模型在各种操作用例中的多功能性和日益重要性,但仍属于 MLOps 的更广泛范畴。最后,AgentOps 明确强调了包含具有特定启发式的生成模型的代理的交互性,并包括工具。所有这些非常专业化的术语的出现突显了该领域的快速发展。MLOps 是一个在行业中广泛使用并得到重视和采纳的已确立术语,因此本章的其余部分将坚持使用 MLOps。在将任何 Agent 或模型投入生产之前,应该首先评估其输出,本章将专注于 LangChain 提供的评估方法。

10.2 LLM 应用评估

对 LLM 进行评估,无论是作为独立实体还是与 Agent 结合都至关重要,因为在实际生产环境中要确保它们能够正确运行并产生可靠结果,这是机器学习生命周期的一个重要组成部分。评估过程决定了模型在有效性、可靠性和效率方面的表现。对 LLM 进行评估的目标是了解其优势和劣势，提高准确性和效率，减少错误，从而最大程度地增强其在解决实际问题中的实用性。此评估过程通常在开发阶段离线进行。离线评估在受控测试条件下提供模型性能的初步评测，包括超参数调整、与同行模型或已建立标准的基准比较等测试。它们是在部署前完善模型的第 1 步。

评估可以确定 LLM 生成的输出是否和任务相关、是否准确和是否有价值。LangChain 中有多种评估 LLM 输出的方式，包括比较链输出、成对字符串比较、字符串距离和嵌入距离。评估结果可通过对输出进行比较来确定首选模型。还可以计算置信区间和 p 值，以评估评估结果的可靠性。LangChain 提供了几种评估 LLM 输出的工具。一种常见的方法是使用 PairwiseStringEvaluator 比较不同模型或提示的输出。这会提示一个评估模型在相同输入下选择两个模型输出，并汇总结果以确定整体上表现更佳的模型，其他评估器允许根据特定标准（如正确性、相关性和简洁性）评估模型输出。CriteriaEvalChain 可以根据自定义或预定义的原则对输出进行评分，无须参考标签，还有其他的一些方法，例如可以通过指定不同的聊天模型（如 ChatGPT）作为评估器来配置评估模型。

10.2.1　比较两个输出

这种评估方法需要一位评估者、一个输入数据集，以及两个或更多的 LLM、链条或 Agent。整个评估过程涉及几个步骤。首先，使用 load_evaluator() 函数加载评估器，确定所需的评估器类型（例如，pairwise_string）。接着，利用 load_dataset() 函数加载输入数据集。随后，通过相应的配置初始化要比较的 LLM、链条或 Agent，包括初始化语言模型和任何其他必需的工具或代理。生成每个模型的输出通常以批处理方式进行，以便提高效率。最后，通过比较每个模型对于每个输入的输出来评估结果。通常采用随机选择的顺序进行比较，以减少位置偏差。以下是来自 LangChain 文档的成对字符串比较的示例，代码如下：

```
//第 10 章/ pairWise.py
from langchain.evaluation import load_evaluator
evaluator = load_evaluator("labeled_pairwise_string")
evaluator.evaluate_string_pairs(
    prediction="there are three dogs",
    prediction_b="4",
    input="how many dogs are in the park? ",
    reference="four",
)
```

评估器的输出如下：

```
{'reasoning': 'Both responses are relevant to the question asked, as they
both provide a numerical answer to the question about the number of dogs in the
park. However, Response A is incorrect according to the reference answer, which
```

```
        states that there are four dogs. Response B, on the other hand, is correct as it
        matches the reference answer. Neither response demonstrates depth of thought, as
        they both simply provide a numerical answer without any additional information or
        context. \n\nBased on these criteria, Response B is the better response.\n',
         'value': 'B',
         'score': 0}
```

评估结果包括一个介于 0 和 1 之间的数值,表示 Agent 的有效性,有时还会附带评估过程的推理和给出分数的理由。在这个例子中,与参考模型相比,两个结果都是错误的。

10.2.2 基于标准的比较

LangChain 提供了几个预定义的评估器,它们各自基于不同的评估标准。这些评估器可以根据特定的评分标准或标准集来评估输出。一些常见的标准包括简洁性、相关性、正确性、连贯性、实用性和争议性。CriteriaEvalChain 允许用户根据自定义或预定义的标准评估模型的输出。它提供了一种验证 LLM 或 Chain 的输出是否符合定义的一组标准的方式。用户可以使用此评估器来评估生成的输出的正确性、相关性、简洁性和其他特性。CriteriaEvalChain 可以配置为使用或不使用参考标签。如果没有参考标签,则评估器依赖于 LLM 的预测答案,并根据指定的标准对其进行评分。如果有参考标签,则评估器会将预测的答案与参考标签进行比较,并确定其是否符合标准。LangChain 中默认使用的评估 LLM 是 GPT-4,但是用户也可以通过指定其他聊天模型(例如 ChatAnthropic 或 ChatOpenAI)和所需的设置(例如温度)来配置评估 LLM。评估器可以通过将 LLM 对象作为参数传递给 load_evaluator() 函数来加载自定义 LLM。LangChain 支持自定义评估标准和预定义原则。用户可以使用包含 criterion_name:criterion_description 对的字典来定义自定义标准。这些标准可根据特定要求或标准来评估输出。以下是一个简单的代码示例:

```
//第 10 章 / criteria.py
custom_criteria = {
    "simplicity": "Is the language straightforward and unpretentious?",
    "clarity": "Are the sentences clear and easy to understand?",
    "precision": "Is the writing precise, with no unnecessary words or details?",
    "truthfulness": "Does the writing feel honest and sincere?",
    "subtext": "Does the writing suggest deeper meanings or themes?",
}
evaluator = load_evaluator("pairwise_string", criteria=custom_criteria)
evaluator.evaluate_string_pairs(
    prediction="Every cheerful household shares a similar rhythm of joy; but
sorrow, in each household, plays a unique, haunting melody.",
    prediction_b="Where one finds a symphony of joy, every domicile of happiness
resounds in harmonious,"
    " identical notes; yet, every abode of despair conducts a dissonant
orchestra, each"
```

```
        " playing an elegy of grief that is peculiar and profound to its own
existence.",
    input="Write some prose about families.",
)
```

运行上述代码可以对两个输出进行非常细致的比较,结果如下:

```
//第 10 章 / criteriaResult.py
{'reasoning': 'Response A is simple, clear, and precise. It uses straightforward
language to convey a deep and sincere message about families. The metaphor of
music is used effectively to suggest deeper meanings about the shared joys and
unique sorrows of families.\n\nResponse B, on the other hand, is less simple and
clear. The language is more complex and pretentious, with phrases like "domicile
of happiness" and "abode of despair" instead of the simpler "household" used in
Response A. The message is similar to that of Response A, but it is less
effectively conveyed due to the unnecessary complexity of the language. \n\
nTherefore, based on the criteria of simplicity, clarity, precision,
truthfulness, and subtext, Response A is the better response.\n\n[[A]]', 'value':
'A', 'score': 1}
```

用户也可以使用 LangChain 中提供的预定义原则。这些原则旨在评估产出的道德、有害和敏感信息。在评估中使用原则可以对生成的文本进行更有针对性的评估。

10.2.3　字符串和语义比较

LangChain 支持字符串比较和距离度量来评估 LLM 的输出。像 Levenshtein 和 Jaro 这样的字符串距离度量提供了预测字符串和参考字符串之间相似性的定量度量。SentenceTransformers 模型则主要用于计算生成文本和期望文本之间的语义相似度的嵌入距离。嵌入距离评估器可以使用基于 GPT-4 或 HuggingFace Embeddings 等嵌入模型,以此来计算预测字符串和参考字符串之间的向量距离。这可以衡量两个字符串之间的语义相似性,并提供有关生成文本质量的见解。以下是 LangChain 文档的一个简单示例代码:

```
from langchain.evaluation import load_evaluator
evaluator = load_evaluator("embedding_distance")
evaluator.evaluate_strings(prediction="I shall go", reference="I shan't go")
```

评估器返回得分 0.0966466944859925。用户可以在 load_evaluator() 调用中的 embeddings 参数中更改所使用的嵌入模型。通常,这比更加旧的字符串距离度量法的效果更好,但这些方法也可用于简单的单元测试和准确性评估。字符串比较评估器将预测的字符串与参考字符串或输入进行比较。字符串距离评估器使用 Levenshtein 或 Jaro 等距离度量方法,以衡量预测字符串与参考字符串之间的相似度。这提供了预测字符串与参考字符串相似程度的定量度量。最后,还有一个 Agent 轨迹评估器,其中使用 evaluate_agent_ trajectory()方法来评估输入、预测和 Agent 轨迹。用户还可以使用 LangSmith 来对 LLM 的性能与数据集进行比较,这将在后面详细讨论。

10.2.4 基准数据集

LangChain 中的 LangSmith 可以评估模型在数据集上的性能。下面通过一个示例来详细介绍。首先,用户需要在 LangSmith 上创建一个账号,网址为 https://smith.langchain.com/。进入整个网址用户可以获取一个 API 密钥,并将其设置为环境变量 LANGCHAIN_API_KEY。这里还可以设置项目 ID 和跟踪的环境变量,代码如下:

```
import os
os.environ["LANGCHAIN_TRACING_V2"] = "true"
os.environ["LANGCHAIN_PROJECT"] = "My Project"
```

这会将 LangChain 配置为记录跟踪。如果不告诉 LangChain 项目 ID,则将记录默认项目。完成此设置后,运行 LangChain 链或 Agent 时,用户将能够在 LangSmith 上看到跟踪,代码如下:

```
from langchain.chat_models import ChatOpenAI
llm = ChatOpenAI()
llm.predict("Hello, world!")
```

我们将在 LangSmith 上看到这样的内容:LangSmith 允许我们在 LangSmith 项目页面上列出到目前为止的所有运行,代码如下:

```
from langsmith import Client
client = Client()
runs = client.list_runs()
print(runs)
```

用户可以列出来自特定项目的运行,也可以按 run_type 列出运行,例如 chain。每次运行都带有输入和输出,代码如下:

```
print(f"inputs: {runs[0].inputs}")
print(f"outputs: {runs[0]. outputs}")
```

用户可以使用 create_example_from_run()函数从现有的 Agent 运行中创建数据集,也可以从其他任何函数创建数据集。下面介绍如何创建包含一组问题的数据集,代码如下:

```
//第 10 章/question_dataset.py
questions = [
    "A ship's parts are replaced over time until no original parts remain. Is it
still the same ship? Why or why not?",  #The Ship of Theseus Paradox
```

```
    "If someone lived their whole life chained in a cave seeing only shadows, how would they react if freed and shown the real world?",   # Plato's Allegory of the Cave
    "Is something good because it is natural, or bad because it is unnatural? Why can this be a faulty argument?",   #Appeal to Nature Fallacy
    "If a coin is flipped 8 times and lands on heads each time, what are the odds it will be tails next flip? Explain your reasoning.",   #Gambler's Fallacy
    "Present two choices as the only options when others exist. Is the statement \"You're either with us or against us\" an example of false dilemma? Why?",   #False Dilemma
    "Do people tend to develop a preference for things simply because they are familiar with them? Does this impact reasoning?",   #Mere Exposure Effect
    "Is it surprising that the universe is suitable for intelligent life since if it weren't, no one would be around to observe it?",   #Anthropic Principle
    "If Theseus' ship is restored by replacing each plank, is it still the same ship? What is identity based on?",   #Theseus' Paradox
    "Does doing one thing really mean that a chain of increasingly negative events will follow? Why is this a problematic argument?",   #Slippery Slope Fallacy
    "Is a claim true because it hasn't been proven false? Why could this impede reasoning?",   #Appeal to Ignorance
]
shared_dataset_name = "Reasoning and Bias"
ds = client.create_dataset(
    dataset_name=shared_dataset_name, description="A few reasoning and cognitive bias questions",
)
for q in questions:
    client.create_example(inputs={"input": q}, dataset_id=ds.id)
```

然后可以在数据集上运行 LLM Agent 或链，代码如下：

```
//第 10 章/ agent_dataset.py
from langchain.chat_models import ChatOpenAI
from langchain.chains import LLMChain
llm = ChatOpenAI(model="gpt-4", temperature=0.0)
def construct_chain():
    return LLMChain.from_string(
        llm,
        template="Help out as best you can.\nQuestion: {input}\nResponse: ",
    )
```

这里使用构造函数来初始化每个输入。为了根据此数据集评估模型性能，需要定义一个评估器，代码如下：

```
//第 10 章/ evaluator.py
from langchain.evaluation import EvaluatorType
```

```
from langchain.smith import RunEvalConfig
evaluation_config = RunEvalConfig(
    evaluators=[
        RunEvalConfig.Criteria({"helpfulness": "Is the response helpful?"}),
        RunEvalConfig. Criteria ({ " insightful": " Is the response carefully thought out?"})
    ]
)
```

然后将数据集和评估器传递给 run_on_dataset()以生成指标和反馈,代码如下:

```
from langchain.smith import run_on_dataset
results = run_on_dataset(
  client=client,
  dataset=dataset,
  llm_factory=lambda: my_agent,
  evaluation=evaluation_config
)
```

类似地,用户可以将数据集和评估器传递给 arun_on_dataset()以异步生成指标和反馈。可以在 LangSmith UI 中查看评估员的反馈,以确定需要改进的地方。在 UI 界面上,用户可以单击这些评估中的任何一个来查看一些细节,其中包括 LLM 的原始答案,代码如下:

```
//第 10 章/ LLMrawdata.py
You are assessing a submitted answer on a given task or input based on a set of criteria. Here is the data:
[BEGIN DATA]
***
[Input]: Is something good because it is natural, or bad because it is unnatural? Why can this be a faulty argument?
***
[Submission]: The argument that something is good because it is natural, or bad because it is unnatural, is often referred to as the "appeal to nature" fallacy. This argument is faulty because it assumes that what is natural is automatically good or beneficial, and what is unnatural is automatically bad or harmful. However, this is not always the case. For example, many natural substances can be harmful or deadly, such as certain plants or animals. Conversely, many unnatural things, such as modern medicine or technology, can greatly benefit our lives. Therefore, whether something is natural or unnatural is not a reliable indicator of its value or harm.
***
[Criteria]: insightful: Is the response carefully thought out?
***
```

```
[END DATA]
Does the submission meet the Criteria? First, write out in a step by step manner
your reasoning about each criterion to be sure that your conclusion is correct.
Avoid simply stating the correct answers at the outset. Then print only the single
character "Y" or "N" (without quotes or punctuation) on its own line corresponding
to the correct answer of whether the submission meets all criteria. At the end,
repeat just the letter again by itself on a new line.
```

得到以下评估：

```
The criterion is whether the response is insightful and carefully thought out.
The submission provides a clear and concise explanation of the "appeal to nature"
fallacy, demonstrating an understanding of the concept. It also provides
examples to illustrate why this argument can be faulty, showing that the
respondent has thought about the question in depth. The response is not just a
simple yes or no, but a detailed explanation that shows careful consideration of
the question.
Therefore, the submission does meet the criterion of being insightful and
carefully thought out.
Y
Y
```

提高一些类型问题的性能的一种方法是执行少量提示。LangSmith 也可以帮助解决这个问题。用户可以在 LangSmith 文档中找到更多示例。

10.3　部署 LLM 应用

鉴于 LLM 在不同领域日益得到广泛应用，了解如何有效地将模型和应用程序部署到生产环境中变得至关重要，其中，部署服务和框架可以帮助克服许多技术问题。生产环境部署 LLM 应用程序或带有生成式 AI 的应用程序有许多不同的方式。将应用程序部署到生产环境需要运维人员对生成式 AI 生态系统进行研究并具备相应的知识，其中包括以下几点。

(1) 模型和 LLM 即服务：LLM 和其他模型直接运行或作为 API 运行在供应商提供的基础设施上。

(2) 模型推理机制：检索增强生成(RAG)、思维树等。

(3) 向量数据库：帮助检索与提示相关的上下文信息。

(4) 提示工程工具：这些工具促进了不需要昂贵微调或敏感数据的上下文学习。

(5) 预训练和微调：用于针对特定任务或领域专用模型。

(6) 提示记录、测试和分析：受到对了解和改进大型语言模型性能的愿望所启发的新兴领域。

(7) 自定义 LLM 堆栈：一套用于塑造和部署基于开源模型构建的解决方案的工具。

本书在前面的章节中依次讨论了模型、模型推理机制、向量数据库及提示和微调。在本章中将专注于研究模型部署的日志记录、监视和定制工具。访问 LLM 的方法无外乎通常使用外部 LLM 提供商或自托管模型。如果使用外部提供商，则计算资源由提供商（如 OpenAI 或 Anthropic）承担，而 LangChain 在其中的作用则是促进业务逻辑的实现。另一方面，使用自托管开源 LLM 则可以显著地降低成本、减少延迟并可以有效地保护隐私问题。一些带有基础设施的工具提供了完整的解决方案。例如，用户可以使用 Chainlit 部署 LangChain Agent，创建类似 ChatGPT 界面的 Chainlit，其中一些关键功能包括中间步骤可视化、元素管理和显示（图像、文本、轮播图等），以及在云端部署。BentoML 是一个框架，可以将机器学习应用容器化，使其成为可以独立运行和扩展的微服务，并自动生成 OpenAPI 和 gRPC 端点。用户还可以将 LangChain 部署到不同的云服务器端点，例如 Azure Machine Learning Online Endpoint。借助 Steamship，LangChain 开发人员可以快速部署其应用，其中包括生产就绪的端点、跨依赖项的水平扩展、应用状态的持久存储、多租户支持等。部署 LLM 常用的服务和框架见表 10-1。

表 10-1 部署 LLM 常用的服务和框架

名 称	描 述	类 型
Streamlit	用于构建和部署 Web 应用的开源 Python 框架	Framework
Gradio	允许用户将模型包装在界面中，并托管在 HuggingFace 上	Framework
Chainlit	构建和部署类似 ChatGPT 的对话式应用	Framework
Apache Beam	用于定义和编排数据处理工作流的工具	Framework
Vercel	用于部署和规模化 Web 应用的平台	Cloud Service
FastAPI	用于构建 API 的 Python Web 框架	Framework
Fly.io	具有自动规模化和全球 CDN 的应用托管平台	Cloud Service
DigitalOcean App Platform	用于构建、部署和扩展应用的平台	Cloud Service
Google Cloud	Cloud Run 等服务，用于托管和扩展容器化应用	Cloud Service
Steamship	用于部署和扩展模型的 ML 基础架构平台	Cloud Service
Langchain-serve	将 LangChain Agent 作为 Web API 提供的工具	Framework
BentoML	模型服务、打包和部署框架	Framework
OpenLLM	为商业 LLM 提供开放的 API	Cloud Service
Databutton	用于构建部署模型工作流的无代码平台	Framework
Azure ML	Azure 上适用于模型的托管 ML 操作服务	Cloud Service

表 10-1 中的所有服务和框架都有各自比较详细的文档，里面都包含了具体的使用案

例,供开发者参考。前面的章节已经展示了使用 Streamlit 和 Gradio 的示例,并讨论了如何将它们部署到 HuggingFace Hub。运行 LLM 应用程序有几个基本的要求:

(1) 可扩展基础设施以处理计算密集型模型和潜在的流量激增问题。

(2) 低延迟以实时提供模型输出。

(3) 持久性存储以管理长时间的对话和应用状态。

(4) API 用于集成到最终用户应用程序和监控。

(5) 日志记录以跟踪指标和模型行为。

保持较高的成本效率在用户量较大且 LLM 服务成本高昂的情况下可能会面临挑战。管理效率的策略包括自行托管模型、根据流量自动扩展资源分配、使用抢占式实例、独立扩展和批量请求以更好地利用 GPU 资源。工具和基础设施的选择决定了这些要求之间的权衡取舍。灵活性和易用性非常重要,因为在实际生产环境中总是希望能够快速迭代,这对于机器学习和 LLM 领域动态性质至关重要。避免被束缚于一种解决方案是至关重要的,这意味着具有灵活性和可扩展性的服务层是关键。模型组合和云服务提供商的选择极大地提升了这种灵活性,基础设施即代码(IaC)工具(如 Terraform、CloudFormation 或 Kubernetes YAML 文件)可以可靠且快速地重新创建基础设施。此外,持续集成和持续交付(CI/CD)流水线可以自动化测试和部署流程,减少错误,并促进更快地进行反馈和迭代。设计强大的 LLM 应用服务可能是一项复杂的任务,需要在评估服务框架时进行权衡和考虑关键因素。利用这些解决方案进行部署,可以让开发人员专注于开发有影响力的 AI 应用,而不是基础设施。本书在前面的内容中提到,LangChain 与几个开源项目和框架,兼容良好,如 Ray Serve、BentoML、OpenLLM、Modal 和 Jina。接下来的部分将带领读者部署一个基于 FastAPI 的聊天服务 Web 服务器。

10.3.1 FastAPI

FastAPI 是部署 Web 服务器非常受欢迎的框架。作为一种现代、高性能的 Web 框架,它被设计为快速、易用和高效地构建 Python 的 API。Lanarky 是一个小型的开源库,用于部署 LLM 应用程序,它提供了对 Flask API 和 Gradio 的便捷封装,可用于 LLM 应用程序的部署。这意味着开发者可以获得 REST API 端点并在浏览器内使数据可视化,而且只需几行代码。

REST API(Representational State Transfer Application Programming Interface)是一组规则和协议,允许不同的软件应用程序在互联网上相互通信。它遵循 REST 的原则,这是一种设计网络化应用程序的架构风格。REST API 使用 HTTP 方法(如 GET、POST、PUT、DELETE)对资源进行操作,通常以标准化格式(如 JSON 或 XML)发送和接收数据。

在 FastAPI 的文档中,有几个示例,包括使用源链进行检索问答(Retrieval QA with Sources Chain)、对话检索应用程序(Conversational Retrieval App)和 Zero Shot Agent。这里将使用 Lanarky 建立一个 Web 服务器,它与 Gradio 集成,其中需要创建一个 ConversationChain 实例,包括 LLM 模型和设置,并定义处理 HTTP 请求的路由。首先导入必要的依赖项,包括用于创建 Web 服务器的 FastAPI,用于与 Gradio 集成的 mount_gradio_app,Langchain 中的 ConversationChain 和 ChatOpenAI 用于处理 LLM 对话,以及其他所需的模块,代码如下:

```
//第 10 章/ Dependencies.py
from fastapi import FastAPI
from lanarky.testing import mount_gradio_app
from langchain import ConversationChain
from langchain.chat_models import ChatOpenAI
from lanarky import LangchainRouter
from starlette.requests import Request
from starlette.templating import Jinja2Templates
```

接下来定义一个 create_chain() 函数,以此来创建 ConversationChain 的实例,并指定 LLM 模型及其设置,代码如下:

```
//第 10 章/ create_chain.py
def create_chain():
    return ConversationChain(
        llm=ChatOpenAI(
            temperature=0,
            streaming=True,
        ),
        verbose=True,
    )
```

然后将链设置为 ConversationChain,代码如下:

```
chain = create_chain()
```

app 变量被分配给 mount_gradio_app,它会创建一个名为 ConversationChainDemo 的 FastAPI 实例,并将其与 Gradio 集成,代码如下:

```
app = mount_gradio_app(FastAPI(title="ConversationChainDemo"))
```

将 templates 变量设置为 Jinja2Templates 类,指定模板所在的目录进行渲染。这指定了网页的显示方式,允许所有类型的自定义,代码如下:

```
templates = Jinja2Templates(directory="webserver/templates")
```

然后使用 FastAPI 装饰器 @app.get 定义用于在根路径处处理 HTTP GET 请求的端点。与此终结点关联的函数返回用于呈现 index.html 模板的模板响应,代码如下:

```
@app.get("/")
```

```
async def get(request: Request):
    return templates.TemplateResponse("index.html", {"request": request})
```

Router 对象被创建为 LangchainRouter 类。该对象负责定义和管理与 ConversationChain 实例相关的路由。用户可以向 Router 添加额外的路由，用于处理基于 JSON 的聊天，甚至可以处理 WebSocket 请求，代码如下：

```
//第 10 章/ langchain_router.py
langchain_router = LangchainRouter(
    langchain_url="/chat", langchain_object=chain, streaming_mode=1
)
langchain_router.add_langchain_api_route(
    "/chat_json", langchain_object=chain, streaming_mode=2
)
langchain_router.add_langchain_api_websocket_route("/ws", langchain_object=
chain)
app.include_router(langchain_router)
```

现在，本例的应用程序就知道如何处理发送到 Router 中定义的指定路由的请求，将它们定向到适当的函数或处理程序进行处理。这里将使用 Uvicorn 运行本例的应用程序。Uvicorn 支持高性能的异步框架，如 FastAPI 和 Starlette。由于其异步性质，它能够处理大量的并发连接，并在重负载下表现良好。可以在终端中运行 Web 服务器，代码如下：

```
uvicorn webserver.chat:app -reload
```

以上命令启动了一个 Web 服务器，在浏览器中用户可以通过本地地址 http://127.0.0.1:8000 查看。这里的 reload 参数特别方便，因为它意味着一旦进行了任何更改，服务器就会自动重新启动。最终完成部署的应用界面如图 10-1 所示。

虽然 Uvicorn 本身不提供内置的负载均衡功能，但它可以与其他工具或技术（如 Nginx 或 HAProxy）协同工作，以在部署设置中实现负载均衡，从而在多个工作进程或实例之间分配传入的客户端请求。将 Uvicorn 与负载均衡器结合使用可实现水平扩展以处理大量流量，缩短客户端响应时间，增强容错能力。10.3.2 节将讲解如何使用 Ray 构建强大且经济高效的生成式 AI 应用程序。下一个例子中包括构建一个简单的搜索引擎，使用 LangChain 进行文本处理，以及使用 Ray 进行扩展、索引和服务。

10.3.2 Ray

Ray 提供了一个灵活的框架，通过在集群间扩展生成式 AI 工作负载，可以应对生产中复杂神经网络的基础设施挑战。Ray 有助于解决一些常见的部署问题，例如低延迟服务、分布式训练和大规模批量推断。Ray 还可以轻松地启动按需微调，或将现有工作负载从单台机器扩展到多台机器。生产环境中需要 Ray 的功能包括以下几个。

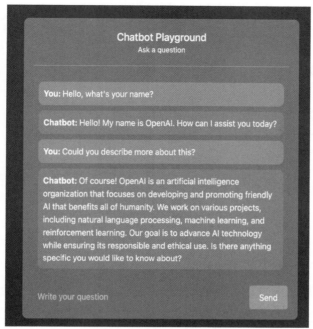

图 10-1 最终完成部署的应用界面

(1) 使用 Ray Train 在 GPU 集群上安排分布式训练作业。
(2) 使用 Ray Serve 以规模化方式部署预训练模型以实现低延迟服务。
(3) 使用 Ray Data 在 CPU 和 GPU 上并行运行 LLM 大规模批量推理。
(4) 编排端到端的生成式 AI 工作流,包括训练、部署和批处理。

本例中将使用 LangChain 和 Ray 构建一个简单的搜索引擎。首先加载和索引 Ray 框架的官方文档,可以快速找到与搜索查询相关的段落,代码如下:

```
//第 10 章/ index_ray_doc.py
#使用 LangChain 加载器加载 Ray 文档
loader = RecursiveUrlLoader("docs.ray.io/en/master/")
docs = loader.load()
#使用 LangChain 拆分器将文档拆分为句子
chunks = text_splitter.create_documents(
    [doc.page_content for doc in docs],
    metadatas=[doc.metadata for doc in docs])
#使用 Transformer 模型将句子嵌入向量中
embeddings = LocalHuggingFaceEmbeddings('multi-qa-mpnet-base-dot-v1')
#通过 LangChain 使用 FAISS 的索引向量
db = FAISS.from_documents(chunks, embeddings)
```

上面的代码通过引入文档、拆分为句子、嵌入句子和索引向量来构建搜索索引。或者,

用户可以通过并行化嵌入步骤来加速索引,代码如下:

```python
//第 10 章/ parallel_embedding.py
#定义分片处理任务
@ray.remote(num_gpus=1)
def process_shard(shard):
    embeddings = LocalHuggingFaceEmbeddings('multi-qa-mpnet-base-dot-v1')
    return FAISS.from_documents(shard, embeddings)
#将块拆分为 8 个分片
shards = np.array_split(chunks, 8)
#并行处理分片
futures = [process_shard.remote(shard) for shard in shards]
results = ray.get(futures)
#合并索引分片
db = results[0]
for result in results[1:]:
    db.merge_from(result)
```

在每个分片上并行运行嵌入可以显著地减少索引时间。使用下面的代码将数据库索引保存到磁盘:

```python
db.save_local(FAISS_INDEX_PATH)
```

FAISS_INDEX_PATH 是任意文件名。这里已将其设置为 faiss_index.db。接下来将使用 Ray Serve 进行搜索查询,代码如下:

```python
//第 10 章/ startray.py
#加载索引和嵌入
db = FAISS.load_local(FAISS_INDEX_PATH)
embedding = LocalHuggingFaceEmbeddings('multi-qa-mpnet-base-dot-v1')
@serve.deployment
class SearchDeployment:
    def __init__(self):
        self.db = db
        self.embedding = embedding

    def __call__(self, request):
        query_embed = self.embedding(request.query_params["query"])
        results = self.db.max_marginal_relevance_search(query_embed)
        return format_results(results)
deployment = SearchDeployment.bind()
#启动服务
serve.run(deployment)
```

运行上面的代码后就可以将搜索查询作为 Web 终结点提供给外部的用户了,运行结果如下：

```
Started a local Ray instance.
View the dashboard at 127.0.0.1:8265
```

下面就可以从 Python 中发送查询请求，首先询问 Ray 框架中不同的组件及这个框架在 LLM 中的应用，代码如下：

```python
//第 10 章/ queryPython.py
import requests
query = "What are the different components of Ray"
        " and how can they help with large language models (LLMs)?"
response = requests.post("http://localhost:8000/", params={"query": query})
print(response.text)
```

结果返回了 Ray 文档链接 https://docs.ray.io/en/latest/ray-overview/use-cases.html。Ray 的仪表盘提供了非常强大的可视性功能，如图 10-2 所示。

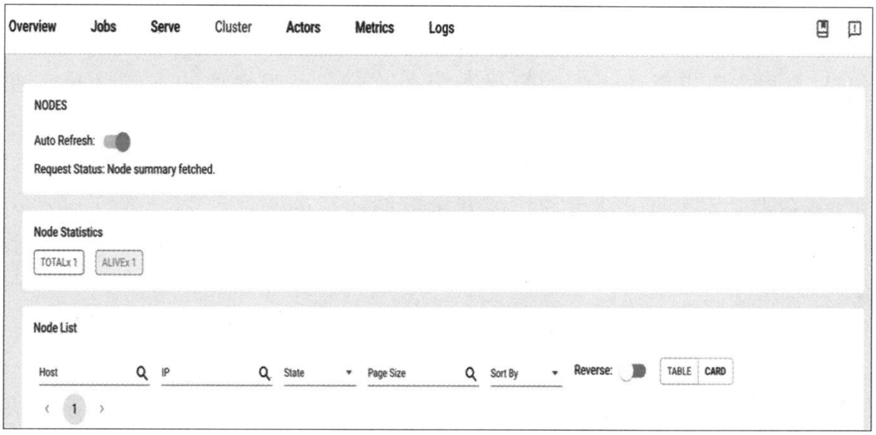

图 10-2　Ray 仪表盘

Ray 的仪表盘非常强大，因为它可以提供大量数据信息。收集数据信息非常简单，因为用户只需在部署对象或者执行器中设置和更新 Counter、Gauge、Histogram 或其他类型的变量。对于时间序列图表，用户可以安装及使用 Prometheus 或 Grafana 服务器。以上就是使用 LangChain 和 Ray 构建的简单语义搜索引擎的例子。随着 LLM 和 LLM 应用程序变得越来越复杂，并且被高度地交织到业务应用的框架中，生产过程中的可观察性和监控变得必不可少，以确保其准确性、效率和可靠性。接下来的部分将重点介绍监控 LLM 的重要性，并着重强调全面监控策略中要跟踪的关键指标。

10.4 监测 LLM 应用

实际生产运营的动态多变性意味着离线评估中所评估的条件几乎无法涵盖 LLM 在生产系统中可能遇到的所有潜在情景，因此，生产环境中需要能够实时监测 LLM 应用，以便持续、实时地进行观察，从而捕捉离线测试无法预料到的异常情况。可观察性保证了能够监测模型与实际输入数据和用户在生产环境中的交互行为和结果，它包括日志记录、跟踪、追踪和警报机制，以确保系统可以健康地运行，可以对性能进行优化，可以及早捕捉漂移等问题。正如前面讨论的那样，LLM 已经成为健康、电子商务和教育等许多应用领域的重要组成部分。

跟踪、追踪和监控是软件运营和管理领域中的 3 个重要概念。尽管它们都与理解和改进系统的性能有关，但它们各自都有着不同的作用。跟踪和追踪是为了保持详细的历史记录以进行分析和调试，而监控则旨在实时观察并立即意识到问题，以确保系统在任何时候都具有最佳的性能。所有这 3 个概念都属于可观察性的范畴。

监控是持续监督系统或应用性能的过程。这可能涉及持续收集和分析与系统健康相关的指标，如内存使用情况、CPU 利用率、网络延迟及整体应用/服务性能（如响应时间）。有效的监控包括建立异常或意外行为的警报系统。监控和可观察性的主要目的是通过实时数据为模型的性能和行为提供信息，以下是实时监控在实际生产环境中的价值。

（1）防止模型漂移：由于输入数据或用户行为特征的变化，模型可能随着时间的推移而退化。定期监控可以早期识别这种情况并采取纠正措施。

（2）性能优化：通过跟踪推断时间、资源使用情况和吞吐量等指标，可以进行调整以提高模型在生产中的效率和效果。

（3）A/B 测试：它有助于比较模型中微小差异可能导致的不同结果，从而帮助决策以改进模型。

（4）故障调试：监控有助于识别运行时可能发生的未预料问题，从而快速解决。

监控策略通常有以下几个关键部分需要考虑。

（1）监控的指标：根据期望的模型性能，定义关键的感兴趣指标，如预测准确性、延迟、吞吐量等。

（2）监控频率：监控频率应根据模型对操作的关键性进行确定，对于高度关键的模型，可能需要接近实时的监控。

（3）日志记录：日志应提供有关 LLM 执行的每个相关操作的详细信息，以便分析人员可以追溯任何异常情况。

（4）警报机制：如果系统检测到异常行为或性能急剧下降，则应该触发警报。

监控 LLM 有多个目的，包括评估模型性能、检测异常或问题、优化资源利用及确保一致且高质量的输出。通过不断评估 LLM 的行为和性能，包括验证、试验发布、解释及可靠的离线评估，企业能够识别潜在风险，保持用户信任，并提供最佳体验。以下是相关的指标。

（1）推理延迟：衡量 LLM 处理请求并生成响应所需的时间。较低的延迟可确保更快速且响应更及时的用户体验。

（2）每秒查询数（Query per Second，QPS）：计算 LLM 在特定时间范围内可以处理的查询或请求数量。监控 QPS 有助于评估可伸缩性并可以对容量进行规划。

（3）每秒生成标记数（Token Per Second，TPS）：跟踪 LLM 生成标记的速率。TPS 指标有助于估算对计算资源的需求和了解模型的效率。

（4）标记使用量：标记数与资源使用率相关，如硬件利用率、延迟和成本。

（5）错误率：监控 LLM 响应中错误或失败的发生情况，确保错误率保持在可接受范围内，以保证输出的质量。

（6）资源利用率：衡量计算资源的消耗，如 CPU、内存和 GPU，以优化资源分配并避免出现瓶颈问题。

（7）模型漂移：通过将 LLM 的输出与基线或实际情况进行比较，检测 LLM 随时间的行为变化情况，确保模型保持准确性并与预期结果一致。

（8）超出分布范围的输入：识别 LLM 训练数据意外分布之外的输入或查询，这可能导致意外或不可靠的输出。

（9）用户反馈指标：监控用户反馈渠道，收集有关用户满意度的见解，确定改进的方向，并验证 LLM 的有效性。

数据科学家和机器学习工程师应使用诸如 LIME 和 SHAP 等模型解释工具来检查模型的陈旧度、错误学习和偏差。如果突然发生最具预测性特征的变化，则可能表明数据泄露。像 AUC 这样的离线指标并不总是与在线转化率的影响相关，因此重要的是找到可靠的离线指标。这些离线指标对业务有实际影响，最好是直接影响系统的单击和购买等指标，然而，需要注意的是，在依赖云服务平台做这些监控时，应仔细研究隐私和数据保护政策。10.4.1 节将研究对 LangChain Agent 轨迹的监控。

10.4.1 跟踪和追踪

跟踪通常指的是记录和管理应用程序或系统中特定操作或一系列操作的信息的过程。例如，在机器学习应用程序或项目中，跟踪可能涉及记录参数、超参数、指标和不同实验或运行结果。它提供了一种记录进展和随时间变化的方式。

追踪是跟踪的一种更专业化形式。它涉及记录软件/系统的执行流程。特别是在单个事务可能跨越多个服务的分布式系统中，追踪有助于保持审计或碎屑信息，即对该请求路径进行详细记录。这种精细的视图使开发人员能够了解各个微服务之间的交互，并通过准确识别事务路径中的问题发生位置来排查延迟或失败等问题。

跟踪 Agent 的轨迹可能会具有挑战性，因为 Agent 具有广泛的行为和生成能力。LangChain 具有用于轨迹跟踪和评估的功能，因此查看 Agent 的轨迹非常简单。用户只需在初始化 Agent 或 LLM 时将 return_intermediate_steps 参数设置为 True。为了具体说明这一功能，接下来将定义一个工具。这里使用 @tool 装饰器，它将使用函数的文档字符串作为工具的描述。第 1 个工具将一个 ping 发送到一个网站网址,并返回传输的包和延迟信息,或者在出现错误时返回错误消息,代码如下:

```
//第 10 章/ pingtool.py
@tool
def ping(url: HttpUrl, return_error: bool) -> str:
    """Ping the fully specified url. Must include https://in the url."""
    hostname = urlparse(str(url)).netloc
    completed_process = subprocess.run(
        ["ping", "-c", "1", hostname], capture_output=True, text=True
    )
    output = completed_process.stdout
    if return_error and completed_process.returncode != 0:
        return completed_process.stderr
    return output]
```

现在,已经设置了一个 Agent,该 Agent 将此工具与 LLM 一起使用,以便在提示的情况下进行调用,代码如下:

```
//第 10 章/ pingtoolagent.py
llm = ChatOpenAI(model="gpt-3.5-turbo-0613", temperature=0)
agent = initialize_agent(
    llm=llm,
    tools=[ping],
    agent=AgentType.OPENAI_MULTI_FUNCTIONS,
    return_intermediate_steps=True,   # IMPORTANT!
)
result = agent("What's the latency like for https://langchain.com? ")
```

上面的代码询问 LangChain 的官网的延迟是多少,Agent 的回复如下:

```
The latency for https://langchain.com is 12.143 ms
```

在 results["intermediate_steps"] 中可以看到有关这个 Agent 操作的所有信息,代码如下:

```
//第 10 章/ intermediatesteps.py
[(_FunctionsAgentAction(tool='ping', tool_input={'url': 'https://langchain.com', 'return_error': False}, log="\nInvoking: `ping` with `{'url': 'https://langchain.com', 'return_error': False}`\n\n\n", message_log=[AIMessage(content='', additional_kwargs={'function_call': {'name': 'tool_selection', 'arguments': '{\n  "actions": [\n    {\n      "action_name": "ping",\n      "action": {\n        "url": "https://langchain.com",\n        "return_error": false\n      }\n    }\n  ]\n}'}}, example=False)]), 'PING langchain.com (192.168.31.93): 56 data bytes\n64 bytes from 192.168.31.93: icmp_seq=0 ttl=249 time=12.143 ms\n\n--- langchain.com ping statistics ---\n1 packets transmitted, 1 packets received, 0.0% packet loss\nround-trip min/avg/max/stddev = 12.143/12.143/12.143/0.000 ms\n')]
```

通过提供对系统的可观察性，LangChain 这种跟踪和评估非常有帮助。如果读者需要详细地了解这方面的知识，可查看 LangChain 文档，此文档详细展示了如何使用轨迹评估器来检查 Agent 生成的完整操作和响应序列，并对 OpenAI 功能 Agent 进行评分。10.4.2 节将探究除了 LangChain 之外的其他一些选项。

10.4.2　可观测性工具

有相当多的工具可以在 LangChain 中通过回调使用。

（1）Argilla：Argilla 是一个开源数据管理平台，可以将用户反馈（人机交互工作流）与提示和响应相结合，从而管理数据集以进行微调。

（2）Portkey：Portkey 增加了基本的 MLOps 功能，例如监控详细指标、跟踪链、缓存和可靠性，从而自动重试 LangChain。

（3）Comet.ml：Comet 提供了强大的 MLOps 功能，用于跟踪实验、比较模型和优化 AI 项目。

（4）LLMonitor：跟踪大量指标，包括成本和使用情况分析，以及跟踪和评估工具。

（5）DeepEval：记录相关性、偏差和毒性等默认指标。还可以帮助测试和监控模型漂移或退化。

（6）Aim：用于 ML 模型的开源可视化和调试平台。它记录组件的输入、输出和序列化状态，从而可以对单个 LangChain 进行可视化检查，并并排比较多个执行。

（7）Splunk：Splunk 的机器学习工具包可以在生产环境中为机器学习模型提供可观测性。

（8）ClearML：一种开源工具，用于自动化训练管道，从研究无缝地过渡到生产。

（9）IBMWatsonOpenScale：一个平台，通过快速识别和解决问题来提供对 AI 运行状况的分析，以帮助降低风险。

（10）DataRobotMLOps：监控和管理模型，以便在问题影响性能之前对其进行检测。

（11）DatadogAPM 集成：此集成允许用户捕获 LangChain 请求、参数、提示完成并可视化 LangChain 操作。用户还可以捕获请求延迟、错误和 Token/成本使用情况等指标。

（12）W&B 跟踪：前面的章节中已经展示了一个 W&B 来监控精细训练的示例，但它也可以跟踪其他指标，以及起到记录和比较提示的作用。

（13）Langfuse：这个开源工具可以方便地监控有关延迟、成本、LangChainAgent 和工具的分数的详细信息。

这些大多数可以非常容易地被集成到 LLM 的流程中。例如，对于 W&B，用户可以通过将 LANGCHAIN_WANDB_TRACING 环境变量设置为 True 来启用跟踪。或者可以使用带有 wandb_tracing_enabled() 的上下文管理器来跟踪特定的代码块。使用 Langfuse 则可以将 langfuse.callback.CallbackHandler() 作为参数传递给 chain.run() 进行调用，其中一些工具是开源的，这些平台的优点在于它允许对隐私很重要的用例进行完全定制和本地部署。例如，Langfuse 是开源的，并提供了自托管选项。开发者可以选择最适合自己项目需求的选项，并按照 LangChain 文档中提供的说明为自己的 Agent 启用跟踪服务。

10.5 LangChain 回调

LangChain 的回调系统是一个强大的工具，允许用户自定义和扩展他们的 LLM 应用程序的功能。回调实际上是事件处理程序，可以订阅应用程序执行的不同阶段，使用户能够执行诸如日志记录、监控、数据流传输或任何其他自定义任务等操作。这种灵活性为用户提供了对应用程序行为的精细控制，使其更容易将 LangChain 集成到各种工作流程和用例中。无论用户需要捕获特定事件、分析模型行为还是集成外部服务，LangChain 的回调系统都可以帮助用户高效地实现目标。

LangChain 提供了一些内置的处理程序，供用户快速使用回调。这些处理程序位于 langchain/callbacks 模块中，其中最基本的处理程序是 StdOutCallbackHandler，它简单地将所有事件记录到标准输出(stdout)。下面是快速请求回调的代码：

```python
//第 10 章/ callbackstart.py
from langchain.callbacks import StdOutCallbackHandler
from langchain.chains import LLMChain
from langchain_openai import OpenAI
from langchain.prompts import PromptTemplate

handler = StdOutCallbackHandler()
llm = OpenAI()
prompt = PromptTemplate.from_template("1 + {number} = ")

#首先构造回调函数:在初始化链时明确设置 StdOutCallbackHandler
chain = LLMChain(llm=llm, prompt=prompt, callbacks=[handler])
chain.run(number=2)

#然后使用 verbose 标志:使用`verbose`标志来达到相同的效果
chain = LLMChain(llm=llm, prompt=prompt, verbose=True)
chain.run(number=2)

#最后请求回调函数:使用请求中的`callbacks`来达到相同的效果
chain = LLMChain(llm=llm, prompt=prompt)
chain.run(number=2, callbacks=[handler])
```

输出的结果如下：

```
> Entering new LLMChain chain...
Prompt after formatting:
1 + 2 =
```

```
> Finished chain.


> Entering new LLMChain chain...
Prompt after formatting:
1 + 2 =

> Finished chain.


> Entering new LLMChain chain...
Prompt after formatting:
1 + 2 =

> Finished chain.


'\n\n3'
```

回调函数参数在大多数对象(链、模型、工具、Agents等)中是可用的,并且有两个不同的位置。构造函数回调是在构造函数中定义的,例如LLMChain(callbacks=[handler],tags=['a-tag']),将用于该对象上的所有调用,并且仅在该对象范围内使用。例如,如果将处理程序传递给LLMChain构造函数,则将不会被附加到该链上的模型中。请求回调是在用于发出请求的run()/apply()方法中定义的,例如chain.run(input, callbacks=[handler]),将仅用于该特定请求及它包含的所有子请求。verbose参数在API的大多数对象(Chains、Models、Tools、Agents等)中作为构造函数参数使用,例如LLMChain(verbose=True),它等效于将ConsoleCallbackHandler传递给该对象和所有子对象的callbacks参数,这对于调试很有用,因为它会将所有事件记录到控制台。在使用这些参数时,构造函数回调最适用于不特定于单个请求的用例,例如日志记录和监视,而请求回调最适用于流式传输等用例,例如希望将单个请求的输出流式传输到特定的Websocket连接或类似用例。

10.5.1 异步回调

在程序开发中,使用异步回调的主要原因在于提高程序的响应性和性能。在异步编程中,经常需要进行非阻塞操作,以确保程序的高效性和响应性。通过AsyncCallbackHandler可以保证回调事件的异步执行,避免了在主线程中阻塞运行循环。这意味着在处理大量请求和事件时,系统可以继续响应其他操作,而不会被阻塞。此外,使用异步回调还可以提高系统的并发性能,允许同时处理多个任务,这在处理大规模请求或高并发环境中尤其有用。如果用户的应用程序需要具备高度的响应性和并发性能,则使用异步回调是一个明智的选择。然而,需要注意的是,如果用户计划使用异步API,则建议使用AsyncCallbackHandler来避免阻塞运行循环。虽然在使用异步方法运行 LLM、链、工具、Agent 时，仍然可以使用同步 CallbackHandler，但在内部它将被调用为 run_in_executor，如果用户的 CallbackHandler 不是线程安全的，则可能会导致问题。下面是使用异步回调的代码：

```python
//第 10 章 / AsyncCallback.py
import asyncio
from typing import Any, Dict, List

from langchain.callbacks.base import AsyncCallbackHandler, BaseCallbackHandler
from langchain.schema import HumanMessage, LLMResult
from langchain_openai import ChatOpenAI


class MyCustomSyncHandler(BaseCallbackHandler):
    def on_llm_new_token(self, token: str, **kwargs) -> None:
        print(f"Sync handler being called in a `thread_pool_executor`: token: {token}")


class MyCustomAsyncHandler(AsyncCallbackHandler):
    """可用于处理来自 LangChain 的回调的异步回调处理程序"""

    async def on_llm_start(
        self, serialized: Dict[str, Any], prompts: List[str], **kwargs: Any
    ) -> None:
        """在链开始运行时运行"""
        print("zzzz....")
        await asyncio.sleep(0.3)
        class_name = serialized["name"]
        print("Hi! I just woke up. Your LLM is starting")

    async def on_llm_end(self, response: LLMResult, **kwargs: Any) -> None:
        """在链结束运行时运行"""
        print("zzzz....")
        await asyncio.sleep(0.3)
        print("Hi! I just woke up. Your LLM is ending")


#如果要启用流式传输，则需要将 `streaming=True` 传递给 ChatModel 构造函数
#此外这里还需要传入一个包含自定义处理程序的列表
chat = ChatOpenAI(
    max_tokens=25,
    streaming=True,
    callbacks=[MyCustomSyncHandler(), MyCustomAsyncHandler()],
)
```

```
await chat.agenerate([[HumanMessage(content="Tell me a joke")]])
```

输出的结果如下：

```
zzzz....
Hi! I just woke up. Your LLM is starting
Sync handler being called in a `thread_pool_executor`: token:
Sync handler being called in a `thread_pool_executor`: token: Why
Sync handler being called in a `thread_pool_executor`: token:  don
Sync handler being called in a `thread_pool_executor`: token: 't
Sync handler being called in a `thread_pool_executor`: token:  scientists
Sync handler being called in a `thread_pool_executor`: token:  trust
Sync handler being called in a `thread_pool_executor`: token:  atoms
Sync handler being called in a `thread_pool_executor`: token: ?
Sync handler being called in a `thread_pool_executor`: token:


Sync handler being called in a `thread_pool_executor`: token: Because
Sync handler being called in a `thread_pool_executor`: token:  they
Sync handler being called in a `thread_pool_executor`: token:  make
Sync handler being called in a `thread_pool_executor`: token:  up
Sync handler being called in a `thread_pool_executor`: token:  everything
Sync handler being called in a `thread_pool_executor`: token: .
Sync handler being called in a `thread_pool_executor`: token:
zzzz....
Hi! I just woke up. Your LLM is ending
```

```
LLMResult(generations=[[ChatGeneration(text="Why don't scientists trust
atoms? \n\nBecause they make up everything.", generation_info=None, message=
AIMessage(content="Why don't scientists trust atoms? \n\nBecause they make up
everything.", additional_kwargs={}, example=False))]], llm_output={'token_
usage': {}, 'model_name': 'gpt-3.5-turbo'})
```

10.5.2 自定义回调处理程序

用户也可以创建一个自定义的处理程序并将其设置到对象上。在下面的示例中将使用自定义处理程序实现流式传输，代码如下：

```
//第 10 章/customHandler.py
from langchain.callbacks.base import BaseCallbackHandler
from langchain.schema import HumanMessage
from langchain_openai import ChatOpenAI
```

```python
class MyCustomHandler(BaseCallbackHandler):
    def on_llm_new_token(self, token: str, **kwargs) -> None:
        print(f" My custom handler,token: {token}")

# 如果要启用流式处理,则需要在 ChatModel 构造函数中传入 `streaming=True`
# 此外,我们还需要传入一个包含自定义处理程序的列表
chat = ChatOpenAI(max_tokens=25, streaming=True, callbacks=[MyCustomHandler()])

chat([HumanMessage(content="告诉我一个笑话")])
```

输出的结果如下:

```
My custom handler, token:
My custom handler, token: Why
My custom handler, token:  don
My custom handler, token: 't
My custom handler, token:  scientists
My custom handler, token:  trust
My custom handler, token:  atoms
My custom handler, token: ?
My custom handler, token:


My custom handler, token: Because
My custom handler, token:  they
My custom handler, token:  make
My custom handler, token:  up
My custom handler, token:  everything
My custom handler, token: .
My custom handler, token:
```

AIMessage(content="Why don't scientists trust atoms? \n\nBecause they make up everything.", additional_kwargs={}, example=False)

10.5.3 记录到文件

下面这个示例演示了如何将日志输出到文件,这在开发和调试过程中非常有用。这里需要使用 FileCallbackHandler,它是一个自定义回调处理程序,用于捕获和记录相关事件。与 StdOutCallbackHandler 不同,它将输出写入文件而不是标准输出。这样就可以在之后轻松地查看和分析日志信息了。此外还使用了 loguru 库,这是一个功能强大的日志记录库,用于记录除了回调处理程序之外的其他输出,确保我们获得全面的日志信息。通过这种方式,可以更轻松地追踪和解决问题,同时保持代码的可维护性和可读性。实现的代码如下:

```python
//第 10 章 / logtofile.py
from langchain.callbacks import FileCallbackHandler
from langchain.chains import LLMChain
from langchain.prompts import PromptTemplate
from langchain_openai import OpenAI
from loguru import logger

logfile = "output.log"

logger.add(logfile, colorize=True, enqueue=True)
handler = FileCallbackHandler(logfile)

llm = OpenAI()
prompt = PromptTemplate.from_template("1 + {number} = ")

#该链将被打印到标准输出(因为 verbose=True)并写入 output.log
#如果 verbose=False,则 FileCallbackHandler 仍将写入 output.log
chain = LLMChain(llm=llm, prompt=prompt, callbacks=[handler], verbose=True)
answer = chain.run(number=2)
logger.info(answer)
```

```
> Entering new LLMChain chain...
Prompt after formatting:
1 + 2 =

> Finished chain.

2023-06-01 18:36:38.929 | INFO     | __main__:<module>:20 -

3
```

现在可以打开 output.log 文件,查看已捕获的输出信息,代码如下:

```
%pip install --upgrade --quiet  ansi2html > /dev/null

//第 10 章 / openfile.py
from ansi2html import Ansi2HTMLConverter
from IPython.display import HTML, display

with open("output.log", "r") as f:
    content = f.read()

conv = Ansi2HTMLConverter()
html = conv.convert(content, full=True)

display(HTML(html))
```

10.5.4　多个回调处理程序

前面的示例在创建对象时使用"callbacks= 参数"传递回调处理程序。在这种情况下，回调处理程序将被限定于该特定对象。然而，在很多情况下，将处理程序传递给运行对象会更有优势。当用户在执行 run 时使用 callbacks 关键字参数传递 CallbackHandlers 时，这些回调将由执行中涉及的所有嵌套对象发出。例如，当将处理程序传递给 Agent 时，它将用于与 Agent 相关的所有回调及 Agent 执行中涉及的所有对象，包括工具、LLMChain 和 LLM。这样就不必手动将处理程序附加到每个单独的嵌套对象上，代码如下：

```python
//第 10 章/ multicallback.py
from typing import Any, Dict, List, Union

from langchain.agents import AgentType, initialize_agent, load_tools
from langchain.callbacks.base import BaseCallbackHandler
from langchain.schema import AgentAction
from langchain_openai import OpenAI


#首先,定义自定义回调处理程序
class MyCustomHandlerOne(BaseCallbackHandler):
    def on_llm_start(
        self, serialized: Dict[str, Any], prompts: List[str], **kwargs: Any
    ) -> Any:
        print(f"on_llm_start {serialized['name']}")

    def on_llm_new_token(self, token: str, **kwargs: Any) -> Any:
        print(f"on_new_token {token}")

    def on_llm_error(
        self, error: Union[Exception, KeyboardInterrupt], **kwargs: Any
    ) -> Any:
        """Run when LLM errors."""

    def on_chain_start(
        self, serialized: Dict[str, Any], inputs: Dict[str, Any], **kwargs: Any
    ) -> Any:
        print(f"on_chain_start {serialized['name']}")

    def on_tool_start(
        self, serialized: Dict[str, Any], input_str: str, **kwargs: Any
    ) -> Any:
        print(f"on_tool_start {serialized['name']}")

    def on_agent_action(self, action: AgentAction, **kwargs: Any) -> Any:
```

```
        print(f"on_agent_action {action}")


class MyCustomHandlerTwo(BaseCallbackHandler):
    def on_llm_start(
        self, serialized: Dict[str, Any], prompts: List[str], **kwargs: Any
    ) -> Any:
        print(f"on_llm_start (I'm the second handler!!) {serialized['name']}")


#实例化处理程序
handler1 = MyCustomHandlerOne()
handler2 = MyCustomHandlerTwo()

#设置 Agent,只有 llm 才会为 handler2 发出回调
llm = OpenAI(temperature=0, streaming=True, callbacks=[handler2])
tools = load_tools(["llm-math"], llm=llm)
agent = initialize_agent(tools, llm, agent=AgentType.ZERO_SHOT_REACT_DESCRIPTION)

#handler1 的回调将由参与其中的每个对象发出
#Agent 执行(llm、llmchain、工具、Agent 执行器)
agent.run("What is 2 raised to the 0.235 power?", callbacks=[handler1])
```

输出的结果如下:

```
on_chain_start AgentExecutor
on_chain_start LLMChain
on_llm_start OpenAI
on_llm_start (I'm the second handler!!) OpenAI
on_new_token  I
on_new_token  need
on_new_token  to
on_new_token  use
on_new_token  a
on_new_token  calculator
on_new_token  to
on_new_token  solve
on_new_token  this
on_new_token  .
on_new_token
Action
on_new_token  :
on_new_token  Calculator
on_new_token
```

```
Action
on_new_token  Input
on_new_token :
on_new_token  2
on_new_token ^
on_new_token 0
on_new_token .
on_new_token 235
on_new_token
on_agent_action AgentAction(tool='Calculator', tool_input='2^0.235', log=' I
need to use a calculator to solve this.\nAction: Calculator\nAction Input: 2^
0.235')
on_tool_start Calculator
on_chain_start LLMMathChain
on_chain_start LLMChain
on_llm_start OpenAI
on_llm_start (I'm the second handler!!) OpenAI
on_new_token
on_new_token ```text
on_new_token

on_new_token 2
on_new_token **
on_new_token 0
on_new_token .
on_new_token 235
on_new_token

on_new_token ```
on_new_token ...
on_new_token num
on_new_token expr
on_new_token .
on_new_token evaluate
on_new_token ("
on_new_token 2
on_new_token **
on_new_token 0
on_new_token .
on_new_token 235
on_new_token ")
on_new_token ...
on_new_token

on_new_token
```

```
on_chain_start LLMChain
on_llm_start OpenAI
on_llm_start (I'm the second handler!!) OpenAI
on_new_token  I
on_new_token  now
on_new_token  know
on_new_token  the
on_new_token  final
on_new_token  answer
on_new_token .
on_new_token
Final
on_new_token  Answer
on_new_token :
on_new_token  1
on_new_token .
on_new_token 17
on_new_token 690
on_new_token 67
on_new_token 372
on_new_token 187
on_new_token 674
on_new_token
```

```
'1.1769067372187674'
```

10.5.5　Token 计算

LangChain 中有一个上下文管理器非常有用，因为它允许用户在执行语言模型任务时可以动态地跟踪和管理 Token 数。这在控制和优化生成文本的过程中非常有用，特别是在涉及 Token 限制的情况下。与回调结合使用时，用户可以在语言模型的执行过程中实时监控标记数，从而更好地控制和管理生成的文本长度，以满足特定需求或限制。这种组合使在处理大型文本生成任务时更加灵活和可控，代码如下：

```
//第 10 章 / tokencount.py
import asyncio

from langchain.callbacks import get_openai_callback
from langchain_openai import OpenAI

llm = OpenAI(temperature=0)
with get_openai_callback() as cb:
    llm("What is the square root of 4?")
```

```python
total_tokens = cb.total_tokens
assert total_tokens > 0

with get_openai_callback() as cb:
    llm("What is the square root of 4?")
    llm("What is the square root of 4?")

assert cb.total_tokens == total_tokens * 2

#用户可以从上下文管理器中启动并发运行
with get_openai_callback() as cb:
    await asyncio.gather(
        *[llm.agenerate(["What is the square root of 4?"]) for _ in range(3)]
    )

assert cb.total_tokens == total_tokens * 3

#上下文管理器是并发安全的
task = asyncio.create_task(llm.agenerate(["What is the square root of 4?"]))
with get_openai_callback() as cb:
    await llm.agenerate(["What is the square root of 4?"])

await task
assert cb.total_tokens == total_tokens
```

10.6　LangSmith

创建可靠的 LLM 应用程序是非常具有挑战性的。前面提到了很多供观测 LLM 的工具和库，LangChain 本身也提供了很多其他工具，但为了使提示、链和 Agent 的性能达到足够的可靠水平以在生产环境中使用，需要一个专业的集成工具。为此，LangChain 团队推出了 LangSmith，它就是为生产环境而生的一个平台工具。

目前 LangChain 默认在后台运行 LangSmith 的追踪功能。对于 Python，可通过设置环境变量实现，在启动虚拟环境或打开 Bash Shell 时会设置这些环境变量，并保持这些设置。LangSmith 的优点是，对 LLM、链条、Agent、工具和检索器的所有调用都将被记录到 LangSmith 中。在大部分时间里用户其实是不查看这些追踪记录的，但一旦需要查看的时候就非常有用。LangSmith 可以用来找出意外的输出结果、Agent 循环的原因、Agent 比预期慢的原因、Agent 使用了多少 Token 等复杂的任务。下面来看一下 LangSmith 在生产环境的各个环节所能扮演的角色。

10.6.1　LangSmith 调试

调试 LLM、链条和 Agent 通常很困难，LangSmith 有助于解决以下痛点问题。

1. 输入输出

LLM 调用通常复杂且难以确定。尽管它们在技术上采用字符串→字符串(或聊天消息→聊天消息)的形式,但看似简单的输入/输出实际上可能会产生误导,因为输入字符串通常由用户输入和辅助功能的组合构成。大多数 LLM 调用的输入是固定模板与输入变量的组合。这些输入变量可能直接来自用户输入或辅助功能(如检索)。当将这些输入变量传递给 LLM 时,它们会被转换为字符串格式,但通常它们不会自然地表示为字符串(可能是列表或文档对象),因此,了解最终进入 LLM 的确切字符串非常重要。这有助于解决格式逻辑错误、对用户输入的意外转换及用户输入的纯粹丢失等问题。

另外,LLM 的输出也是如此。通常,LLM 的输出在技术上是一个字符串,但该字符串可能包含一些结构(JSON、YAML),因此了解确切的输出有助于确定是否需要以不同的解析方式进行解析。LangSmith 提供了 LLM 调用的确切输入/输出的直观可视化,让用户能够轻松理解它们。

2. 提示词和输出的关系

设想遇到这样一个问题:某次当发现输出错误时,开发者进入 LangSmith 查看具体情况。发现某个问题的 LLM 调用后看到了确切的输入内容。这时开发者必然想尝试更改输入中的一个单词或者词语来观察输出的变化。

这个问题经常会发生,因此 LangSmith 界面构建了一个游乐场。在检查 LLM 调用时,可以单击"在游乐场中打开"按钮来使用这个游乐场,在里面可以无限次地修改提示并重新运行以观察输出结果的变化,目前这个功能只支持 OpenAI 和 Anthropic 模型,未来计划将其功能扩展到更多类型的 LLM、链条、Agent 和检索器。

3. 追溯事件

在复杂的链条和代理中,通常很难理解底层的运作机制。例如,进行了哪些调用?顺序是怎样的?每个调用的输入和输出是什么?LangSmith 内置的跟踪功能提供了一种可视化工具,可以帮助澄清这些序列。这个工具对于理解复杂和冗长的链条和 Agent 至关重要。对于链条,它可以揭示调用的顺序及它们是如何相互作用的。对于 Agent,由于调用序列是非确定性的,它有助于可视化特定运行的具体序列,而这是事先无法知道的。

4. 延迟和 Token 数

如果一个链条的运行时间比预期长,就需要确定原因。通过跟踪每个步骤的延迟,LangSmith 允许用户识别并可能消除链条中最慢的部分。构建和原型设计 LLM 应用可能成本高昂。LangSmith 追踪了链条的总 Token 使用量及每个步骤的 Token 使用量。这使用户容易识别链条中可能成本较高的部分。

10.6.2 LangSmith 样本收集

大多数时候,进行调试是因为应用程序中发生了不良或意外的结果。这些失败是宝贵的数据样本。通过识别链条可能的失败原因并监控这些失败,可以针对这些已知问题测试未来的链条版本。

为什么这样做如此重要？在构建 LLM 应用时，通常从没有任何数据集就开始构建是常见的。这是 LLM 的一部分强大之处，即它们可以进行零样本学习，但这也有很大的缺点，因为当调整提示时意味着没有任何示例可以用来对比标准化更改。LangSmith 通过为每次运行包含一个"添加到数据集"按钮来解决这个问题，使将输入/输出示例添加到所选数据集变得容易。可以在将示例添加到数据集之前编辑它们，包括编辑预期结果，这对于不良样本尤其有用。

这个功能在嵌套链条的每个步骤都可用，从而可以为端到端链条、中间链条（例如 LLM 链条）或仅仅是 LLM 或聊天模型添加示例。端到端链的示例非常适合测试整体流程，而单一、模块化的 LLM 链条或 LLM/聊天模型示例对于测试简单可修改的组件比较有帮助。

10.6.3　LangSmith 测试评估

前面的章节提到起初对于评估工作主要采用手动和临时的方法。用户输入不同的数据，观察结果如何，然而，在某个时刻应用程序可能会表现出色，这时就希望更加严格地测试变化。这里可以使用之前提到的构建数据集。或者可以花一些时间手工构建一个小型数据集。对于这些情况，LangSmith 简化了数据集的上传过程。

一旦拥有了一个数据集，接下来要如何使用它来测试提示或链条的变化？最基本的方法是运行链条对数据点进行处理并可视化输出。尽管技术在不断进步，但仍然没有什么可以替代亲自查看输出结果。目前，需要在客户端对数据点运行链条。LangSmith 客户端可以轻松地下载数据集，然后对其运行链条，将结果记录到与数据集关联的新项目中，然后用户可以对其进行审查。LangSmith 已经简化了在 Web 应用程序中直接分配反馈并将其标记为正确或不正确的操作，同时显示每个测试项目的汇总统计信息。此外，LangSmith 还使评估这些运行更加容易。为此 LangSmith 在开源的 LangChain 库中添加了一组评估器。这些评估器可以在启动测试运行时进行指定，并在测试运行完成后对结果进行评估。

10.6.4　LangSmith 人工评估

自动评估指标对于获得模型行为的一致性和可扩展信息非常有帮助，但人工审查仍然不可或缺，以确保应用程序的最高质量和可靠性。LangSmith 通过注释队列轻松地实现了手动审查和注释运行，允许根据模型类型或自动评估分数等标准选择运行并排队等待人工审查。审阅者可以迅速查看输入、输出和任何现有标签，然后添加自己的反馈。这种方法常用于评估自动评估器难以处理的主观品质，如创造力或幽默感，并对经过自动评估的一部分运行进行抽样和验证，以确保自动度量仍可靠且可捕获正确信息。所有通过队列进行的注释都分配给源运行的"反馈"，以便稍后轻松地进行过滤和分析，还可以快速编辑示例并将其添加到数据集中，以扩展评估集的范围或对模型进行微调以提高质量或降低成本。

10.6.5　LangSmith 监控

最后阶段，用户的 LLM 应用程序最终可能准备好投入生产。与之前用于调试的方式相似，LangSmith 还可以用于监视应用程序。用户可以记录所有的迹象，以及可视化延迟和 Token 使用统计信息，并在问题出现时进行故障排除。每个运行还可以分配字符串标签或键值元数据，允许附加关联标识或 AB 测试变体，并相应地筛选运行。

LangSmith 还实现了将反馈与运行程序关联的可能性。这意味着如果应用程序上有一个"赞/踩"按钮，则可以使用它将反馈记录到 LangSmith。这可用于随时间跟踪性能并准确定位性能较差的数据点，随后可以将其添加到用于未来测试的数据集中，这与调试模式的方法相似。

10.6.6 LangSmith 实战演示

本示例将带领读者完整地了解 LangSmith 的各种使用方法。

正如前面所讨论的，LangChain 使开发 LLM 应用程序和 Agent 程序变得轻而易举，然而，将 LLM 应用程序投入生产可能会更加困难，开发者通常需要对提示、链条及其他组件进行深度定制和迭代，以打造高质量的产品。LangSmith 是一个统一的平台，用于调试、测试和监视用户的 LLM 应用程序，它可以快速调试新的链、Agent 或工具、可视化组件之间的关系，评估不同提示和 LLM，多次运行链以确保质量达标，以及在捕获使用迹象并使用 LLM 或分析管道生成见解等方面提供便利。

在开始运行本例之前，用户需要到 https://smith.langchain.com/ 去创建一个账号，然后在左下角会有个生成 API 密钥的按钮，用户需要生成自己的 API 密钥。

1. 将运行记录到 LangSmith

首先，在配置环境变量之前，需要确保自己的 LangChain 环境已正确设置，以便开始记录迹象。要实现这一点，需要将 LANGCHAIN_TRACING_V2 环境变量设置为 True，这将启用追踪功能。此外，用户可以使用 LANGCHAIN_PROJECT 环境变量指定要记录到哪个项目中。如果未设置该环境变量，则 LangChain 会将运行情况记录到默认项目中。重要的是，如果指定的项目不存在，则 LangChain 将自动创建该项目，使整个记录过程更加流畅。此外，还需要设置 LANGCHAIN_ENDPOINT 和 LANGCHAIN_API_KEY 环境变量，以确保与 LangChain 的通信和身份验证是安全的和有效的。这些环境变量的配置步骤是确保能够顺利记录和管理 LangChain 迹象的重要一步。

如果要获得有关其他设置追踪的方式的更多信息，则可参阅 LangSmith 文档。需要注意，为了运行以下示例，用户必须设置自己的 OPENAI_API_KEY 环境变量。API 密钥只能在首次创建时访问，因此需要妥善保管。此外，用户还可以在 Python 中使用上下文管理器。

首先安装必要的库，并且设置环境变量，代码如下：

```
%pip install -U langchain langsmith langchainhub --quiet
%pip install openai tiktoken pandas duckduckgo-search --quiet
```

```
//第 10 章/ environmentVariable.py
import os
```

```
from uuid import uuid4

unique_id = uuid4().hex[0:8]
os.environ["LANGCHAIN_TRACING_V2"] = "true"
os.environ["LANGCHAIN_PROJECT"] = f"Tracing Walkthrough - {unique_id}"
os.environ["LANGCHAIN_ENDPOINT"] = "https://api.smith.langchain.com"
os.environ["LANGCHAIN_API_KEY"] = "用户自己的 LangChain API Key"   #更新自己的
#LangChain API Key


os.environ["OPENAI_API_KEY"] = "用户自己的 OpenAI API Key"
```

然后创建 LangSmith 客户端以与 API 交互，代码如下：

```
from langsmith import Client

client = Client()
```

接下来创建一个 LangChain 组件，并将运行日志记录到平台上。在此示例中将创建一个 ReAct 风格的 Agent，该 Agent 可以访问常规搜索工具（DuckDuckGo），Agent 的提示词可以在这个 Hub 中查（https://smith.langchain.com/hub/wfh/langsmith-agent-prompt），代码如下：

```
//第 10 章/ langChainComponent.py
from langchain import hub
from langchain.agents import AgentExecutor
from langchain.agents.format_scratchpad import format_to_openai_function_messages
from langchain.agents.output_parsers import OpenAIFunctionsAgentOutputParser
from langchain.tools import DuckDuckGoSearchResults
from langchain_community.chat_models import ChatOpenAI
from langchain_community.tools.convert_to_openai import format_tool_to_openai_function

#获取此提示的最新版本
prompt = hub.pull("wfh/langsmith-agent-prompt:latest")

llm = ChatOpenAI(
    model="gpt-3.5-turbo-16k",
    temperature=0,
)

tools = [
    DuckDuckGoSearchResults(
        name="duck_duck_go"
```

```
    ),  #使用DuckDuckGo进行一般互联网搜索
]

llm_with_tools = llm.bind(functions=[format_tool_to_openai_function(t) for t in tools])

runnable_agent = (
    {
        "input": lambda x: x["input"],
        "agent_scratchpad": lambda x: format_to_openai_function_messages(
            x["intermediate_steps"]
        ),
    }
    | prompt
    | llm_with_tools
    | OpenAIFunctionsAgentOutputParser()
)

agent_executor = AgentExecutor(
    agent=runnable_agent, tools=tools, handle_parsing_errors=True
)
```

为了减少延迟，这里在多个输入上并发运行 Agent。在后台将运行日志记录到 LangSmith，因此执行延迟不受影响，代码如下：

```
//第 10 章/ input.py
inputs = [
    "什么是 LangChain？",
    "什么是 LangSmith？",
    "Llama-v2 是什么时候发布的？",
    "LangChain 什么时候发布了 Hub？",
]

results = agent_executor.batch([{"input": x} for x in inputs], return_exceptions=True)
```

如果用户自己已成功设置环境，则 Agent 跟踪信息将显示在应用的"项目"部分中，如图 10-3 所示。

从图 10-3 来看，Agent 没有有效地使用外部工具。接下来是评估环节。

2. 评估 Agent

除了允许记录运行情况外，LangSmith 还允许用户测试和评估 LLM 应用程序。接下来将利用 LangSmith 创建一个基准数据集，并在一个 Agent 上运行 AI 辅助评估器。将按照以下几个步骤进行：创建数据集、初始化一个新的 Agent 以进行基准测试、配置评估器以

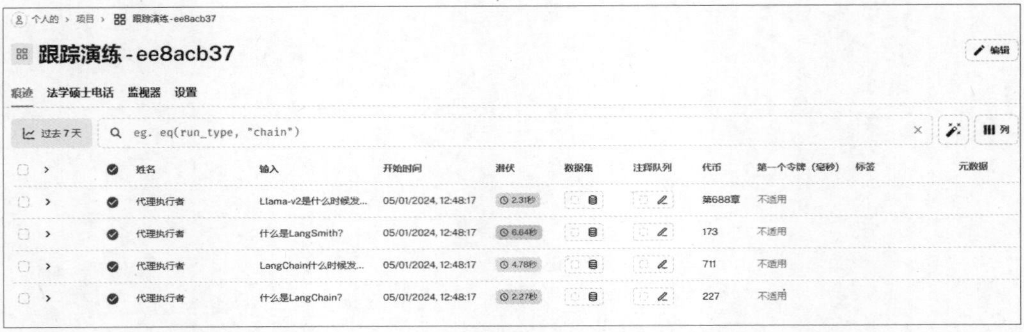

图 10-3　LangSmith 中的 Agent 跟踪信息

对 Agent 的输出进行评分，最后在数据集上运行 Agent 并评估结果。

首先创建 LangSmith 数据集。以下使用 LangSmith 客户端根据上述输入问题和标签列表创建数据集。稍后将使用这些数据来衡量新 Agent 的性能。数据集是一组示例，它们实际上就是可以用作应用程序测试用例的输入-输出对。有关数据集的更多信息，包括如何从 CSV 文件或其他文件创建数据集，以及如何在平台上创建数据集，用户可以参阅 LangSmith 文档。

创建数据集的代码如下：

```
outputs = [
    "LangChain 是一个使用大型语言模型构建应用程序的开源框架。它也是建立 LangSmith 的公司的名称。",
    "LangSmith 是一个统一的平台，用于调试、测试和监控由 LangChain 驱动的语言模型应用程序和代理。",
    "2023 年 7 月 18 日",
    "LangSmith 菜谱是一个 GitHub 存储库，其中包含如何使用 LangSmith 来调试、评估和监控大型语言模型驱动应用程序的详细示例。",
    "2023 年 9 月 5 日",
]

dataset_name = f"agent-qa-{unique_id}"

dataset = client.create_dataset(
    dataset_name,
    description="LangSmith 文档中的问题示例数据集。",
)

for query, answer in zip(inputs, outputs):
    client.create_example(
        inputs={"input": query}, outputs={"output": answer}, dataset_id=dataset.id
    )
```

接下来初始化一个新的 Agent 以进行基准测试。LangSmith 允许用户评估任何 LLM、链条、Agent，甚至是自定义函数。会话 Agent 是有状态的（它们有记忆）；为了确保此状态不会在数据集运行之间共享，这里将传入一个 chain_factory 函数（也称为构造函数），以便对每个调用进行初始化。在这种情况下将测试一个使用 OpenAI 函数调用端点的代理，代码如下：

```python
//第 10 章 / openai_agent.py
from langchain import hub
from langchain.agents import AgentExecutor, AgentType, initialize_agent, load_tools
from langchain.agents.format_scratchpad import format_to_openai_function_messages
from langchain.agents.output_parsers import OpenAIFunctionsAgentOutputParser
from langchain_community.chat_models import ChatOpenAI
from langchain_community.tools.convert_to_openai import format_tool_to_openai_function


# 由于链条可以是有状态的(例如它们可以有内存)，所以提供了
# 一种为数据集中的每行初始化新链条的方法，这已经完成了
# 通过传入一个工厂函数，该函数为每行返回一个新链条
def agent_factory(prompt):
    llm_with_tools = llm.bind(
        functions=[format_tool_to_openai_function(t) for t in tools]
    )
    runnable_agent = (
        {
            "input": lambda x: x["input"],
            "agent_scratchpad": lambda x: format_to_openai_function_messages(
                x["intermediate_steps"]
            ),
        }
        | prompt
        | llm_with_tools
        | OpenAIFunctionsAgentOutputParser()
    )
    return AgentExecutor(agent=runnable_agent, tools=tools, handle_parsing_errors=True)
```

然后到了配置评估的环节。手动在用户界面中比较链条的结果虽然有效，但可能会耗费大量时间。使用自动化指标和 AI 辅助反馈来评估组件性能可能会很有帮助。下面创建一些预先实现的运行评估器，需要执行以下操作：将结果与基准标签进行比较、使用嵌入距离测量语义相似性、使用自定义标准以无参考方式评估 Agent 响应的不同方面。如果用户需要更深入地了解如何为用例选择适当的评估器及如何创建自定义评估器，则可参考

LangSmith 文档，代码如下：

```python
//第 10 章/ configure_evaluation.py
from langchain.evaluation import EvaluatorType
from langchain.smith import RunEvalConfig

evaluation_config = RunEvalConfig(
    #评估器可以是评估器类型(例如,qa、criteria、embedding_distance 等)或该评估器的
    #配置
    evaluators=[
        #通过参考答案衡量 QA 响应是否"正确"
        #也可以通过原始字符串 qa 进行选择
        EvaluatorType.QA,
        #测量输出与参考答案之间的嵌入距离
        #相当于 EvalConfig.EmbeddingDistance(embeddings=OpenAIEmbeddings())
        EvaluatorType.EMBEDDING_DISTANCE,
        #输出的评分是否满足所述标准
        #可以选择默认标准,例如 helpfulness,也可以提供自己的标准
        RunEvalConfig.LabeledCriteria("helpfulness"),
        #LabeledScoreString 评估器在 1~10 分值范围内输出分数
        #可以使用默认标准或编写自己的评分标准
        RunEvalConfig.LabeledScoreString(
            {
                "accuracy": """
Score 1: 答案与参考答案完全无关。
Score 3: 答案有轻微相关性,但与参考答案不符。
Score 5: 答案有中度相关性,但存在不准确之处。
Score 7: 答案与参考答案相符,但存在轻微错误或遗漏。
Score 10: 答案完全准确,与参考答案完全一致。"""
            },
            normalize_by=10,
        ),
    ],
    #还可以在此处添加自定义的 StringEvaluator 或 RunEvaluator 对象,这些对象将被自动
    #应用于每个预测
    #查看文档以获取示例
    custom_evaluators=[],
)
```

最后一步是运行 Agent 和评估器。使用 run_on_dataset(或异步 arun_on_dataset)函数评估模型。这里有 3 个环节：

(1) 从指定的数据集中获取示例行。
(2) 在每个示例上运行 Agent(或任何自定义函数)。
(3) 将评估器应用于生成的运行跟踪和相应的参考示例,以生成自动反馈。

运行结果将在 LangSmith 应用程序中可见,代码如下：

```python
from langchain import hub

# 这里将测试此版本的提示
prompt = hub.pull("wfh/langsmith-agent-prompt:798e7324")
```

```python
// 第 10 章 / run_agent_evaluator.py
import functools

from langchain.smith import (
    arun_on_dataset,
    run_on_dataset,
)

chain_results = run_on_dataset(
    dataset_name=dataset_name,
    llm_or_chain_factory=functools.partial(agent_factory, prompt=prompt),
    evaluation=evaluation_config,
    verbose=True,
    client=client,
    project_name=f"runnable-agent-test-5d466cbc-{unique_id}",
    tags=[
        "testing-Notebook",
        "prompt:5d466cbc",
    ],  # 可选,为生成的链运行添加标签
)

# 有时,由于解析问题、不兼容的工具输入等,代理会出错
# 这些错误会被记录为警告,并在追踪 UI 中捕获为错误
```

运行完成后可以得到如下的结果:

```
View the evaluation results for project 'runnable-agent-test-5d466cbc-bf2162aa' at:
https://smith. langchain. com/o/ebbaf2eb - 769b - 4505 - aca2 - d11de10372a4/
projects/p/0c3d22fa-f8b0-4608-b086-2187c18361a5
 [>                                                  ] 0/5[--------------------
------------------------------>] 5/5
Eval quantiles:
                            0.25      0.5       0.75      mean      mode
embedding_cosine_distance   0.086614  0.118841  0.183672  0.151444  0.050158
correctness                 0.000000  0.500000  1.000000  0.500000  0.000000
score_string:accuracy       0.775000  1.000000  1.000000  0.775000  1.000000
helpfulness                 0.750000  1.000000  1.000000  0.750000  1.000000


Chain failed for example 54b4fce8-4492-409d-94af-708f51698b39 with inputs
{'input': '谁训练了 Llama-v2? '}
Error Type: TypeError, Message: DuckDuckGoSearchResults. _ run ( ) got an
unexpected keyword argument 'arg1'
```

用户可以通过单击上面代码输出中的 URL 或导航到 LangSmith "agent-qa-{unique_id}" 数据集中的"测试和数据集"页面来查看如图 10-4 所示的结果。

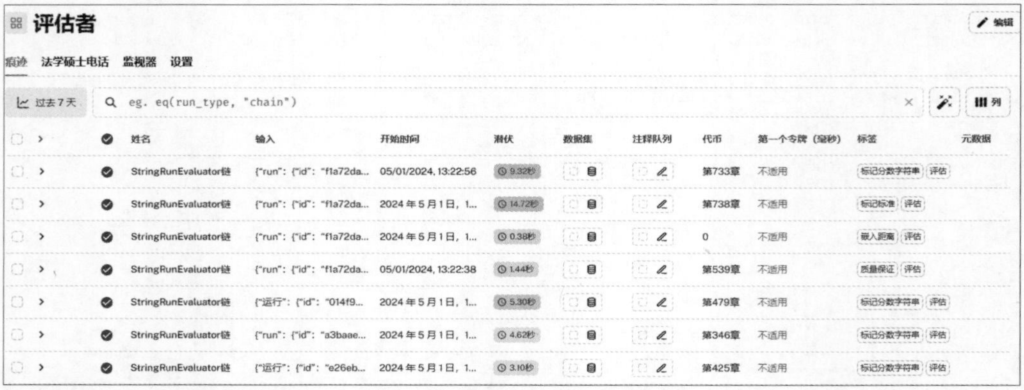

图 10-4　新的 Agent 评估结果

图 10-4 显示了新运行和所选评估器记录的反馈。用户还可以输入下面的代码以表格格式浏览结果摘要，代码如下：

```
chain_results.to_dataframe()
```

最后一步，用户可以导出数据集和评估结果。LangSmith 允许用户直接在 Web 应用程序中将数据导出为 CSV 或 JSONL 等常见格式。同时还可以使用客户端获取运行日志以进行进一步分析，存储在自己的数据库中或与他人共享。接下来从评估运行中获取运行跟踪记录，代码如下：

```
runs = client.list_runs(project_name=chain_results["project_name"], execution_order=1)
```

```
client.read_project(project_name=chain_results["project_name"]).feedback_stats
```

10.7　总　结

在生产环境中成功部署 LLM 和其他生成式 AI 模型是一项复杂但易于管理的任务，需要仔细考虑众多因素。开发者需要解决与数据质量、偏见、道德、法规遵从性、可解释性、资源要求及持续监控和维护等相关的问题。LLM 的评估是评估其性能和质量的重要一步。LangChain 支持模型之间的比较评估，根据标准检查输出、简单的字符串匹配和语义相似度指标。这些提供了对模型质量、准确性和适当生成的不同见解。系统评估是确保 LLM 产生有用、相关和合理的输出的关键。监控 LLM 是部署和维护这些复杂系统的一个重要方面。随着LLM在各种应用中的日益普及，确保其性能、有效性和可靠性至关重要。本章讨论了监控LLM的重要性，强调了要跟踪的关键指标，以制定全面的监控策略，并举例说明了如何在实践中跟踪指标。LangSmith提供了强大的功能来跟踪、进行基准测试和优化使用LangChain构建的LLM，其自动化评估器、指标和可视化有助于加速LLM的开发和验证。

第 11 章 生成式 AI 的未来展望

本书到此为止已经讨论了用于构建应用程序的 LLM 的各个方面,带领读者探讨了 LLM 和图像模型用于内容创作、工具使用、Agent 策略、带检索增强生成的语义搜索,以及通过提示和微调对模型进行条件设置等。此外还创建了一些简单的应用,例如为开发人员和数据科学家开发的应用。本章将讨论生成式 AI 未来的发展方向。过去一年中,AI 领域的进展速度急剧加快,像 DALL-E、Midjourney 和 ChatGPT 等突破性成果的出现产生了令人震惊的结果。这些生成式 AI 模型可以创建逼真的图像,撰写文章和代码,并具有超越大多数人类的对话能力。生成式 AI 初创公司的风险投资在 2022 年飙升,几乎与前五年的总投资相匹配。最近,Salesforce 和 Accenture 等主要参与者已经对生成式 AI 做出了数十亿美元的重大承诺。为特定用例和特定场景定制基础 LLM 被视为真正的价值创造机会。在技术层面上,像 ChatGPT 这样的生成模型通常作为黑匣子运行,其决策过程具有有限的透明度。模型可解释性的缺乏使人们难以完全理解模型的行为或控制其输出。人们还担心可能由于不完美的训练数据而产生潜在偏见。在实际层面上,生成模型需要大量的计算资源进行训练和部署。对于许多组织来讲,获得有效利用这些 AI 系统的基础设施仍然是一道障碍。积极的一面是,AI 可以使技能民主化,允许业余人士在设计、写作等方面产生专业质量的输出。企业方面则可以从更快、更便宜、按需提供的工作中受益。

然而,人们对于工作岗位流失存在比较大的担忧,特别是像平面设计师、律师和医生这样的专业中产阶级岗位。他们的工作正在被自动化,而低技能的工人则学会利用 AI 作为超能力。更令人担忧的是,AI 可能被恐怖分子、犯罪分子用于宣传和提高影响力,例如实时制作的 Deepfakes 可以高相似度地模拟任何人的画面和声音,带来巨大的欺诈风险,通过承认风险、促进开放讨论和实施谨慎的政策,可以激发 AI 的更大潜力并为人类所用。

11.1 当前的生成式 AI

近年来,生成式 AI 模型在跨越文本、图像、音频和视频等多种形式的人类内容生成方面取得了新的里程碑。领先的模型,如 OpenAI 的 GPT-4 和 DALL-E 2,谷歌的 Imagen 和

Parti，以及 Anthropic 的 Claude，在语言生成方面展现出令人印象深刻的流畅度，同时具备创造性的视觉艺术。在 2022 年到 2023 年，这些模型取得了巨大的进步。生成模型以前只能产生几乎不连贯的文本或模糊的图像，但现在我们看到了高质量的三维模型、视频，以及生成连贯且在语境中相关的散文和对话，甚至可以与人类的流畅水平相媲美或超越人类。这些 AI 模型利用庞大的数据集和计算规模，使它们能够捕捉复杂的语言模式，展现对世界知识的细致理解，翻译文本，总结内容，回答自然语言问题，创造吸引人的视觉艺术作品，并获得描述图像的能力。仿佛神奇般地，AI 生成的输出模仿了人类的创造，例如绘制原创艺术作品、创作诗歌、生成人类级别的散文，甚至对来自不同来源的信息进行汇总和综合，但是，生成模型既有优势也有弱点。与人类认知相比，模型仍然存在不足，包括频繁生成具有合理但不正确或荒谬陈述的问题。幻觉现象显示出缺乏现实基础，因为它们基于数据中的模式而不是对真实世界的理解。此外，模型在进行数学、逻辑或因果推理方面存在困难。它们很容易被复杂的推理问题所困扰，这可能会限制它们在某些工作领域的适用性。对于预测的解释性及模型本身缺乏可解释性，也构成了黑匣子问题，阻碍了故障排除工作，并且在期望的参数范围内控制模型行为仍然具有挑战性。AI 模型可能会展现出有害的意外偏见，这带来了重大的道德问题，这个问题主要是由训练数据本身存在的偏见所导致的。偏见问题不仅会扭曲输出，还可能传播和放大社会的不平等。将当前 LLM 与人类认知的关键优势和不足进行可视化对比，见表 11-1。

表 11-1 LLM 的优缺点对比

LLM 的优点	LLM 的缺点
语言流利度：能够生成语法连贯、上下文相关的散文和对话。GPT-4 能够生成人类水平的散文	事实准确性：LLM 经常会生成看似合理但不正确或荒谬的陈述。缺乏现实基础
知识综合：能够对来自不同来源的信息进行聚合和呈现	逻辑推理：无法进行数学、逻辑或因果推理。容易被复杂的推理问题所迷惑
创意输出：能够生成富有想象力的原创文本、艺术作品、音乐，反映人类的聪明才智。例如 Claude 能够写诗，DALL-E 2 能原创艺术作品	可控性：难以将模型行为约束在所需参数内。可能会表现出有害的意外偏见
	偏见：传播和放大训练数据中存在的社会偏见的潜力。引发伦理问题
	透明度：模型预测缺乏可解释性。黑匣子问题限制了故障排除

虽然生成式 AI 的性能已经取得了长足的进步，但它的短板需要尽快得到解决，以便这些技术在未来可有效地发挥作用。尽管存在一些问题，但是如果生成 AI 模型得到负责任的开发和监管，则它们的巨大潜力预示着一个令人兴奋的未来。下面探讨一下由于这些短板带来的技术挑战。

尽管取得了迅速的进展，但要安全、负责地实现生成式 AI 的全部潜力，仍存在重大的技术障碍。正如前文所述，生成式 AI 模型尽管取得了可观的进展，但仍在应对重大的技术挑战,需要克服这些挑战,以允许其安全、负责地发挥其全部潜力。在前几章讨论了其中一些问题及可能的解决方案。一些挑战及解决这些挑战的技术方法见表11-2。

表11-2 一些挑战及解决这些挑战的技术方法

挑 战	描 述	解 决 方 案
逼真和多样化的内容生成	现有模型在逻辑一致性和事实合理性方面存在困难。生成重复、乏味的样本,缺乏人类的细微差别	从人类反馈中强化学习 数据增强和综合技术 模块化领域知识
输出质量控制	缺乏可靠的约束生成内容属性的机制。模型偶尔会产生有害的、有偏见的或荒谬的结果	受约束的优化目标 审核系统 中断和纠正技术
避免偏见	模型无意中放大了训练数据中存在的社会偏见。开发减少偏见的技术仍然很困难	平衡且具有代表性的训练数据 偏差缓解算法 持续的测试和审计
事实准确性	无法对客观事实进行推理限制了实际应用的可靠性。常识和物理学中的接地模型是一个悬而未决的问题	整合知识库 混合神经符号架构 检索增强生成
可解释性	大型神经网络的不透明行为为排除故障或偏差带来了障碍,因此需要可解释的 AI 技术	模型内省技术 概念归因方法 简化的模型架构
数据隐私	收集和处理海量数据集会带来有关同意、匿名化、访问控制和数据滥用的挑战	差分隐私和安全的多方计算 合成数据生成
延迟和计算	部署大型模型需要大量的计算资源,从而延迟了许多应用程序所需的实时交互性	将蒸馏建模为更小的外形尺寸 优化的推理引擎 专用 AI 硬件加速器
数据许可	组织可能需要获得商业许可才能使用现有数据集或构建定制数据集来训练生成模型。这可能是一个复杂且耗时的过程	开源和合成数据

　　LLM 生成的内容经常受到现实感和多样性的限制。虽然它们展示了模仿人类语言和创造力的令人印象深刻的能力,但在生成逻辑一致且事实可信的内容方面仍然存在缺陷。它们的输出经常缺乏人类的细微差别,变得重复和单调。潜在的解决方案包括从人类反馈中进行强化学习,以改善连贯性和细微差别,控制数据增强和综合技术,以及结合模块化领域知识的架构。另一个关键障碍是输出质量的控制。尽管经过严格的训练和开发,现有的 AI 机制在可靠地限制生成内容的特性方面仍然存在不足。这导致不时产生可能有害、有偏见或完全荒谬的内容,对它们更广泛的接受和应用构成风险。有希望的方法包括约束优化目标、人在循环中的审查系统,以及在生成过程中中断和纠正模型输出的技术。

　　偏见确实是这些 AI 模型的一个主要问题,因为它们频繁且无意地放大了训练数据中存在的社会偏见。开发纠正这类偏见的技术仍然是一个复杂的问题。诸如平衡和代表性训练数据、偏见缓解算法及公平性的持续测试和审计等策略旨在解决这一问题。这些 AI 模型无法推理客观真理的能力明显限制了它们在现实应用中的可靠性。将这些模型基于常识和物理学的基础是AI社区仍在努力解决的一个未解问题。混合神经符号结构、知识库的整合和检索增强生成提供了前进的方向。

AI的黑匣子特性提出了另一个复杂的挑战：可解释性。大型神经网络的不透明行为为解决故障或偏见设置了障碍，这强调了更透明的AI技术的必要性。模型内省、概念归因方法和简化的模型架构可能提供解决方案。此外，数据隐私问题因广泛数据的收集和处理而变得突出。这一方面带来了关于同意、匿名化、访问控制和数据滥用的挑战。差分隐私、安全多方计算、合成数据生成和联邦学习等技术可能有助于应对隐私风险。最后，部署这些庞大模型需要大量的计算资源，从而导致显著的延迟和计算问题。这可能延迟许多应用所需的实时交互，表明提高效率至关重要。解决方案包括将模型压缩为更小的形态因素、优化推理引擎和专用的AI硬件加速器。

11.2 未来的能力

当前大型模型的训练计算时间翻倍周期约为8个月，超过了摩尔定律（集成电路上晶体管密度每18个月增加一倍）或洛克定律（诸如GPU和TPU等硬件成本每4年减半）等扩展规律。著名机器学习系统随时间的训练计算如图11-1所示[20]。

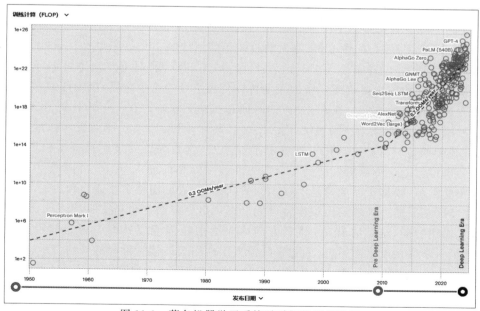

图11-1　著名机器学习系统随时间的训练计算

正如本书第1章所讨论的，大型系统的参数大小增长速度与训练计算相似，这意味着如果这种增长持续下去，人们则可能会看到更大更昂贵的系统。基于经验得出的扩展规律预测了大型语言模型的性能,这些规律基于训练预算、数据集大小和参数数量。这可能意味着高度强大的系统将集中在大型科技公司手中。由 Kaplan 及其同事提出的 KM 扩展规律,通过对不同数据大小、模型大小和训练计算进行实证分析和模型性能拟合得出,呈现出幂律关系,表明模型性能与模型大小、数据集大小和训练计算等因素之间存在着强烈的依赖关系。由谷歌 DeepMind 团队开发的 Chinchilla 扩展规律,涉及对更广泛范围的模型大小和数据大小进行实验,建议将计算预算最优地分配给模型大小和数据大小,这可以通过在约束条件下优化特定损失函数来确定。

然而,未来的 AI 模型的发展可能更多地取决于数据效率和模型质量,而不是单纯的规模大小。尽管巨型模型吸引了媒体关注,但计算能力和能源限制了模型的无节制增长。未来将会看到庞大的通用模型与较小、可访问的专业化细分模型共存,这些模型提供更快、更便宜的训练、维护和推理。已经有研究表明,规模较小的专业化模型可以表现出非常高的性能。最近出现了一些模型,例如 phi-1[1],其参数数量约为 10 亿个,尽管规模较小,但在评估基准上仍然实现了高精度。作者指出,改善数据质量可以极大地改变扩展规律的形态。更多的研究表明,模型可以大幅缩小,相应带来的精度损失很少[21],这支持了模型训练和访问的大众化的论点。此外,诸如迁移学习、蒸馏和提示技术等技术可以使较小的模型利用大型基础模型的能力,而无须复制它们的成本。为了弥补局限性,像搜索引擎和计算器这样的工具已经被整合到 Agent 和多步推理策略中,插件和扩展可能越来越多地用于扩展能力。

AI 训练成本的降低是由于不同的因素造成的。根据 ARK 投资管理有限责任公司的数据,AI 的训练成本每年约下降 70%。最近由 Mosaic ML 发布的 AI 训练工具可以将语言模型训练至类似 GPT-3 的性能,成本大约是两年前估计的 460 万美元的十分之一。这将促进更多的模型训练,但真正的进步不光来自训练成本的降低,而更多地来自训练制度、数据质量和新颖的架构。在由资源丰富的大型科技公司主导的竞争之后,负责任的、经济的创新可能成为优先考虑的方向。在 3~5 年的时间范围内(2025—2027 年),计算和人才可用性的限制可能会显著缓解,削弱了中心化的壁垒。具体来讲,如果云计算成本如预期般下降,并且 AI 技能通过教育和自动化工具变得更加普及,自我训练的定制化语言模型可能对许多公司变得可行。这可以更好地满足个性化和数据隐私的需求,然而,一些能力,例如在特定环境中的学习,根据扩展规律是不可预测的,只有在大型模型中才能出现。进一步推测认为,即使在更多数据上训练的巨型模型可能会表现出更多的行为和技能,极端扩展最终可能产生人工通用智能,这意味着 AI 将具有与人类智力相当或超越人类智力的推理能力,然而,从神经科学的角度来看,目前的技术阶段人工通用智能接管世界的威胁似乎被高度夸大了[22]。如果需要达到和人类对话基本相近的程度,则目前 LLM 主要还有以下的一些障碍。

(1)缺乏具体、嵌入式信息:当前的 LLM 纯粹依赖文本数据进行训练,而无法获得丰富的多模态输入,这些输入使人类能够对物理世界形成常识性推理。这种缺乏基于实际经验的学习是发展人类水平智能的主要障碍。

(2)与生物大脑不同的架构:诸如 GPT-4 之类的模型采用的是相对简单的堆叠变换器架构,缺乏人类脑中被认为支持意识和一般推理的复杂循环和分层结构。

（3）能力狭窄：现有模型仍然专门针对特定领域（如文本），在灵活性、因果推理、规划、社交技能和一般问题解决等智能方面存在不足。这可能会随着工具使用增加或模型的根本性改变而发生改变。

（4）社交能力和意图最小：当前AI系统没有内在的动机、社交智能或超出其训练目标的意图。担心恶意目标或对支配的欲望似乎是没有根据的。

（5）有限的现实世界知识：尽管摄取了庞大的数据集，大型模型的事实知识和常识仍然与人类相比非常有限。这影响了在物理世界中的适用性。

（6）数据驱动的局限性：依赖于来自训练数据的模式识别，而不是结构化知识，使对新情况的可靠泛化变得困难。

鉴于这些论点，如今AI迅速演变为恶意超级智能的风险似乎不太可能。尽管如此，随着AI能力的持续提升，审慎地长期关注安全和伦理问题仍然是明智的选择，但对即将发生的世界主导权被夺的担忧，缺乏来自神经科学或当前模型能力的证据支持，因此，对必然即将到来的人工通用智能的宣称缺乏严格的证据支持，然而，技术进步的速度确实令人担忧，人类可能被淘汰，工作可能被取代，这可能进一步加剧经济阶层的分化。与过去的实物自动化不同，生成式AI威胁到先前被认为不会被自动化取代的认知工作类别，例如设计类的工作。以合乎道德和公平的方式管理这种劳动力过渡将需要远见和规划。此外，围绕AI是否应该创作反映人类状况的艺术、文学或音乐作品，存在着哲学上的辩论。接下来将探讨AI的社会影响。

11.3 AI的社会影响

高性能生成式AI的出现可能在未来几年中改变社会的许多方面。随着生成式模型的不断发展并为企业和创意项目增加价值，生成式AI将在跨领域的技术和人际交互中塑造未来。尽管它们的广泛应用为企业和个人带来了众多的好处和机遇，但由此带来的对各个领域的AI模型依赖程度增加而引发的伦理和社会关切问题必须得到解决。

如果能够慎重部署，则生成式AI在个人、社会和工业领域都有巨大的潜在益处。在个人层面上，这些模型可以增强创造力和生产力，提高对诸如医疗保健、教育和金融等服务的可获得性。通过大型模型获取知识，它们可以帮助学生学习或者辅助专业人士通过综合专业知识做出决策。作为虚拟助手，它们提供即时定制信息以便于日常任务的完成。通过自动化机械化的任务，它们可能使人类释放出更多时间，从而从事更高价值的工作，提高经济产出。经济上，生产率的提高很可能会导致某些工作类别的大规模变革。新的产业和工作可能会出现以支持AI系统，因此慎重考虑并应对这些变化至关重要。

随着模型的不断改进和运行成本的降低，这可能会引发生成式AI和LLM应用大规模地扩展到新领域。除了硬件成本的降低之外，根据Wright定律，随着每次AI系统的累积翻倍，成本可能会下降10%~30%。这个成本曲线反映了代码、工具和技术的重复使用所带来的效率提升。随着成本的降低，应用范围的扩大，形成了良性循环，进一步推动了成本的降低。这将引发一个循环,即更高的效率驱动更广泛的使用,进而推动更高的效率。

Wright 定律,也被称为经验曲线效应,在经济和商业领域是一个观察结果,指出对于许多产品来讲,每当累积生产量翻倍时,成本会按固定百分比下降。它表明每当生产量翻倍时,成本都会稳定地降低(通常为 10%~30%)。这个定律得名自 Theodore Paul Wright,他是一位美国飞机工程师,于 1936 年首次观察到这种趋势,当时他在研究飞机生产成本。Wright 观察到随着机体总产量的翻倍,生产飞机所需劳动力成本每次都会稳定地减少 10%~15%。这种关联可以用数学方式进行表达,如式 11-1 所示。

$$C_x = C_1 x^{\log_2 b} \tag{11-1}$$

其中,C_1 代表生产第一单位的成本,C_x 代表生产第 x 单位的成本,而 b 则是进步比率,据估计在许多行业中介于 0.75~0.9。Wright 定律背后的逻辑是随着生产的增加,工人通过实践、标准化工作流程和改进工具与流程,在制造产品方面变得更加高效。公司还会找到优化供应链、物流和资源利用的方法以降低成本。

在工业领域,这些模型带来了广泛的机遇,可以增强人类能力并重塑工作流程。在内容生产方面,生成式 AI 可以比人类更快地起草营销活动或新闻报道的初稿,从而实现更大程度的创造力和定制化。对于开发人员,自动生成的代码和快速迭代可以加速软件构建。研究人员可以快速综合论文中的发现以推动科学进步。生成式 AI 还可以为消费者提供规模化的个性化体验。推荐内容可以针对个体进行定制。跨不同领域和地理位置的营销可以进行定制。总体而言,这些模型可以提高从工业设计到供应链等各个行业的生产力。

就技术传播而言,存在两种主要场景。在第 1 种场景中,每家公司或个人都使用其专有数据训练自己定制的模型,然而,公司的开发人员需要较为丰富的 AI/机器学习专业知识才能正确地开发、训练和部署这些系统,目前这方面的人才仍然稀缺且昂贵。计算成本也极其高昂,专门的硬件(如 GPU 集群)的成本巨大,只有大型公司才能负担得起。当模型基于敏感信息进行训练时,还存在着数据隐私合规方面的进一步风险。

如果能够克服围绕专业知识、计算要求和数据隐私的障碍,则经过微调以适应组织特定目标和数据的个性化 LLM 能够通过自动化例行任务提供针对具体业务的见解,显著提升生产力和效率,然而,不足之处在于基于小型私有数据集训练的模型可能缺乏基于程度不同的公共语料库训练的模型的泛化能力。集中式和自服务模型可以共存,为不同的使用案例提供服务。短期内,由于资源的优势,大型科技公司在提供面向特定行业的精细调整服务方面具有优势,但随着定制化和隐私需求的增加,越来越多的内部训练可能会出现。在降低成本、传播专业知识和解决稳健性挑战方面的快速创新将决定集中优势的持续时间。这些领域的快速创新有利于消除优势,但围绕主流框架、数据集和模型的平台效应可能使当前领导者继续集中。

如果出现简化和自动化 AI 开发的强大工具,则定制的生成式模型甚至可能适用于地方政府、社区组织和个人以解决超局部的挑战。尽管目前集中的大型科技公司受益于规模经济,但来自较小实体的分布式创新可以释放生成式 AI 在社会各个领域的全部潜力。最后,生成式 AI 的出现与我们如何生产和消费创意作品的更广泛变革相交汇。互联网已经培育了一种混合文化,其中派生作品和协作内容创作是司空见惯的。当 AI 模型通过重新组合现有材料生成新的作品时,它们符合迭代、集体生产的混合文化原则,然而,生成式模型合成和重新利用受版权保护内容的规模提出了棘手的法律问题。目前的检测机制无法以高于随机水平发现由生成式 AI 创作的内容。这引发了围绕作者身份和版权法的广泛争议,所以接下来看一下目前 AI 给创意行业带来的影响。

11.3.1　AI 和创意行业

游戏和娱乐行业正在利用生成式 AI 打造独具沉浸式的用户体验。自动化创意任务带来的效率大幅提升,这可能会增加在线休闲时间。生成式 AI 可以让机器通过学习模式和示例生成新颖原创的内容,如艺术品、音乐和文学作品。这对创意产业具有重要意义,因为它可以提升创作效率并可能创造新的收入来源。它也为媒体、电影和广告开启了新的个性化、动态内容创作,然而,在完全部署之前,生成式内容需要对质量进行控制,确保准确性并消除偏见。对于媒体、电影和广告,AI 解锁了新的个性化、动态内容创作尺度。在新闻报道方面,利用海量数据集进行自动化文章生成可以释放记者的时间,让他们更专注于更复杂的调查报道。通过提高效率和多样性,AI 生成的内容(AIGC)在改变媒体制作和传递方面发挥着日益增长的作用。

在新闻报道领域,文本生成工具自动化了传统由人类记者完成的写作任务,极大地提高了生产率,同时保持了及时性。例如,像美联社这样的媒体机构每年会利用 AIGC 生成成千上万篇报道。像洛杉矶时报的"地震机器人" Quakebot 这样的机器记者可以迅速撰写突发新闻的文章,其中聊天机器人可以创建个性化的一句新闻摘要。AIGC 还实现了 AI 新闻主播,通过模仿人类的外貌和从文本输入的语音,与真实主播共同主持节目。中国新华社的虚拟主播就是一个例子,它以不同角度呈现新闻节目,营造出沉浸式效果。AIGC 正在改变从剧本创作到后期制作的电影制作。AI 剧本创作工具分析数据以生成优化的剧本。视觉效果团队将 AI 增强的数字环境和实景拍摄进行融合,创造出沉浸式视觉效果。Deep Fake 技术以逼真的方式重新制作或复活角色。AI 还可以进行自动生成字幕,甚至通过对大量音频样本进行训练可以预测无声电影中的对话。这扩大了通过字幕进行无障碍访问,并重新创建与场景同步的配音。在后期制作中,AI 色彩分级和编辑工具(如 Colourlab.AI 和 Descript)简化了色彩校正等流程。

在广告领域,AIGC 释放了有效、定制的广告创意和个性化的新潜力。AI 生成的内容允许广告商以规模化的方式创建针对个体消费者的个性化、引人入胜的广告。像创意广告系统(CAS)和个性化广告文案智能生成系统(SGS-PAC)这样的平台利用数据自动生成针对特定用户需求和兴趣的广告宣传语。AI 还可以协助广告创意和设计,像 Vinci 这样的工具可以从产品图像和口号生成定制的吸引人的海报,而像 Brandmark.io 这样的公司可以根据用户偏好生成各种标志。GAN 技术自动化产品列表生成并加上关键词以进行有效的点对点营销。合成广告制作也在逐渐兴起,它能够实现高度个性化和可扩展的广告宣传,极大地节省了时间。在音乐领域,像谷歌的 Magenta、IBM 的 Watson Beat 或 Sony CSL 的 Flow

Machine 等工具可以生成原创的旋律。AIVA 同样可以根据用户调整的参数创建独特的作品。LANDR 的 AI 母带制作利用机器学习处理和改善音频质量。在视觉艺术方面，Midjourney 利用神经网络生成启发式图像，可以启动绘画项目。艺术家们利用它的输出创作出获奖作品。DeepDream 算法将图案添加到图像上，创作出迷幻的艺术作品。GAN 可以生成符合特定风格的抽象绘画。AI 绘画修复可以分析艺术作品进行数字修复和恢复。

动画工具（如 Adobe 的 Character Animator 或 Anthropic 的 Claude）可以帮助生成定制的角色、场景和动作序列，为非专业人员打开了动画潜力。ControlNet 通过增加输出变化性的约束来指导扩散模型。对于所有这些应用程序，先进的 AI 通过生成式内容和数据驱动的洞察力扩展了创意的可能性，然而，值得注意的是在正式部署上线 AI 模型之前，生成式内容需要对质量进行控制，确保准确性并消除偏见。对于广告业务，合乎道德和法律使用消费者数据和人工监督仍然非常重要。在 AI 全面到来的时代，正确合理评判人类艺术家、开发人员和训练数据的贡献仍然是一个持续的挑战。

11.3.2 AI 和社会经济

生成式 AI 和其他技术的投入运用有望加速提高生产力，从而部分弥补就业增长下降，并促进整体经济增长。假设能源和计算可持续地扩展，将生成式 AI 整合到商业流程中所带来的巨大生产力增益，似乎可能在未来十年推动许多任务实现自动化，然而，这种转变可能会对劳动力市场产生冲击，因此需要进行调整。麦肯锡公司的研究估计到 2030 年至 2060 年，当前 30%～50% 的工作活动可能会被自动化。生成式 AI 可能到 2030 年每年为全球生产力增加 6 万～8 万亿美元，使 AI 对经济影响的预估增加了 15%～40%。

另外，Tyna Eloundou 及其同事的研究表明，大约 80% 的美国工人的至少 10% 的工作任务受到 LLM 的影响，而 19% 的人从事的工作可能有超过 50% 受到影响。这种影响涵盖了各个工资水平，高工资工作面临更大的曝光。只有 15% 的美国工人的所有任务可以通过 LLM 单独完成，但是如果是通过 LLM 驱动的软件工作，则这一比例增加到了 47%～56%，这显示了辅助技术的重大影响。从地理角度来看，与生成式 AI 相关的外部私人投资主要来自科技巨头和风险投资公司，主要集中在北美地区，反映了该地区目前对整体 AI 投资景观的主导地位。总部位于美国的与生成式 AI 相关的公司在 2020 年至 2022 年期间筹集了约 80 亿美元，占同期此类公司总投资额的 75%。

11.3.3 AI 和教育

一个可能发生的未来情景是，个性化 AI 导师和指导者的兴起可能会让与 AI 驱动经济相关的高需求技能的教育变得更加普及化。在教育领域，生成式 AI 已经在改变我们的教学和学习方式。像 ChatGPT 这样的工具可以自动生成个性化课程和为个别学生定制内容。这通过自动化重复的教学任务大幅减轻了教师的工作负担。AI 导师可以对学生的写作作业提供实时反馈，释放教师的精力去关注更复杂的技能。由生成式 AI 驱动的虚拟模拟还可以创造引人入胜的、量身定制的学习体验，以满足不同学习者的需求和兴趣，然而，随着这些技术的发展，对持续存在的偏见和传播错误信息的风险需要进一步研究。

知识的加速增长和科学发现的过时意味着培养孩子的好奇心驱动学习应着重于发展涉及启动和维持好奇心的认知机制，例如对知识空缺的认识和使用适当策略来解决它们。虽然为每名学生量身定制的 AI 导师可能会提高学习成果和参与度，但是较差的学校可能会被落下，从而加剧不平等。政府应该促进平等获取，以防止生成式 AI 成为富裕阶层的特权。为所有学生提供机会仍然至关重要。如果实施得当，则个性化的 AI 教育可以使任何有学习动力的人获得重要的技能。交互式 AI 助手可根据学生的优势、需求和兴趣调整课程，使学习高效、有趣且公平，但是需要解决关于获取、偏见和社会化方面的挑战问题。

11.3.4　AI 和就业

假设能源和计算能够可持续发展，将生成式 AI 整合到业务流程中所带来的巨大生产力提升，似乎可能在未来十年内引发许多任务的自动化。这种转变可能会扰乱劳动力市场，需要做出调整。AI 所带来的自动化可能会在短期内取代许多行政、客户服务、写作、法律和创意工作，而农业和建筑等行业可能几乎不受影响，然而，过去的工业革命最终导致了新型的工作和产业，尽管工作人员的转变过程艰难。相同的动态很可能在 AI 自动化方面长期发挥作用。这将在某种程度上影响绝大多数职业，尽管有些职业受到的影响更为严重。

生成式 AI 分析和生成自然语言内容的能力可能会显著增加对跨多个白领职业的沟通、协作和报告等活动的自动化潜力，然而，仍然无法确定整个工作岗位是否会被取代。过去的技术创新最终创造了新的工作类型，即使这个转变过程很困难。根据麦肯锡公司的研究，在生成式 AI 用例能够提供的价值中，约有 75% 分布在 4 个领域：客户运营、市场营销、软件工程和研发。突出的例子包括生成式 AI 支持与客户的互动、为营销和销售生成创意内容，以及根据自然语言提示起草计算机代码等众多任务。语言模型和生成式 AI 潜在地扰乱和自动化了传统上由人类执行的各个行业内的任务。下面是对未来不同职业的一些预测：

（1）初级软件工程师可能会得到 AI 编码助手的增强或替代。

（2）利用 AI 客服代理提供的数据洞察分析师和顾问可能被对话式 AI 所取代。

（3）技术撰稿人和新闻工作者可以得到 AI 内容生成的帮助。

（4）教师利用 AI 进行课程准备和个性化辅导。

（5）法律助理利用 AI 进行摘要和文件审查。

（6）图形设计师将通过 AI 图像生成得到更强大的支持，然而，图像创作和操作能够让更多人使用也可能对工资产生影响。

（7）对于高级软件工程师来讲，需求仍然旺盛，他们开发专门的 AI 解决方案和系统。

（8）数据科学家可能会转向验证、调试和最大化 AI 系统价值的工作，而不是构建预测模型。

（9）程序员将越来越多地编写工具以辅助 AI 开发。

（10）新兴职位如提示工程师已开始出现。

AI可以执行涉及自然语言处理、内容创作甚至复杂创意工作的某些任务,效率高、错误少于人类。技能较低的人可能能够从事更高技能的工作,而高技能的人可能面临较少的工作机会。例如,法律助理使用模板文件并填写必要信息以满足客户需求。配备了广泛法律文件、法规、大学课程、期刊、新闻文章和法院案例知识的AI能比法律助理更好地完成这项任务。结果可能是对初级律师起草工作需求的潜在减少,法律助理使用AI软件传达客户特定需求。

软件开发人员和数据科学家都可以从大型语言模型的潜力中受益,但必须仔细考虑其能力和局限性以实现最佳使用。对于初级开发人员和数据科学家,大型语言模型可以自动化例行任务,提供基本解决方案,并减少错误,通过释放更多时间进行更复杂的工作以加速学习,然而,仅依赖AI存在阻碍更深层次技术增长的风险,因此应将大型语言模型视为辅助工具,同时积极发展实际专业技能。高级开发人员和数据科学家拥有领域知识和超越当前AI能力的问题解决能力。虽然自动化标准解决方案可能节省一些时间,但他们的专业知识对引导AI工具、确保可靠且可扩展的结果至关重要。AI生产力的激增意味着公司急需AI人才,并且在招聘和留住这些人才方面竞争激烈。保护AI系统免受攻击的网络安全专家也会越来越受欢迎。此外,随着AI系统的普及,AI伦理、法规和公共政策等领域的工作也可能增加,因此,吸引和培养这样的人才对公司在这个快速变化的环境中保持相关性至关重要。

在音乐方面,AI正在协助音乐家进行创作,从创作歌词和旋律到数字化音频的母带处理和增强。生成艺术工具允许视觉艺术家尝试根据其独特风格定制的绘画作品。2023年3月高盛公司的一项研究指出,行政和法律角色处于最高风险之中。他们估计约三分之二的现有工作将面临AI自动化的风险,并得出结论,生成AI工具可能影响全球3亿个全职工作岗位,占当前劳动力的20%以上。采用的速度是一个重要的未知数。麦肯锡分析师估计自动化将吸收60%~70%的员工工时,因此在2030年至2060年间,大约有一半的当今工作活动可能被自动化。根据普华永道的说法,到2030年代中期,多达30%的工作可能是可自动化的,但实际的应用取决于许多难以预测的因素,例如法规、社会认可和再培训政策等。在知识工作领域,如软件和应用程序开发,已经开始看到这种转变的影响。

生成AI已被用于简化从初始代码生成到图像编辑和设计的任务。它减少了开发人员和设计师的重复手工作业,使他们能够将精力集中在更高价值的创新上,然而,在自动生成输出中进行细致的监控和迭代仍然至关重要。大规模的工作自动化可能会导致劳动力需求的重大转变,引起职业方面的重大变化,并迫使员工掌握新技能。由于它们的能力基本上是为了认知,生成AI可能对知识工作产生的影响最大,特别是涉及决策制定和合作的活动,而这些活动之前的自动化潜力最小。在过去,自动化的影响在中低收入的中间五分位数最大,进一步而言,生成AI可能对高薪工作的活动产生最大影响。相当多的工人将需要在他们目前的职业或新职业中进行重大转变。他们还需要支持实现新工作活动的过渡。管理这种变化将需要政策引导,通过再培训计划、创造就业激励和便携式福利来减少被裁员工的困难。如果能够管理工人转型和其他风险,则生成AI可能会对经济增长做出重要贡献,并支持更加可持续、包容的世界,人们将从重复性工作中解放出来。如果AI自动化带来的效率收益得到良好的再投资,则长远来看可能会创造新的产业和就业机会,但是,顺利的劳动力转变将需要政策引导和临时员工培训。总之,虽然在近期可能会有一些工作被 AI 取代,但 AI 可能只是自动化特定活动而不是消除整个职业。不过可以肯定的是,数据科学家和程序员等技术专家将继续是开发 AI 工具和实现其全面商业潜力的关键。

11.3.5　AI 和其他行业

1. 法律

LLM 这样的生成模型可以自动化常规的法律任务,例如合同审查、文件生成和简要准备。它们还可以实现更快速、全面的法律研究和分析,其他应用包括用通俗语言解释复杂的法律概念及利用案例数据预测诉讼结果,然而,考虑到透明度、公平性和问责制等方面的因素,负责任和道德的使用仍然至关重要。总体来讲,正确实施的 AI 工具承诺提高法律生产力和司法准入,同时需要对可靠性和道德问题进行持续审查。

2. 制造业

在汽车行业中,AI 被用于生成用于模拟的三维环境,并协助汽车的开发。此外,生成式 AI 被用于使用合成数据进行自动驾驶汽车的道路测试。这些模型还可以处理物体信息以理解周围环境,通过对话了解人类意图,对人类输入生成自然语言回应,并创建操控计划以协助人类完成各种任务。

3. 医药

能够准确预测基因序列的物理特性的 AI 模型将代表医学上的重大突破,并可能对社会产生深远影响。它可以进一步加速药物发现,实现更早的疾病预测和预防,提供对复杂疾病更深入的理解,然而,这也带来了围绕基因工程的重大道德关切,并可能加剧社会不平等。目前已经有新技术采用神经网络降低长读 DNA 测序的错误率[23],根据 ARK Investment Management 2023 年的报告,在短期内像这样的技术已经可以以不到 1000 美元的价格提供第一份高质量的完整长读基因组,这意味着大规模的基因到表达模型可能也就不远了。

4. 军事

世界各国军队正在投资研究开发致命的自主武器系统(Lethal Autonomous Weapons Systems,LAWS)。机器人和无人机可以在没有任何人类监督的情况下识别目标并使用致命武器。机器能够处理信息并比人类反应更快,因为它们没有情感,然而,这引发了重大的道德问题。让机器决定是否夺取生命跨越了令人不安的界限,即使具备先进的 AI 技术,战争中的复杂因素需要人类判断,如平民与战斗人员的区分。如果随意部署这种 AI 武器系统,则完全自主的致命武器将代表着放弃对生死决策的控制,这是令人极度担忧的。一旦完全独立释放,自主杀手机器人的行为将无法预测或约束。

11.3.6　AI 和网络安全

AI 在网络安全领域是一把双刃剑。虽然它能够实现规模化的检测,但自动化使传播复杂且个性化的宣传变得更加容易。AI 可能对网络安全构成威胁,它增加了对误导信息的脆弱性，以及利用生成式黑客和社会工程学进行网络攻击。与 AI 技术相关的威胁包括微观定位和深度伪造。强大的 AI 可以在心理层面对用户进行个性化误导，从而促成隐蔽的操纵，避开了广泛的审查。大数据和 AI 可能被用来利用心理漏洞，并渗透在线论坛以攻击和传播阴谋论。虚假信息已经转变成了一个多方面的现象，涉及有偏见的信息、操纵、宣传及意图影响政治行为。例如，在 COVID-19 大流行期间，虚假信息和信息疫情的传播已经成为一个主要挑战。AI 有潜力影响公众对选举、战争或外国势力等话题的观点。它还可以生成虚假的音频/视频内容来损害名誉并制造混乱。国家和非国家行为者正在将这些能力武器化，用于宣传、损害名誉和制造混乱。政党、政府、犯罪团伙甚至法律系统都可以利用 AI 发起诉讼并获取金钱。这可能会在各个领域产生深远的影响。互联网用户中的大部分人可能在不访问外部网站的情况下获得所需的信息。大型企业成为信息的守门人并控制舆论的风险存在，他们能够有效地限制某些行动或观点。谨慎的治理和数字素养是建立抵御力的关键。尽管没有单一的解决方案，但促进负责任的 AI 发展的集体努力可以帮助世界各个社会应对新兴威胁。

11.4 应用难题探索

当前，负责任地利用生成式 AI 涉及解决一系列实际的法律、伦理和监管问题。

（1）法律：就生成式 AI 生成的内容而言，版权法仍存在模糊不清之处。产出物归属于谁，是模型的创建者、训练数据的贡献者，还是最终用户？在训练中复制受版权保护的数据也引发了有关公平使用的争议，这需要明确地进行界定。

（2）数据保护：收集、处理和存储训练先进模型所需的庞大数据集会带来数据隐私和安全风险。确保同意、匿名性和安全获取的治理模式至关重要。

（3）监督与监管：目前对 AI 监管的呼声越来越高，要求监督以确保先进 AI 系统的非歧视性、准确性和责任性，但这需要灵活的政策，平衡创新与风险，而非沉重的官僚主义。

（4）伦理：引导朝着有益成果方向发展的框架至关重要。通过专注于透明度、可解释性和人类监督的设计实践，将伦理融入其中有助于建立信任。

总体而言，政策制定者、研究人员和公民社会之间的主动合作对解决围绕权利、伦理和治理方面尚未解决的问题至关重要。在设定务实的监管措施的同时，生成式模型可以充分发挥其潜力，同时降低其危害性，但公共利益必须始终作为指引 AI 进步的指南。对算法透明度的需求与日俱增。这意味着科技公司和开发者应该公开其系统的源代码和内部运作方式，然而，这些公司和开发者对此持抵制态度，他们认为披露内部信息会损害他们的竞争优势。另一方面开源模型将继续蓬勃发展，欧盟和其他一些国家的地方立法将推动 AI 的透明使用。

AI 偏见的后果包括由 AI 系统偏见决策可能对个人或群体造成的潜在危害。将伦理培训纳入计算机科学课程可以帮助减少 AI 代码中的偏见。通过教导开发人员如何构建符合道德设计原则的应用程序，可以将偏见嵌入代码中的可能性降至最低。为了走上正确的道路，组织需要优先考虑透明度、问责制和监管措施，以防止其AI系统存在偏见。AI偏见的预防是许多组织的长期优先任务，但是，在没有立法推动的情况下，引入这一变革可能需要时间。例如，欧盟国家的地方立法像欧洲委员会关于AI监管规则的提案，将促进更加道德的语言和图像使用。目前德国关于虚假信息的法律规定了平台删除虚假信息和仇恨言论的24h时间限制，但对于大型和小型平台来讲都是不切实际的。此外，较小平台的资源有限，因此使它们无法监管所有内容成为现实。此外，在线平台不应该单独拥有决定什么是真相的权威，因为这可能导致过度的审查制度。社会需要更加微妙的政策来平衡言论自由、问责制和技术多样性平台的执行可行性。政府、公民社会、学术界和产业之间更广泛的合作可以制定更有效的框架来打击虚假信息，同时保护用户的权利。

11.5 写在最后

新一代生成式AI模型的到来带来了众多引人入胜的机遇和空前的进步，然而其中也夹杂着许多不确定性。正如本书所讨论的，近年来AI取得了许多突破，但随之而来的挑战仍然存在，主要涉及这些模型的精度、理性能力、可控性及根深蒂固的偏见。虽然未来可能出现超智能AI的宏伟声明似乎有些夸大，但连续不断的趋势预示着几十年内会出现复杂的能力。在个人层面上，生成式内容的大量增加引发了人们对于误导信息、学术抄袭及在线空间中的冒充行为的合理担忧。随着这些模型在模仿人类表达方面变得更加娴熟，人们可能难以分辨人类生成的和AI生成的内容，从而可能产生新形式的欺骗。人们还担心生成式模型加剧了社交媒体成瘾，因为它们能够制作无穷无尽的定制内容。

从社会角度来看，飞速发展所带来的不安源于对于人类被淘汰和工作岗位流失的担忧，这可能会进一步分化经济阶层。与过去的物理自动化不同，生成式AI威胁到了以往认为不会被自动化取代的认知型工作类别。在道德和公平管理这种工作力量转变方面，需要有远见和规划。围绕AI是否应该创作反映人类状况的艺术、文学或音乐作品，也存在着哲学上的争议。对于公司而言，尚未建立起有效的治理框架来规范可接受的使用案例。生成式模型放大了各种滥用的风险，从制作深度伪造视频到生成不安全的医疗建议。内容许可和知识产权方面的法律问题也应运而生。虽然这些模型可以提高业务生产力，但质量控制和偏见缓解会增加额外成本。

目前，大型科技公司主导着生成式AI的研究和发展，但更小的实体最终可能会从这些技术中获得最多的利益。随着计算、数据存储和AI人才成本的降低，针对特定模型的定制预训练对于中小型公司而言可能会变得可行。与依赖大型科技公司的通用模型不同，针对特定数据集微调的定制生成式AI可能会更好地满足独特需求。初创公司和非营利组织通常擅长快速迭代，为特定领域构建尖端解决方案。通过降低成本可以让这些专注的参与者训练出性能卓越的模型，超越通用系统的能力。关于生成式AI工具的饱和度问题已经出现，因为使用基础模型相对容易。定制模型和工具的个性化将会创造价值，但目前尚不清楚谁将会获得最大的利益。

尽管当前市场在炒作 AI 技术，但投资者鉴于较低的估值及对 2021 年 AI 繁荣/衰退周期后的怀疑正在冷静思考。2021 年的人工智能繁荣/衰退周期指的是人工智能初创公司领域投资和增长的急剧加速，随后在 2022 年由于预期未能实现和估值下降而出现市场冷却和稳定。

下面是一个对这段时期的简要的总结。

（1）繁荣阶段（2020—2021 年）：人们对于提供计算机视觉、自然语言处理、机器人技术和机器学习平台等创新能力的人工智能初创公司表现出巨大的兴趣，并进行了大规模投资。据 Pitchbook 统计，2021 年全球人工智能初创公司的总融资额达到创纪录的 730 多亿美元。在此期间，数百家人工智能初创公司成立并获得了资金支持。

（2）衰退阶段（2022 年）：2022 年，市场经历了一轮修正，人工智能初创公司的估值大幅下降，远低于 2021 年的高点。诸如 Anthropic 和 Cohere 等多家知名人工智能初创公司面临了估值降低。许多投资者变得更加谨慎和选择性地对人工智能初创公司提供资金支持。更广泛的科技行业市场调整也促成了这轮衰退。

（3）一些关键因素：过度炒作、不切实际的增长预期、2021 年历史上的高估值及更广泛的经济状况都为繁荣/衰退周期做出了贡献。这一周期沿袭了以往互联网和区块链等领域所见的经典模式。

展望未来几十年，或许最深层次的挑战在于伦理问题。随着 AI 被委托处理更重要的决策，与人类价值观的一致性变得至关重要。虽然准确性、推理能力、可控性和减少偏见仍然是技术上的重点，但其他重点应该包括加强模型的稳健性，促进透明度，并确保与人类价值观的一致性。为了最大化利益，公司需要确保在开发中有人类监督、多样性和透明度。决策者可能需要实施防止滥用的防护措施，并为随着活动转移而需要转变的工作者提供支持。通过负责任的实施，生成式 AI 可以推动更加繁荣的社会的增长性、创造力和可访问性。早期解决潜在风险问题并确保利益的公正分配将培养出与利益相关者之间的信任感，这些信任感包括以下几方面。

（1）AI 与人类共生：与其追求彻底的自动化，更有利的系统应该将人类的创造力与 AI 的高效生产能力相融合和互补。这样的混合模式将确保最佳的监督。

（2）促进获取与包容：对于消除差异的扩大，关键在于资源的平等获取、相关教育和涉及 AI 的无数机会。代表性和多样性应该得到优先考虑。

（3）预防措施与风险管理：通过跨学科的见解持续评估新出现的能力是为了避免未来发生危险的必要步骤，然而，过度的忧虑不应妨碍潜在的进展。

（4）维护民主规范：与单一实体强加的单方面法令相比，协作讨论、共同努力和达成妥协必将更有建设性地定义 AI 未来的进程。公共利益必须优先考虑。

尽管未来 AI 的能力仍不确定，但积极的治理对于引导这些技术朝着公平、善意的方向发展至关重要。研究人员、政策制定者和民间社会围绕透明度、问责制和道德等问题的合作，有助于将新兴创新与共同的人类价值观相一致。AI 的最终目标应该是赋予人类潜能，而不仅仅是技术上的进步。

参 考 文 献

扫描下方二维码可查询参考文献。

图 书 推 荐

书　名	作　者
Diffusion AI 绘图模型构造与训练实战	李福林
图像识别——深度学习模型理论与实战	于浩文
HuggingFace 自然语言处理详解——基于 BERT 中文模型的任务实战	李福林
动手学推荐系统——基于 PyTorch 的算法实现(微课视频版)	於方仁
TensorFlow 计算机视觉原理与实战	欧阳鹏程、任浩然
自然语言处理——原理、方法与应用	王志立、雷鹏斌、吴宇凡
人工智能算法——原理、技巧及应用	韩龙、张娜、汝洪芳
跟我一起学机器学习	王成、黄晓辉
深度强化学习理论与实践	龙强、章胜
Java+OpenCV 高效入门	姚利民
Java+OpenCV 案例佳作选	姚利民
计算机视觉——基于 OpenCV 与 TensorFlow 的深度学习方法	余海林、翟中华
深度学习——理论、方法与 PyTorch 实践	翟中华、孟翔宇
Flink 原理深入与编程实战——Scala+Java(微课视频版)	辛立伟
Spark 原理深入与编程实战(微课视频版)	辛立伟、张帆、张会娟
PySpark 原理深入与编程实战(微课视频版)	辛立伟、辛雨桐
Python 预测分析与机器学习	王沁晨
Python 人工智能——原理、实践及应用	杨博雄 等
Python 深度学习	王志立
编程改变生活——用 Python 提升你的能力(基础篇·微课视频版)	邢世通
编程改变生活——用 Python 提升你的能力(进阶篇·微课视频版)	邢世通
编程改变生活——用 PySide6/PyQt6 创建 GUI 程序(基础篇·微课视频版)	邢世通
编程改变生活——用 PySide6/PyQt6 创建 GUI 程序(进阶篇·微课视频版)	邢世通
Python 量化交易实战——使用 vn.py 构建交易系统	欧阳鹏程
Python 从入门到全栈开发	钱超
Python 全栈开发——基础入门	夏正东
Python 全栈开发——高阶编程	夏正东
Python 全栈开发——数据分析	夏正东
Python 编程与科学计算(微课视频版)	李志远、黄化人、姚明菊 等
Python 游戏编程项目开发实战	李志远
Python 数据分析实战——从 Excel 轻松入门 Pandas	曾贤志
Python 概率统计	李爽
Python 数据分析从 0 到 1	邓立文、俞心宇、牛瑶
Python Web 数据分析可视化——基于 Django 框架的开发实战	韩伟、赵盼
Python 玩转数学问题——轻松学习 NumPy、SciPy 和 Matplotlib	张骞
AR Foundation 增强现实开发实战(ARKit 版)	汪祥春
AR Foundation 增强现实开发实战(ARCore 版)	汪祥春
ARKit 原生开发入门精粹——RealityKit + Swift + SwiftUI	汪祥春
HoloLens 2 开发入门精要——基于 Unity 和 MRTK	汪祥春
Octave GUI 开发实战	于红博
Octave AR 应用实战	于红博
HarmonyOS 移动应用开发(ArkTS 版)	刘安战、余雨萍、陈争艳 等

续表

书 名	作 者
openEuler 操作系统管理入门	陈争艳、刘安战、贾玉祥 等
JavaScript 修炼之路	张云鹏、戚爱斌
深度探索 Vue.js——原理剖析与实战应用	张云鹏
前端三剑客——HTML5＋CSS3＋JavaScript 从入门到实战	贾志杰
剑指大前端全栈工程师	贾志杰、史广、赵东彦
HarmonyOS 应用开发实战（JavaScript 版）	徐礼文
HarmonyOS 原子化服务卡片原理与实战	李洋
鸿蒙操作系统开发入门经典	徐礼文
鸿蒙应用程序开发	董昱
鸿蒙操作系统应用开发实践	陈美汝、郑森文、武延军、吴敬征
HarmonyOS 移动应用开发	刘安战、余雨萍、李勇军 等
HarmonyOS App 开发从 0 到 1	张诏添、李凯杰
从数据科学看懂数字化转型——数据如何改变世界	刘通
JavaScript 基础语法详解	张旭乾
5G 核心网原理与实践	易飞、何宇、刘子琦
恶意代码逆向分析基础详解	刘晓阳
深度探索 Go 语言——对象模型与 runtime 的原理、特性及应用	封幼林
深入理解 Go 语言	刘丹冰
Vue＋Spring Boot 前后端分离开发实战	贾志杰
Spring Boot 3.0 开发实战	李西明、陈立为
Flutter 组件精讲与实战	赵龙
Flutter 组件详解与实战	［加］王浩然（Bradley Wang）
Dart 语言实战——基于 Flutter 框架的程序开发（第 2 版）	亢少军
Dart 语言实战——基于 Angular 框架的 Web 开发	刘仕文
IntelliJ IDEA 软件开发与应用	乔国辉
FFmpeg 入门详解——音视频原理及应用	梅会东
FFmpeg 入门详解——SDK 二次开发与直播美颜原理及应用	梅会东
FFmpeg 入门详解——流媒体直播原理及应用	梅会东
FFmpeg 入门详解——命令行与音视频特效原理及应用	梅会东
FFmpeg 入门详解——音视频流媒体播放器原理及应用	梅会东
Power Query M 函数应用技巧与实战	邹慧
Pandas 通关实战	黄福星
深入浅出 Power Query M 语言	黄福星
深入浅出 DAX——Excel Power Pivot 和 Power BI 高效数据分析	黄福星
从 Excel 到 Python 数据分析：Pandas、xlwings、openpyxl、Matplotlib 的交互与应用	黄福星
云原生开发实践	高尚衡
云计算管理配置与实战	杨昌家
虚拟化 KVM 极速入门	陈涛
虚拟化 KVM 进阶实践	陈涛
Octave 程序设计	于红博